Advances in Experimental Medicine and Biology

Volume 796

For further volumes:
http://www.springer.com/series/5584

Marta Filizola
Editor

G Protein-Coupled Receptors - Modeling and Simulation

Springer

Editor
Marta Filizola
Department of Structural and Chemical Biology
Icahn School of Medicine at Mount Sinai
New York, NY
USA

ISSN 0065-2598 ISSN 2214-8019 (electronic)
ISBN 978-94-007-7422-3 ISBN 978-94-007-7423-0 (eBook)
DOI 10.1007/978-94-007-7423-0
Springer Dordrecht Heidelberg New York London

Library of Congress Control Number: 2013951640

Printed on acid-free paper

Springer is part of Springer Science+Business Media (www.springer.com)

Preface

G protein-coupled receptors (GPCRs) are membrane proteins of significant interest in pharmaceutical research owing to their involvement in several important biological processes, including those leading to some serious medical conditions. In spite of focused research, progress towards the discovery of effective therapeutics for GPCRs has long been hampered by the lack of high-resolution structural information about these receptors. Although the number of high-resolution crystal structures of GPCRs has grown significantly in the past few years, the information they provide is limited. Not only are we still far from a comprehensive structural coverage of the GPCR superfamily, but the available structures refer to static, heavily engineered, and generally inactive conformational states of receptor subtypes stripped out of their natural lipid environment. Furthermore, it has become increasingly clear that a full understanding of GPCR structure and function requires dynamic information at a level of detail that is likely to require integration of experimental and computational approaches.

Significant progress has been made over the past decade in the development and application of computational approaches to the large family of GPCRs. A dedicated book that discusses in depth this important topic is lacking but strongly needed owing to: (a) the critical (but sometimes unappreciated) impact that these computational approaches have on understanding the molecular mechanisms underlying the physiological function of GPCRs in support of rational drug discovery, (b) the recent advances in theory, hardware, and software, and (c) the potential for much-improved applications using newly available experimentally-derived structural and dynamic information on GPCRs. Thus, I sought the help of experts with an established reputation in the development and/or application of computational methods to GPCRs, and asked them to contribute their state-of-the-art views on modeling and simulation of this important family of membrane proteins. I am indebted to the highly distinguished authors of this book for agreeing to participate in this project and to provide chapters for four different sessions: a first one describing the impact of currently available GPCR crystal structures on structural modeling of other receptor subtypes, a second one reporting on critical insights from simulations, a third one focusing on recent progress in rational ligand discovery and mathematical modeling, and a fourth one providing an overview of bioinformatics tools and resources that are available for GPCRs.

Heartfelt thanks also go to the several anonymous reviewers of the chapters, and to Thijs van Vlijmen from Springer for the opportunity he offered me to edit this volume. I believe this book adds a unique facet to the "Advances in Experimental Medicine and Biology" series, and I hope the reader will find it both fascinating and of enduring interest.

New York, NY, USA
June 3, 2013

Marta Filizola, Ph.D.

Contents

Part I

Progress in Structural Modeling of GPCRs

The GPCR Crystallography Boom: Providing an Invaluable Source of Structural Information and Expanding the Scope of Homology Modeling

1

Stefano Costanzi and Keyun Wang

Abstract

G protein-coupled receptors (GPCRs) are integral membrane proteins of high pharmaceutical interest. Until relatively recently, their structures have been particularly elusive, and rhodopsin has been for many years the only member of the superfamily with experimentally elucidated structures. However, a number of recent technical and scientific advancements made the determination of GPCR structures more feasible, thus leading to the solution of the structures of several receptors. Besides providing direct structural information, these experimental GPCR structures also provide templates for the construction of GPCR models. In depth studies have been performed to probe the accuracy of these models, in particular with respect to the interactions with their ligands, and to assess their applicability the rational discovery of GPCR modulators. Given the current state of the art and the pace of the field, the future of GPCR structural studies is likely to be characterized by a landscape populated by an increasingly higher number of experimental and theoretical structures.

Keywords

Rhodopsin • Homology • Crystal structures • Docking • 3D Models

1.1 Introduction

For the scientific community involved in the study of G protein-coupled receptors (GPCRs), 2012 was a special year, since the Nobel Prize in Chemistry was awarded to Robert Lefkowitz and Brian Kobilka for their pioneering and revolutionary studies that led to our current understanding of the structure and the functioning of these notable receptor proteins.

GPCRs form a superfamily that encompasses a large number of receptors embedded on the lipid bilayer of the plasma membrane of animal cells as well as other eukaryotic cells, including yeast. In light of their characteristic topology, which features one polypeptide chain that spans seven times the membrane with seven α-helices, they are also known as seven transmembrane

S. Costanzi (✉) • K. Wang
Department of Chemistry and Center for Behavioral Neuroscience, American University, Washington, DC 20016, USA
e-mail: costanzi@american.edu

M. Filizola (ed.), *G Protein-Coupled Receptors - Modeling and Simulation*, Advances in Experimental Medicine and Biology 796, DOI 10.1007/978-94-007-7423-0_1,

(7TM) receptors (Pierce et al. 2002). These receptors function as cellular sensors for stimuli provided by extracellular first messengers, which can be either endogenous or exogenous substances. The first category encompasses a wide variety of signaling molecules secreted by the organism, among which are neurotransmitters or hormones. Conversely, the second category includes environmental substances that are detected by the organism, for instance molecules responsible for odor or taste. Beyond chemical first messengers, some GPCRs are also stimulated by physical messengers such as light photons. Upon activation, GPCRs couple with an array of direct intracellular partners, chiefly G proteins and arrestins, and transduce the stimulus into biochemical signals involving complex pathways (Pierce et al. 2002).

As stated by Robert Lefkowitz in an insightful personal retrospective on his career devoted to the study of seven transmembrane receptors, GPCR signaling ultimately regulates virtually all known physiological pathways (Lefkowitz 2007a). Thus, the superfamily of GPCRs provides a wealth of targets for pharmacological intervention. As a result, the modulation of GPCR signaling is the mechanism of action of many of the currently approved drugs (Overington et al. 2006). Some of these drugs act directly on the receptors, either by stimulating their activity, as in the case of agonists of the μ opioid receptors used for the treatment of pain (Spetea and Schmidhammer 2012), or by preventing their stimulation by natural agonists, as in the case of blockers of the $P2Y_{12}$ receptor used for the prevention of platelet aggregation (Jacobson et al. 2011). Conversely, other drugs alter GPCR signaling through indirect means, for instance by preventing the reuptake of the natural agonists of the receptors, as in the case of the selective serotonin reuptake inhibitors (SSRIs) used for the treatment of depression (Donati and Rasenick 2003), or by altering the activity of the enzymes responsible for the synthesis of these molecules, as in the case of the inhibitors of the angiotensin-converting enzyme used for the treatment of hypertension (Rosskopf and Michel 2008).

Experimentally determined three-dimensional structures of drug targets are very powerful tools for drug discovery, as they provide a direct insight into the structure-function relationships of the targets and a platform for the discovery of novel modulators of their activity (Congreve et al. 2011; Congreve and Marshall 2009; Mason et al. 2012). However, being integral membrane proteins, GPCRs are not easy targets for structural biology studies. The first experimentally determined three-dimensional (3D) GPCR structures, first in the form of a projection map (Schertler et al. 1993), and then of 3D X-ray crystallographic structures (Palczewski et al. 2000), were determined for rhodopsin, a photoreceptor activated by light photons. Rhodopsin is a rather peculiar GPCR, which features a covalently bound 11-*cis*-retinal molecule that acts as an inverse agonist. Light photons isomerize the ligand to its all-*trans* form and consequently trigger activation of the receptor. However, despite these unique characteristics, rhodopsin shares its main structural features with the entire GPCR superfamily. Historically, the first direct evidence that, beyond rhodopsin, a seven transmembrane domain topology was actually shared by all GPCRs emerged in the 1980s, when the laboratories of Lefkowitz and Strader cloned the β_2 adrenergic receptor and detected a striking similarity with the hydropathicity profile shown by rhodopsin (Dixon et al. 1986; Dohlman et al. 1987). As Lefkowitz recalled in his personal retrospective, prior to these discoveries, nobody had thought of any possible relationships between rhodopsin and other receptors (Lefkowitz 2007a, b).

We now know that rhodopsin is just one of the members of the very large superfamily of G protein-coupled receptors (Fredriksson et al. 2003) (Gloriam et al. 2007). In humans, the superfamily counts over 800 distinct members that, on the basis of sequence similarity, can be divided into five different families: the glutamate family (G), also known as class C or family III; the rhodopsin family (R), which is by far the largest and is also known as class A or family I; the adhesion family (A) and the secretin family (S), which taken together are also known as class B or family 2; and the frizzled (F)/taste2 family (Fredriksson et al. 2003; Gloriam et al. 2007).

As reviewed by Costanzi et al., rhodopsin remained for several years the only GPCR with experimentally elucidated three-dimensional structures (Costanzi et al. 2009). Beyond providing direct information on the light-activated receptor, these structures also superseded those of rhodopsin as the template of choice for the study of other members of the superfamily, in particular those belonging to class A (Audet and Bouvier 2012; Costanzi et al. 2009). However, in light of the vast pharmaceutical interest revolving around GPCRs, enormous efforts were put toward the crystallization of other members of the superfamily. In 2007, these efforts culminated in a number of breakthroughs that yielded the solution of the X-ray structure of the β_2-adrenergic receptors (Cherezov et al. 2007; Rasmussen et al. 2007; Rosenbaum et al. 2007). Thanks to these advances, the field of GPCR crystallography is now in full blossom and, at the time of a recent survey study conducted by Jacobson and Costanzi during the summer of 2012, the Protein Data Bank (www.rcsb.org) enlisted structures for a total of 73 structures for 15 distinct receptors (Jacobson and Costanzi 2012). At the time of this writing, structures for two additional receptors, namely the NTSR1 neurotensin receptor 1 (NTSR1) (White et al. 2012) and the protease-activated receptor 1 (PAR1) (Zhang et al. 2012), have been published. Moreover, another milestone in the field of GPCR structural studies has been reached with the determination of the structure of the chemokine CXCR1 receptor through nuclear magnetic resonance (NMR) spectroscopy (Park et al. 2012). Given the fast pace of the field, many more structures are very likely to be solved in the near future.

1.2 There Once Was Rhodopsin

Since the 1970s, Rhodopsin has been the object of biochemical investigations intended to unveil its amino acid sequence, facilitated by the fact that the protein could be obtained in high quantities from the retinas of cows (Hargrave 2001). The first complete amino acid sequence of rhodopsin was published in 1982 and 1983 by the laboratories of Ovchinnikov and Hargrave (Hargrave et al. 1983; Ovchinnikov et al. 1982). The hydropathicity profile inferred from the sequence indicated a seven transmembrane topology. Moreover, Hargrave and coworkers had previously determined that the C-terminus of the receptor was located in the cytoplasm. Hence, taken together, these data led to the drawing of the first bi-dimensional model of rhodopsin as a serpentine originating in the extracellular milieu, spanning seven times the plasma membrane with seven α-helical domains connected by three extracellular and three intracellular loops, and terminating in the cytoplasm (Hargrave et al. 1983) – see Fig. 1.1.

The first direct experimental insights into the three-dimensional structure of rhodopsin were provided by Schertler and coworkers, who subjected the receptor to electron crystallography

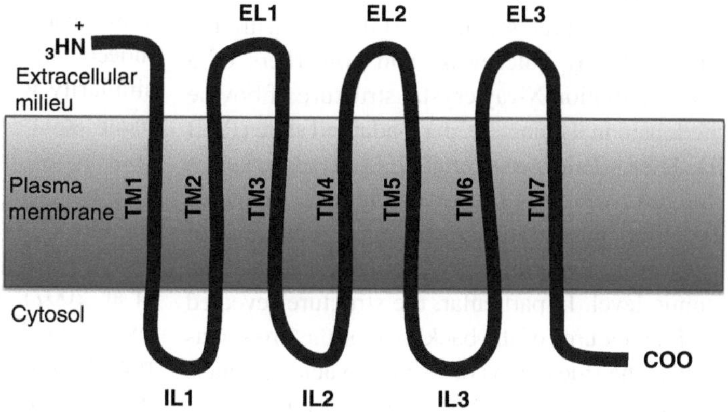

Fig. 1.1 Schematic representation of Rhodopsin as a membrane spanning serpentine. Abbreviations: *TM* transmembrane domain, *IL* intracellular loop, *EL* extracellular loop

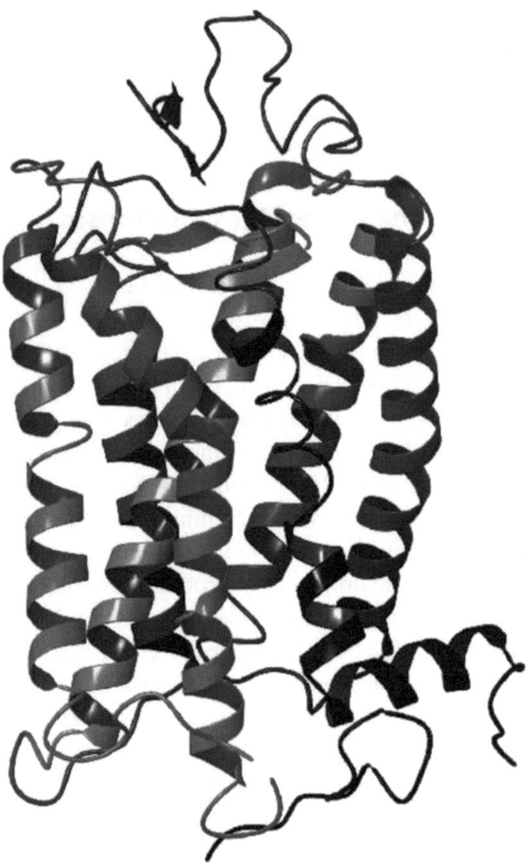

Fig. 1.2 The three-dimensional structure of bovine rhodopsin (PDB ID: 1F88), showing the backbone of the receptor schematically represented as a ribbon

and obtained a 2D projection map (Schertler et al. 1993). On the basis of this map, Baldwin subsequently built a molecular model of the receptor that allocated the individual TMs to the various peaks (Baldwin 1993).

The year 2000 brought another leap forward for the structural studies of GPCRs, with the publication by Palczewski and coworkers of a 2.8 Å resolution X-ray crystal structure of bovine rhodopsin in its inactive, dark-adapted state (PDB ID: 1F88) (Palczewski et al. 2000). Palczewski's structure, which is shown in Fig. 1.2, provided the first three-dimensional representation of rhodopsin experimentally elucidated at the atomic level. In particular, the structure revealed the fine features of the backbone of the protein as well as the side chains of its amino acid residues. Importantly, not only the α-helical bundle, but also part of the termini and almost the entirety of the intracellular and extracellular loops that interconnect the transmembrane domains were visible in the structure. In this regard, the structure unveiled that the second extracellular loop, which connects the fourth and fifth transmembrane domains, in rhodopsin assumes a β-hairpin conformation and obstructs the access to the helical bundle from the extracellular milieu like a closed lid, thus creating a closed binding pocket for retinal. As we discuss later in the article, subsequent X-ray crystallography studies highlighted that the extracellular domains in general, and the second extracellular loop in particular, are highly variable structures that adopt different conformations in different members of the GPCR family.

1.3 2007: The Year of Change

Twenty years after the discovery of the homology between rhodopsin and the β adrenergic receptors (Dixon et al. 1986; Dohlman et al. 1987) and 7 years after the solution of the crystal structure of rhodopsin (Palczewski et al. 2000), crystal structures were solved for members of the β adrenergic receptor family (Fig. 1.3) (Cherezov et al. 2007; Day et al. 2007; Rasmussen et al. 2007; Rosenbaum et al. 2007). This milestone ultimately sanctioned the structural homology between rhodopsin and the β adrenergic receptors and, more in general, the entire superfamily of GPCRs postulated 20 years earlier by Lefkowitz and coworkers (Dixon et al. 1986; Dohlman et al. 1987). Just as the experiments conducted in the 1980s had suggested, the β adrenergic receptors indeed share a very high degree of structural similarity with rhodopsin. In particular, the root mean square deviations (RMSD) of the atomic coordinates between the Cα atoms of the amino acid residues located in the membrane spanning regions of the two receptors amounts to a mere about 2.7 Å (Cherezov et al. 2007; Rosenbaum et al. 2007). Moreover, a significant overlap between the ligands co-crystallized with the two receptors is noticeable (Cherezov et al. 2007; Rosenbaum et al. 2007).

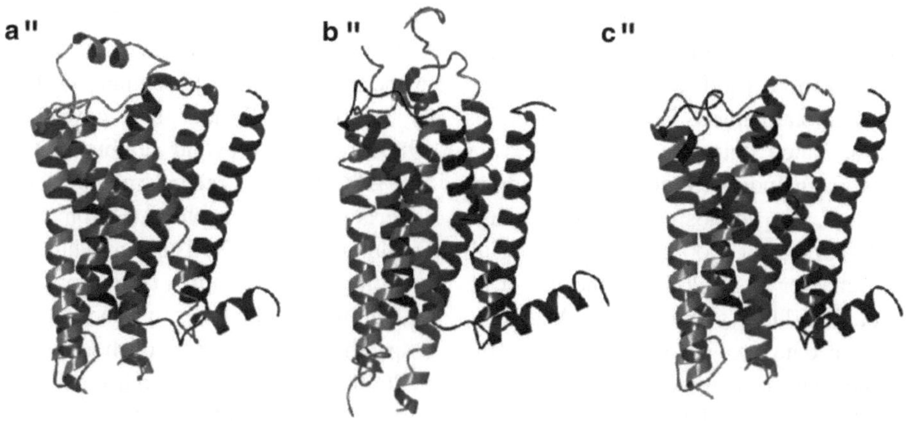

Fig. 1.3 The three-dimensional structures of (**a**) the human β_2-adrenergic receptor (PDB ID: 2RH1) and (**b**) the adenosine A_{2A} receptor (PDB ID: 3EML). (**c**) Our model of the adenosine A_{2A} receptor constructed on the basis of the β_2 adrenergic receptor for the first blind assessment of GPCR modeling and docking (Michino et al. 2009). The figure shows the backbone of the three receptors schematically represented as a ribbon

The solution of the structures of the adrenergic receptors marked the beginning of a new era. The technical expedient that made the solution of these structures attainable, which include the fusion of GPCRs with easily crystallizable proteins, the use of antibodies, and the introduction of stabilizing mutations, provided general tools for the crystallization of other members of the superfamily as well (Hanson and Stevens 2009; Salon et al. 2011; Steyaert and Kobilka 2011; Tate 2012). As a result, in 2008 it became the turn of a member of the adenosine receptor family, namely the A_{2A} receptor, to be solved through X-ray crystallography (Fig. 1.3) (Jaakola et al. 2008). The solution of the structure of various other receptors belonging to the large rhodopsin family rapidly followed, with a continuous flow that keeps making the list of experimentally solved GPCRs longer (Jacobson and Costanzi 2012).

Taken together, the currently solved GPCR structures indicate that the members of the superfamily, or at least those belonging to the rhodopsin family, share a high degree of structural similarity, which is particularly pronounced within the helical bundle. However, it also highlighted the fact that each receptor has its own peculiarity. Specifically, it is now evident that for all the crystallized receptors the ligand binding cavity lies within the heptahelical bundle, near its opening toward the extracellular space. However, the various ligands co-crystallized with their respective receptors occupy different regions of this cavity, more or less overlapping with each other. Moreover, in some cases different ligands were found to bind to the same receptor by exploiting different regions of the binding cavity. This situation is particularly evident when comparing the binding of a small molecule antagonist and a much larger cyclic peptide antagonist to the CXCR4 chemokine receptor, which show a complete lack of overlap. For pictorial views of the ligand binding cavities of the crystallized GPCRs and a comparison of ligand binding modes, see a recent review by Jacobson and Costanzi (2012).

It is also now evident that the extracellular regions assume different conformations in the various receptor families. This high variability is also found in the second extracellular loop, which in most receptors is the extracellular domain that shows more contacts with the bound ligand. Notably, however, all the structures of receptors naturally activated by peptides that were solved to date, including chemokine, opioid, and nociceptin receptors, featured a very similar second extracellular loop, which adopted a particularly solvent exposed β-hairpin conformation that kept the binding cavity wide open and ready for the binding of large ligands (Granier et al. 2012; Manglik et al. 2012; Thompson et al. 2012; Wu et al. 2010, 2012).

Most of the crystal structure of GPCRs have been solved solely in the inactive state, which is believed to be more rigid and, therefore, more amenable to crystallization. However, structures that reflect activated states have been solved for some receptors, including rhodopsin, the β adrenergic receptors, the adenosine receptors (Choe et al. 2011; Park et al. 2008; Rasmussen et al. 2011a, b; Scheerer et al. 2008; Standfuss et al. 2011; Xu et al. 2011). As reviewed by Audet and Bouvier, these data provide significant insights into the molecular mechanisms at the basis of the activation of GPCRs (Audet and Bouvier 2012). Most notable is the structure that was solved by Kobilka and coworkers for the β2 adrenergic receptor in its activated state, in complex with an agonist and an activated G protein heterotrimer (Rasmussen et al. 2011b). This structure, which captures a GPCR in the act of signaling, arguably represents the current pinnacle of the structural studies of GPCRs.

1.4 Assessing the Accuracy of Homology Models in Light of the Corresponding Crystal Structures

Besides unveiling the three-dimensional topology of the protein for which they have been determined, experimental structures also provide templates for the construction of homology models of homologous proteins. This application is particularly prominent in the field of GPCRs, given the size of the superfamily, the pharmaceutical relevance of its members and the paucity of direct structural information. Prior to the solution of the structures of rhodopsin, earlier GPCR modeling studies were based on the structure of bacteriorhodopsin, the structure of which was first obtained in the mid 1970s through electron microscopy (Ballesteros and Weinstein 1992; Henderson and Unwin 1975; IJzerman et al. 1992, 1994; Oliveira et al. 1993; Pardo et al. 1992). Bacteriorhodopsin, however is not a GPCR but a proton pump. Moreover, despite featuring seven transmembrane domains, it shows a significantly different three-dimensional topology (Audet and Bouvier 2012; Okada and Palczewski 2001). Conversely, as we mentioned in the introduction, the seminal work conducted by Lefkowitz, Strader and coworkers in the 1980s revealed beyond doubts that rhodopsin and GPCRs were indeed members of a single superfamily of receptors (Dixon et al. 1986; Dohlman et al. 1987). Thus, in light of this proven homology, the structures of rhodopsin published by Baldwin on the basis of Schertler's projection maps, the X-ray structure of rhodopsin published in 2000 and a number of additional subsequently solved X-ray structures of the same receptor, became the templates of choice for the modeling of GPCRs and maintained this status for several years, until the structures of additional members of the superfamily became available (Audet and Bouvier 2012; Costanzi et al. 2009).

Currently, due to the explosion of GPCR crystallography, a variety of templates are available for the construction of GPCR homology models. Ultimately, the success of any homology modeling effort rests in the use of a suitable template and the accuracy of the sequence alignment. The presence of high sequence similarity and the detection of shared features such as common sequence motifs, putatively shared disulfide bridges and a similar distribution of proline and glycine residues, which notoriously affects the shape of α-helices, are criteria commonly used for template selection. Beyond the selection of single templates, the use of multiple templates to model different regions of a target receptor have been proposed as well (Costanzi 2012; Mobarec et al. 2009; Worth et al. 2009). For a detailed methodological article on the procedures that typically followed by our group for the construction of homology modeling of GPCRs, the reader is referred to an article recently published in the journal "Methods in Molecular Biology" (Costanzi 2012).

For many years a direct and straightforward analysis of the accuracy of GPCR models was not possible, since experimental structures were available only for rhodopsin. However, such comparisons became suddenly possible with the availability of multiple templates yielded by the breakthroughs of 2007. In fact, it became possible to build a model of an experimentally-solved receptor on the basis of a different template, to then compare the similarity between the model and the experimental structure.

In this context, our group studied the feasibility of the construction of molecular models of GPCRs in complex with their ligands through homology modeling followed by molecular docking (Costanzi 2010). In particular, soon after the unveiling of the crystal structure of the β_2 adrenergic receptor in complex with a potent inverse agonist, a comparison of this experimental structure with models of the same receptor-ligand complex obtained through rhodopsin-based homology modeling followed by molecular docking was published (Costanzi 2008). The encouraging results of these comparative studies were confirmed by our subsequent success in the first blind assessment of GPCR modeling and docking, which was conducted in concomitance with the solution of the crystal structure of the adenosine A_{2A} receptors (Fig. 1.3) (Jaakola et al. 2008; Michino et al. 2009). This controlled assessment was organized to gage how closely receptor-ligand complexes obtained through homology modeling and molecular docking would resemble the experimental structure itself. Importantly, the assessment was conducted in a blind manner, i.e. the models were submitted to the evaluators before the unveiling of the experimental structure. Taken together, our models of the β_2 adrenergic receptor (Costanzi 2008) as well as the models of the adenosine A_{2A} receptor submitted to the blind assessment (Michino et al. 2009) illustrated that reasonably accurate models of the complexes can be constructed. The most accurate models resulted to be those that incorporated experimental derived from site-directed mutagenesis, with average root mean square deviations (RMSD) from the atomic coordinates of the experimental structures ranging from 1.7 Å (for the β_2 adrenergic receptor) to 2.8 Å (for the adenosine A_{2A} receptor) for the ligands and 2.7 Å (for the β_2 adrenergic receptor) to 3.4 Å (for the adenosine A_{2A} receptor) for the residues surrounding the binding pocket. When the models are built for the study of receptor-ligand interactions of for drug-discovery purposes, it is of particular importance that the side chains of the residues lining the binding cavity be modeled with accuracy (Costanzi 2010; Mobarec et al. 2009). Thus, whenever the rotameric state of a residue thought to be in close proximity to the ligand cannot be guessed from the template, the identification and incorporation of experimentally derived constraints that can guide the selection of the conformation of its side chain may yield a markedly more accurate model.

Concerning the overall structure of the receptor, at the level of the transmembrane domains, which demonstrated a high degree of structural conservation, homology models usually show a significant structural accuracy, with average RMSD values below 3 Å. Conversely, they are substantially less accurate at the level of the extracellular and intracellular domains, especially the long ones. This is not surprising, in light of the low degree of sequence and structural conservation shown by these regions. As mentioned, the second extracellular loop is often of particular importance for ligand recognition, since it lines the interhelical binding cavity for many GPCRs. The homology modeling of this domain, which assumes different conformations in different receptor families, is not always feasible. However, the modeling process is assisted by the fact that, in most GPCRs, this domain is endowed with a characteristic disulfide bridge that links it to TM3. The portion of the loop that shows the most contacts with the ligand is usually the segment downstream of this bridge. This is typically a relatively short stretch of residues, and hence, is suited for *de novo* modeling, which typically yields more accurate results than homology modeling. Particularly encouraging results in this direction have been obtained by Friesner and coworkers, who demonstrated the possibility of reconstructing through *de novo* modeling loops previously expunged from GPCR structures (Goldfeld et al. 2011, 2012).

A second blind assessment intended to probe the status of GPCR modeling and docking was organized in 2010, in coordination with the experimental elucidation of the structures of the dopamine D3 receptor and the chemokine CXCR4 receptor through X-ray crystallography

(Chien et al. 2010; Kufareva et al. 2011; Wu et al. 2010). This exercise highlighted that the accuracy of the predictions depend dramatically on the similarity of the target receptor with the templates. In particular, while the structures of receptors that share a substantial sequence similarity with the template used for their modeling are generally endowed with a significant level of accuracy, predicting the structures of receptors that are more distant from the available templates is a particularly difficult problem, especially in the absence of experimentally derived constraints (Kufareva et al. 2011).

The ability of experimental structures and homology models of GPCRs to serve as a platform for virtual screening campaigns intended to identify active compounds out of large databases of molecules has been probed through a number of controlled experiments that measured the ability of these structures to retrieve a pool of known ligands dispersed within a larger set of inactive compounds, a non-comprehensive list of which is provided in the reference (Cavasotto 2011; Cavasotto et al. 2008; Katritch et al. 2010; Phatak et al. 2010; Vilar et al. 2010, 2011) – for a review of the application of virtual screening to GPCRs, see Chapter 18 of the book "G Protein-Coupled Receptors: From Structure to Function" published by the Royal Society of Chemistry (Costanzi 2011). These studies highlighted that accurate models are indeed capable of prioritizing binders versus non-binders, albeit not as well as crystal structures. Not surprisingly, the models tended to perform according to their level of structural accuracy. For instance, a study conducted by Katrich and coworkers on the basis of the A_{2A} models submitted to the blind assessment revealed very good performance levels for the three most structurally accurate models (Costanzi, Katritch/Abagyan and Lam/Abagyan) (Katritch et al. 2010). Conversely, less accurate models performed very poorly.

Not surprisingly, virtual screening campaigns based on homology models seem to perform better when the models are built on the basis of closely rather than distantly related templates. Accordingly, a recent study conducted by Shoichet and coworkers revealed that a homology model of the dopamine D_3 receptor constructed on the basis of a close homologue, namely the β_2 adrenergic receptor, was as effective in a prospective virtual screening as the crystal structure of the same receptor (Carlsson et al. 2011). Conversely, a model of the CXCR4 receptor built on the basis of a more distant homologue performed poorly in comparison to the crystal structure of the same receptor (Mysinger et al. 2012).

1.5 Summary and Conclusions

Due to the pharmaceutical interest that revolves around them, G protein-coupled receptors (GPCRs) have always been the object of intense efforts toward their structural determination. However, until relatively recently their structures proved to be elusive. In this context, rhodopsin has been for years the only GPCR with an experimentally solved structure and has been heavily used to construct models of the other members of the superfamily through a technique known as homology modeling. Thanks to a number of technical and scientific advancements, now the solution of GPCR structures through X-ray crystallography and, more recently, NMR is more feasible. Thus, structures of several GPCRs have been recently disclosed. These structures, besides providing direct structural information on the receptor in question, also provide a larger basis of templates for the construction of more accurate models of those receptors that have not yet been experimentally solved yet. The availability of multiple templates has also offered the possibility of performing accurate studies intended to probe the strengths and the limits of homology modeling and molecular docking applied to homology models. Moreover, controlled virtual screening experiments have been conducted to investigate the extent of the applicability of homology models to molecular recognition campaigns and to compare it to that shown by experimental structures. The current pace of structural biology and the rapid advancements of computational chemistry suggest a future landscape populated by increasingly more numerous experimental and theoretical GPCR structures. The former and the latter, especially when supported by experimental data and built on the basis of closely related templates, will provide a solid basis for the rational discovery of novel modulators of GPCR activity.

References

Audet M, Bouvier M (2012) Restructuring G-protein-coupled receptor activation. Cell 151:14–23

Baldwin J (1993) The probable arrangement of the helices in G protein-coupled receptors. EMBO J 12:1693–1703

Ballesteros JA, Weinstein H (1992) Analysis and refinement of criteria for predicting the structure and relative orientations of transmembranal helical domains. Biophys J 62:107–109

Carlsson J, Coleman RG, Setola V, Irwin JJ, Fan H, Schlessinger A, Sali A, Roth BL, Shoichet BK (2011) Ligand discovery from a dopamine D(3) receptor homology model and crystal structure. Nat Chem Biol 7:769–778

Cavasotto CN (2011) Homology models in docking and high-throughput docking. Curr Top Med Chem 11:1528–1534

Cavasotto CN, Orry AJ, Murgolo NJ, Czarniecki MF, Kocsi SA, Hawes BE, O'Neill KA, Hine H, Burton MS, Voigt JH et al (2008) Discovery of novel chemotypes to a G-protein-coupled receptor through ligand-steered homology modeling and structure-based virtual screening. J Med Chem 51:581–588

Cherezov V, Rosenbaum D, Hanson M, Rasmussen S, Thian F, Kobilka T, Choi H, Kuhn P, Weis W, Kobilka B et al (2007) High-resolution crystal structure of an engineered human beta2-adrenergic G protein-coupled receptor. Science 318:1258–1265

Chien EY, Liu W, Zhao Q, Katritch V, Han GW, Hanson MA, Shi L, Newman AH, Javitch JA, Cherezov V et al (2010) Structure of the human dopamine D3 receptor in complex with a D2/D3 selective antagonist. Science 330:1091–1095

Choe HW, Kim YJ, Park JH, Morizumi T, Pai EF, Krauss N, Hofmann KP, Scheerer P, Ernst OP (2011) Crystal structure of metarhodopsin II. Nature 471:651–655

Congreve M, Marshall F (2009) The impact of GPCR structures on pharmacology and structure-based drug design. Br J Pharmacol 159:986–996

Congreve M, Langmead CJ, Mason JS, Marshall FH (2011) Progress in structure based drug design for G protein-coupled receptors. J Med Chem 54:4283–4311

Costanzi S (2008) On the applicability of GPCR homology models to computer-aided drug discovery: a comparison between in silico and crystal structures of the beta2-adrenergic receptor. J Med Chem 51: 2907–2914

Costanzi S (2010) Modeling G protein-coupled receptors: a concrete possibility. Chim Oggi 28:26–31

Costanzi S (2011) Chapter 18. Structure-based virtual screening for ligands of G protein-coupled receptors. In: Giraldo J, Pin J-P (eds) G protein-coupled receptors: from structure to function. The Royal Society of Chemistry, London and Cambridge, pp 359–374

Costanzi S (2012) Homology modeling of class a G protein-coupled receptors. Methods Mol Biol 857:259–279

Costanzi S, Siegel J, Tikhonova I, Jacobson K (2009) Rhodopsin and the others: a historical perspective on structural studies of G protein-coupled receptors. Curr Pharm Des 15:3994–4002

Day P, Rasmussen S, Parnot C, Fung J, Masood A, Kobilka T, Yao X, Choi H, Weis W, Rohrer D et al (2007) A monoclonal antibody for G protein-coupled receptor crystallography. Nat Methods 4: 927–929

Dixon R, Kobilka B, Strader D, Benovic J, Dohlman H, Frielle T, Bolanowski M, Bennett C, Rands E, Diehl R et al (1986) Cloning of the gene and cDNA for mammalian beta-adrenergic receptor and homology with rhodopsin. Nature 321:75–79

Dohlman H, Bouvier M, Benovic J, Caron M, Lefkowitz R (1987) The multiple membrane spanning topography of the beta 2-adrenergic receptor. Localization of the sites of binding, glycosylation, and regulatory phosphorylation by limited proteolysis. J Biol Chem 262:14282–14288

Donati RJ, Rasenick MM (2003) G protein signaling and the molecular basis of antidepressant action. Life Sci 73:1–17

Fredriksson R, Lagerström M, Lundin L, Schiöth H (2003) The G-protein-coupled receptors in the human genome form five main families. Phylogenetic analysis, paralogon groups, and fingerprints. Mol Pharmacol 63:1256–1272

Gloriam D, Fredriksson R, Schiöth H (2007) The G protein-coupled receptor subset of the rat genome. BMC Genomics 8:338

Goldfeld DA, Zhu K, Beuming T, Friesner RA (2011) Successful prediction of the intra- and extracellular loops of four G-protein-coupled receptors. Proc Natl Acad Sci U S A 108:8275–8280

Goldfeld DA, Zhu K, Beuming T, Friesner RA (2012) Loop prediction for a GPCR homology model: algorithms and results. Proteins 81:214–228

Granier S, Manglik A, Kruse AC, Kobilka TS, Thian FS, Weis WI, Kobilka BK (2012) Structure of the delta-opioid receptor bound to naltrindole. Nature 485:400–404

Hanson MA, Stevens RC (2009) Discovery of new GPCR biology: one receptor structure at a time. Structure 17:8–14

Hargrave PA (2001) Rhodopsin structure, function, and topography the Friedenwald lecture. Invest Ophthalmol Vis Sci 42:3–9

Hargrave P, McDowell J, Curtis D, Wang J, Juszczak E, Fong S, Rao J, Argos P (1983) The structure of bovine rhodopsin. Biophys Struct Mech 9:235–244

Henderson R, Unwin PN (1975) Three-dimensional model of purple membrane obtained by electron microscopy. Nature 257:28–32

IJzerman A, Van Galen P, Jacobson K (1992) Molecular modeling of adenosine receptors. I. The ligand binding site on the A1 receptor. Drug Des Discov 9:49–67

IJzerman AP, van der Wenden EM, Van Galen PJ, Jacobson KA (1994) Molecular modeling of adenosine receptors. The ligand binding site on the rat adenosine A2A receptor. Eur J Pharmacol 268:95–104

Jaakola V, Griffith M, Hanson M, Cherezov V, Chien E, Lane J, Ijzerman A, Stevens R (2008) The 2.6 angstrom crystal structure of a human A2A adenosine receptor bound to an antagonist. Science 322: 1211–1217

Jacobson KA, Costanzi S (2012) New insights for drug design from the x-ray crystallographic structures of g-protein-coupled receptors. Mol Pharmacol 82: 361–371

Jacobson KA, Deflorian F, Mishra S, Costanzi S (2011) Pharmacochemistry of the platelet purinergic receptors. Purinergic Signal 7:305–324

Katritch V, Rueda M, Lam P, Yeager M, Abagyan R (2010) GPCR 3D homology models for ligand screening: lessons learned from blind predictions of adenosine A2a receptor complex. Proteins 78:197–211

Kufareva I, Rueda M, Katritch V, Stevens RC, Abagyan R (2011) Status of GPCR modeling and docking as reflected by community-wide GPCR Dock 2010 assessment. Structure 19:1108–1126

Lefkowitz RJ (2007a) Seven transmembrane receptors: a brief personal retrospective. Biochim Biophys Acta 1768:748–755

Lefkowitz RJ (2007b) Seven transmembrane receptors: something old, something new. Acta Physiol (Oxf) 190:9–19

Manglik A, Kruse AC, Kobilka TS, Thian FS, Mathiesen JM, Sunahara RK, Pardo L, Weis WI, Kobilka BK, Granier S (2012) Crystal structure of the mu-opioid receptor bound to a morphinan antagonist. Nature 485:321–326

Mason JS, Bortolato A, Congreve M, Marshall FH (2012) New insights from structural biology into the druggability of G protein-coupled receptors. Trends Pharmacol Sci 33:249–260

Michino M, Abola E, 2008 Participants G, Brooks Cr, Dixon J, Moult J, Stevens R (2009) Community-wide assessment of GPCR structure modelling and ligand docking: GPCR Dock 2008. Nat Rev Drug Discov 8:455–463

Mobarec J, Sanchez R, Filizola M (2009) Modern homology modeling of G-protein coupled receptors: which structural template to use? J Med Chem 52:5207–5216

Mysinger MM, Weiss DR, Ziarek JJ, Gravel S, Doak AK, Karpiak J, Heveker N, Shoichet BK, Volkman BF (2012) Structure-based ligand discovery for the protein-protein interface of chemokine receptor CXCR4. Proc Natl Acad Sci U S A 109:5517–5522

Okada T, Palczewski K (2001) Crystal structure of rhodopsin: implications for vision and beyond. Curr Opin Struct Biol 11:420–426

Oliveira L, Paiva ACM, Vriend G (1993) A common motif in G-protein-coupled 7 transmembrane helix receptors. J Comput Aided Mol Des 7:649–658

Ovchinnikov Y, Abdulaev N, Feigina M, Artamonov I, Zolotarev A, Kostina M, Bogachuk A, Miroshnikov A, Martinov V, Kudelin A (1982) The complete amino-acid-sequence of visual rhodopsin. Bioorganicheskaya Khimiya 8:1011–1014

Overington JP, Al-Lazikani B, Hopkins AL (2006) How many drug targets are there? Nat Rev Drug Discov 5:993–996

Palczewski K, Kumasaka T, Hori T, Behnke C, Motoshima H, Fox B, Le Trong I, Teller D, Okada T, Stenkamp R et al (2000) Crystal structure of rhodopsin: a G protein-coupled receptor. Science 289:739–745

Pardo L, Ballesteros JA, Osman R, Weinstein H (1992) On the use of the transmembrane domain of bacteriorhodopsin as a template for modeling the three-dimensional structure of guanine nucleotide-binding regulatory protein-coupled receptors. Proc Natl Acad Sci U S A 89:4009–4012

Park JH, Scheerer P, Hofmann KP, Choe HW, Ernst OP (2008) Crystal structure of the ligand-free G-protein-coupled receptor opsin. Nature 454:183–187

Park SH, Das BB, Casagrande F, Tian Y, Nothnagel HJ, Chu M, Kiefer H, Maier K, De Angelis AA, Marassi FM et al (2012) Structure of the chemokine receptor CXCR1 in phospholipid bilayers. Nature 491: 779–783

Phatak SS, Gatica EA, Cavasotto CN (2010) Ligand-steered modeling and docking: a benchmarking study in class a G-protein-coupled receptors. J Chem Inf Model 50:2119–2128

Pierce K, Premont R, Lefkowitz R (2002) Seven-transmembrane receptors. Nat Rev Mol Cell Biol 3:639–650

Rasmussen S, Choi H, Rosenbaum D, Kobilka T, Thian F, Edwards P, Burghammer M, Ratnala V, Sanishvili R, Fischetti R et al (2007) Crystal structure of the human beta2 adrenergic G-protein-coupled receptor. Nature 450:383–387

Rasmussen SG, Choi HJ, Fung JJ, Pardon E, Casarosa P, Chae PS, Devree BT, Rosenbaum DM, Thian FS, Kobilka TS et al (2011a) Structure of a nanobody-stabilized active state of the beta(2) adrenoceptor. Nature 469:175–180

Rasmussen SG, DeVree BT, Zou Y, Kruse AC, Chung KY, Kobilka TS, Thian FS, Chae PS, Pardon E, Calinski D et al (2011b) Crystal structure of the beta2 adrenergic receptor-Gs protein complex. Nature 477:549–555

Rosenbaum D, Cherezov V, Hanson M, Rasmussen S, Thian F, Kobilka T, Choi H, Yao X, Weis W, Stevens R et al (2007) GPCR engineering yields high-resolution structural insights into beta2-adrenergic receptor function. Science 318:1266–1273

Rosskopf D, Michel MC (2008) Pharmacogenomics of G protein-coupled receptor ligands in cardiovascular medicine. Pharmacol Rev 60:513–535

Salon JA, Lodowski DT, Palczewski K (2011) The significance of G protein-coupled receptor crystallography for drug discovery. Pharmacol Rev 63:901–937

Scheerer P, Park JH, Hildebrand PW, Kim YJ, Krauss N, Choe HW, Hofmann KP, Ernst OP (2008) Crystal structure of opsin in its G-protein-interacting conformation. Nature 455:497–502

Schertler G, Villa C, Henderson R (1993) Projection structure of rhodopsin. Nature 362:770–772

Spetea M, Schmidhammer H (2012) Recent advances in the development of 14-alkoxy substituted morphinans as potent and safer opioid analgesics. Curr Med Chem 19:2442–2457

Standfuss J, Edwards PC, D'Antona A, Fransen M, Xie G, Oprian DD, Schertler GF (2011) The structural basis of agonist-induced activation in constitutively active rhodopsin. Nature 471:656–660

Steyaert J, Kobilka BK (2011) Nanobody stabilization of G protein-coupled receptor conformational states. Curr Opin Struct Biol 21:567–572

Tate CG (2012) A crystal clear solution for determining G-protein-coupled receptor structures. Trends Biochem Sci 37:343–352

Thompson AA, Liu W, Chun E, Katritch V, Wu H, Vardy E, Huang XP, Trapella C, Guerrini R, Calo G et al (2012) Structure of the nociceptin/orphanin FQ receptor in complex with a peptide mimetic. Nature 485:395–399

Vilar S, Ferino G, Phatak SS, Berk B, Cavasotto CN, Costanzi S (2010) Docking-based virtual screening for ligands of G protein-coupled receptors: Not only crystal structures but also in silico models. J Mol Graph Model. doi:10.1016/j.jmgm.2010.1011.1005

Vilar S, Karpiak J, Berk B, Costanzi S (2011) In silico analysis of the binding of agonists and blockers to the beta2-adrenergic receptor. J Mol Graph Model 29:809–817

White JF, Noinaj N, Shibata Y, Love J, Kloss B, Xu F, Gvozdenovic-Jeremic J, Shah P, Shiloach J, Tate CG et al (2012) Structure of the agonist-bound neurotensin receptor. Nature 490:508–513

Worth C, Kleinau G, Krause G (2009) Comparative sequence and structural analyses of G-protein-coupled receptor crystal structures and implications for molecular models. PLoS One 4:e7011

Wu B, Chien EY, Mol CD, Fenalti G, Liu W, Katritch V, Abagyan R, Brooun A, Wells P, Bi FC et al (2010) Structures of the CXCR4 chemokine GPCR with small-molecule and cyclic peptide antagonists. Science 330:1066–1071

Wu H, Wacker D, Mileni M, Katritch V, Han GW, Vardy E, Liu W, Thompson AA, Huang XP, Carroll FI et al (2012) Structure of the human kappa-opioid receptor in complex with JDTic. Nature 485:327–332

Xu F, Wu H, Katritch V, Han GW, Jacobson KA, Gao ZG, Cherezov V, Stevens RC (2011) Structure of an agonist-bound human A2A adenosine receptor. Science 332:322–327

Zhang C, Srinivasan Y, Arlow DH, Fung JJ, Palmer D, Zheng Y, Green HF, Pandey A, Dror RO, Shaw DE et al (2012) High-resolution crystal structure of human protease-activated receptor 1. Nature 492: 387–392

Modeling of G Protein-Coupled Receptors Using Crystal Structures: From Monomers to Signaling Complexes

2

Angel Gonzalez, Arnau Cordomí, Minos Matsoukas, Julian Zachmann, and Leonardo Pardo

Abstract

G protein–coupled receptors constitute a large and functionally diverse family of transmembrane proteins. They are fundamental in the transfer of extracellular stimuli to intracellular signaling pathways and are among the most targeted proteins in drug discovery. Recent advances in crystallization methods have permitted to resolve the molecular structure of several members of the family. This chapter focuses on the impact of these structures in the use of homology modeling techniques for building three-dimensional models of homologous G protein–coupled receptors, higher order oligomers, and their complexes with ligands and signaling proteins.

Keywords

Transmembrane helices • Homology modeling • Conformation • Ligand binding • Activation mechanism

2.1 Introduction

Membrane receptors coupled to guanine nucleotide-binding proteins (commonly known as G protein-coupled receptors, GPCRs) comprise one of the widest and most adaptable families of cellular sensors, as they are able to mediate a wide range of transmembrane signal transduction processes (Kristiansen 2004). GPCRs are present in almost every eukaryotic organism, including fungi and plants. They are highly diversified in mammalian genomes with current estimates of about 1,000 genes (2–3 % of the human proteome) (Fredriksson and Schioth 2005). GPCRs transduce sensory signals of external origin such as odors, pheromones, or tastes; and endogenous signals such as neurotransmitters, (neuro)peptides, proteases, glycoprotein hormones, purine ligands and ions, among others. The response is operated through second messenger cascades controlled by different heterotrimeric guanine nucleotide-binding proteins (G-proteins) coupled at their intracellular regions (Oldham and Hamm 2008). Due to their relevance to cellular physiology (Smit et al. 2007) and their accessibility from the extracellular environment, membrane proteins

A. Gonzalez • A. Cordomí • M. Matsoukas • J. Zachmann • L. Pardo (✉)
Laboratori de Medicina Computacional, Unitat de Bioestadística, Facultat de Medicina, Universitat Autònoma de Barcelona, 08193 Bellaterra, Spain
e-mail: Leonardo.Pardo@uab.es

M. Filizola (ed.), *G Protein-Coupled Receptors - Modeling and Simulation*, Advances in Experimental Medicine and Biology 796, DOI 10.1007/978-94-007-7423-0_2,

represent a significant portion of therapeutic drug targets (Arinaminpathy et al. 2009; Imming et al. 2006).

2.2 The Structure of G Protein-Coupled Receptors

Significant advances in crystallization of GPCRs (Day et al. 2007; Serrano-Vega et al. 2008) have permitted to elucidate the crystal structures of many receptors (Table 2.1) (see (Katritch et al. 2012, 2013) for recent reviews). All these structures share the common architecture of seven plasma membrane-spanning (or transmembrane) domains (TMs, which also terms this family of proteins as 7TM receptors) connected to each other with three extracellular (ECL) and three intracellular loops (ICL), a disulphide bridge between ECL 2 and TM 3, and a cytoplasmic C–terminus containing an α-helix (Hx8) parallel to the cell membrane. In addition, GPCRs contain an extracellular N-terminal region (N-terminus) and an intracellular C-terminal tail (C-tail).

2.3 Homology Modeling of G Protein-Coupled Receptors

Because of the limited high-resolution structural information on GPCRs, computational techniques to predict their structure from the amino acid sequence are a valuable tool (Pieper et al. 2013). Recently, de novo techniques using evolutionary constraints have been applied to predict 3D structures of TM proteins (Hopf et al. 2012). However, homology models of proteins with unknown experimental structure can also be built from homologous proteins of known structure and similar sequence (templates). This method is based on the fact that in homologous proteins, structure is more conserved than sequence. Thus, in general, homologous proteins with a sequence identity above 35 % have a similar 3D structure (Krissinel and Henrick 2004). Because membrane proteins contain only two types of folds in their TM domains, α-helix bundles and β-barrels, a significant set of membrane proteins maintains a strong conservation of the TM structure even at low sequence identity (<20 %) (Olivella et al. 2013).

The GPCR family is not an exception. All crystal structures preserve analogous secondary/tertiary structures at the seven-helical-bundle domain (Fig. 2.1) despite the percentage of sequence identity in the TM segments is very low (Mobarec et al. 2009; Gonzalez et al. 2012). Structure conservation in the GPCR family is associated, in contrast to other proteins, to the presence of at least one highly conserved amino acid in each helix (Mirzadegan et al. 2003): N in TM1 (present in 98 % of the sequences), D in TM2 (93 %), R in TM3 (95 %), W in TM4 (96 %), P in TM5 (76 %), P in TM6 (98 %), and P in TM7 (93 %). This feature was used by Ballesteros and Weinstein (1995) to define a general numbering scheme consisting of two numbers: the first (1 through 7) corresponds to the helix in which the amino acid of interest is located; the second indicates its position relative to the most conserved residue in the helix, arbitrarily assigned to 50. Significantly, the position of these highly conserved amino acids in each helix is the same in the superimposition of the currently available crystal structures (Fig. 2.1). This finding validates the use of these amino acids as reference points in TM sequence alignments (instead of the common procedure of using substitution matrices and fast sequence similarity search algorithms) (see red box in Fig. 2.2), and in the construction of homology models of GPCRs with unknown structure (de la Fuente et al. 2010; Blattermann et al. 2012).

2.4 The Conformation of Transmembrane Helices in G Protein-Coupled Receptors

Figure 2.1 shows the superimposition of the TM domain of representative crystal structures. Clearly, the structure of the cytoplasmic part is highly conserved. This structural conservation correlates with the fact that most conserved residues are clustered in the central and

Table 2.1 Crystal structures of G protein coupled receptors

Receptor	Ligand	PDB	Reference
Rhodopsin			
Bovine Rhodopsin (bRho)	11-*cis* retinal	1F88, 1GZM	Palczewski et al. (2000) and Li et al. (2004)
Squid Rhodopsin (sRho)	11-*cis* retinal	2Z73	Murakami and Kouyama (2008)
Opsin		3CAP	Park et al. (2008)
Opsin + transducin peptide		3DQB	Scheerer et al. (2008)
Constitutively active rhodopsin		2X72	Standfuss et al. (2011)
Metarhodopsin II	11-*trans* retinal	3PXO	Choe et al. (2011)
Metarhodopsin II + transducin peptide	11-*trans* retinal	3PQR	Choe et al. (2011)
Biogenic amine receptors			
β_1-adrenergic (β_1AR)	Cyanopindolol	2VT4	Warne et al. (2008)
β_1AR	Isoprenaline	2Y03	Warne et al. (2011)
β_1AR homo-oligomer		4GPO	Huang et al. (2013)
β_2-adrenergic (β_2AR)	Carazolol	2RH1	Cherezov et al. (2007) and Rosenbaum et al. (2007)
β_2AR + nanobody	BI-167107	3POG	Rasmussen et al. (2011a)
β_2AR + Gs	BI-167107	3SN6	Rasmussen et al. (2011b)
Dopamine D3 (D3R)	Eticlopride	3PBL	Chien et al. (2010)
Histamine H_1 (H_1R)	Doxepin	3RZE	Shimamura et al. (2011)
Muscarinic M_2 (M_2R)	3-quinuclidinyl-benzilate	3UON	Haga et al. (2012)
Muscarinic M_3 (M_3R)	Tiotropium	4DAJ	Kruse et al. (2012)
Serotonin $5HT_{1B}$ ($5HT_{1B}$R)	Ergotamine	4IAR	Wang et al. (2013a)
Serotonin $5HT_{1B}$ ($5HT_{2B}$R)	Ergotamine	4IB4	Wacker et al. (2013)
Nucleotide			
Adenosine A_{2A} (A_{2A}R)	ZM241385	3EML	Jaakola et al. (2008)
A_{2A}R	UK-432097	3QAK	Xu et al. (2011)
A_{2A}R + Na^+	ZM241385	4EIY	Liu et al. (2012)
Peptide receptors			
CXCR4	CVX15	3OE0	Wu et al. (2010)
CXCR4	IT1t	3ODU	Wu et al. (2010)
μ-opioid (μ-OR)	β-funaltrexamine	4DKL	Manglik et al. (2012)
κ-opioid (κ-OR)	JDTic	4DJH	Wu et al. (2012)
δ-opioid (δ-OR)	Naltrindole	4EJ4	Granier et al. (2012)
Nociceptin/orphanin FQ	C-24	4EA3	Thompson et al. (2012)
Neurotensin1 (NTSR1)	Neurotensin (8–13)	4GRV	White et al. (2012)
Protease-activated receptor 1 (PAR1)	Vorapaxar	3VW7	Zhang et al. (2012)
Lipid			
Sphingosine S1P ($S1P_1$R)	ML056	3V2Y	Hanson et al. (2012)
Frizzled (class F)			
Smoothened (SMO)	LY2940680	4JKV	Wang et al. (2013b)

intracellular regions of the receptor (Mirzadegan et al. 2003). In contrast, there is a low degree of sequence conservation among different GPCRs at their extracellular domains. Accordingly, the structure of the extracellular part of TM helices is more divergent. We have previously suggested that GPCRs, during their evolution, have evolved to adjust the structural characteristics of their cognate ligands, by customizing a preserved scaffold (7TM receptors) through *conformational plasticity* (Deupi et al. 2007). We use this term to describe the structural differences

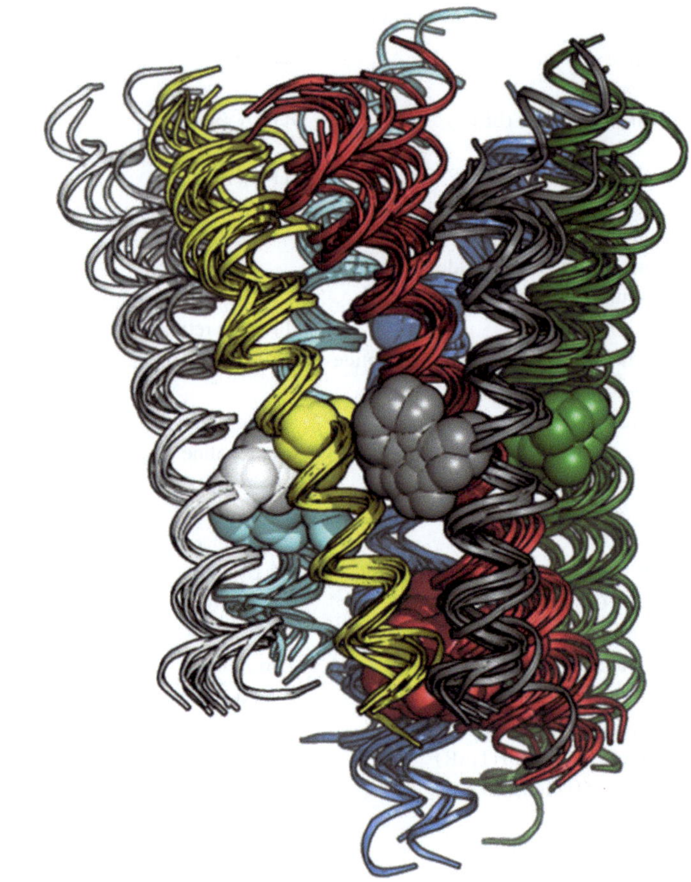

Fig. 2.1 Comparison of the TM bundle of GPCRs. The structures of bRho (PDB code 1U19), β_2R (2RH1), D_3R (3PBL), H_1R (3RZE), M_2R, $5HT_{1b}$R (4IAR), A_{2A}R (4EIY), CXCR4 (3ODU), μOR (4DKL), NTSR1 (4GRV), PAR1 (3VW7), and $S1P_1$R (3V2Y) are shown. The colour code of the helices is TMs 1 in *white*, 2 in *yellow*, 3 in *red*, 4 in *grey*, 5 in *green*, 6 in *dark blue*, and 7 in *light blue*. The highly conserved N1.50 (in *white*), D2.50 (in *yellow*), R3.50 (in *red*), W4.50 (in *grey*), P5.50 (in *green*), P6.50 (in *dark blue*), and P7.50 (in *light blue*) are shown as spheres

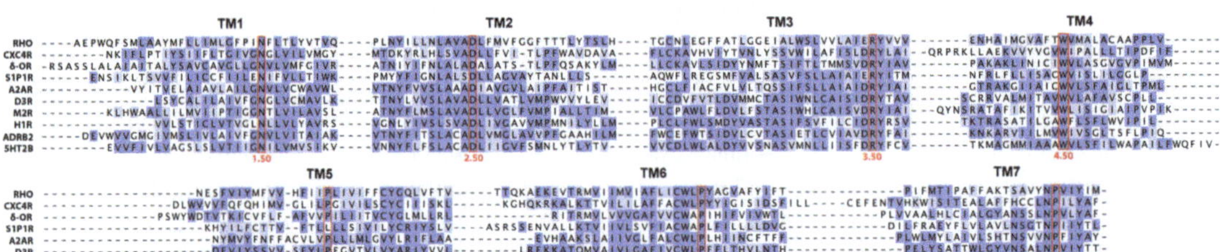

Fig. 2.2 Sequence alignments of TMs 1–7 of GPCRs with known structures. The highly conserved amino acids in each helix, used as reference points in TM sequence alignments are boxed in *red*

among different receptor subfamilies within the extracellular side, near the binding site crevices, responsible for recognition and selectivity of diverse ligands.

Moreover, comparison among the crystal structures of GPCRs revealed backbone anomalies, in the form of kinks and bulges, in the majority of TM helices. These non-canonical elements are frequent in TM proteins, modulating the polytopic membrane protein architecture (Riek et al. 2001). Deviations from the regular α-helical context have been associated to prolines (Von Heijne 1991), glycines (Senes et al. 2000), serines and threonines (Deupi et al. 2004, 2010), or to the insertion or deletions (indels) of residues within the TMs (Deville et al. 2009). Moreover, specific intra- and interhelical interactions involving polar side

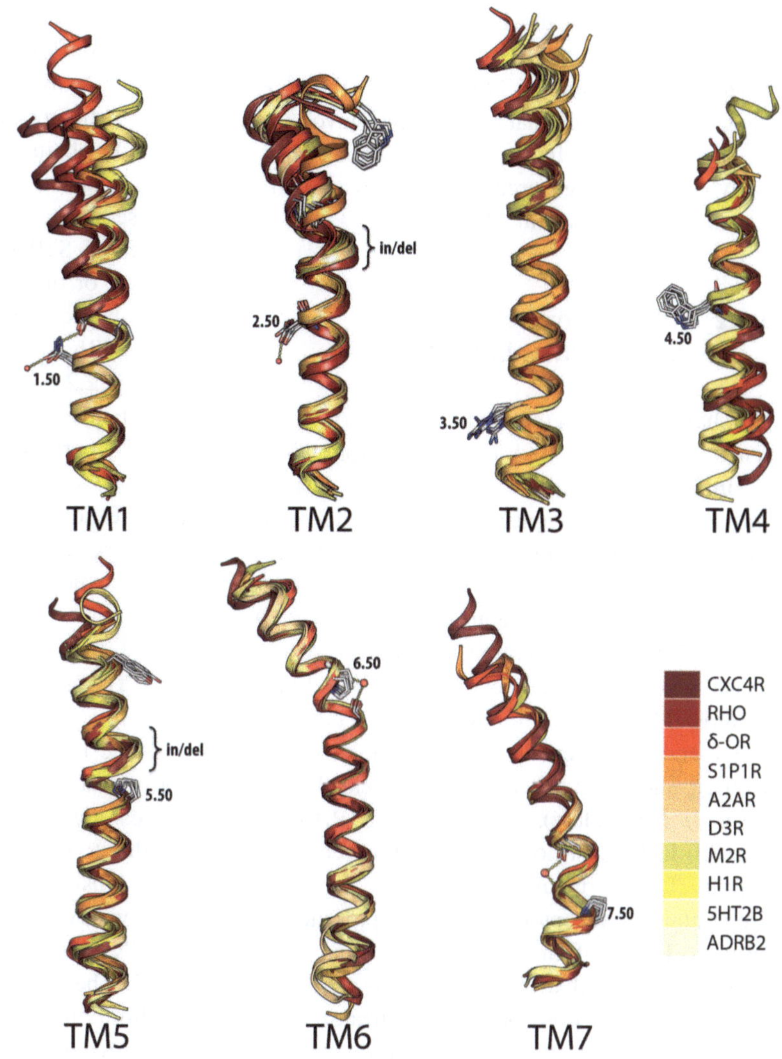

Fig. 2.3 Comparison of individual TM helices in representative structures of the inactive state of GPCRs. The most conserved domain of the helices is superimposed. The crystal structures of GPCRs revealed backbone anomalies, in the form of kinks and bulges, in the majority of TM helices

chains, backbone carbonyls, disulphide bridges and, in some cases, structural water molecules embedded in the TM bundle (Pardo et al. 2007) also cause these distortions. Here we present a detailed analysis of these distortions and their implication in modeling other GPCRs. Figure 2.3 shows the superimposition of the more conserved part of individual TM helices in representative structures of the inactive state of GPCRs.

2.4.1 Transmembrane Helix 1

The extracellular region of TM1 displays a bending propensity in some of the crystal structures (Fig. 2.3). It appears shifted towards the central axis of the receptor in Rho, $A_{2A}R$, CXCR4, opioid receptors, NTSR1, and PAR1. The major displacement of TM1 corresponds to CXCR4 due to the formation of a disulphide bond between the C28 in the N-terminal region and $C274^{7.25}$ in TM7 (Wu et al. 2010). In contrast, TM1 is pointing outside of the bundle in biogenic amine receptors. The highly conserved $N^{1.50}$ (97 % in class A non-olfactory GPCRs) most probably influences the packing of the TM bundle (see Fig. 2.3) since its $N_{\delta 2}$-H atoms act as hydrogen bond donors in the interaction with the backbone carbonyl oxygen of residues at positions 1.46 and 7.46, linking TMs 1 and 7. Moreover, $O_{\delta 1}$ of

$N^{1.50}$ interacts with the highly conserved $D^{2.50}$ (in 92 % of the sequences), via a conserved water molecule, linking TMs 1 and 2. Previous studies have shown that interactions involving a polar Asn side chain provide a strong thermodynamic driving force for membrane helix association (Choma et al. 2000).

2.4.2 Transmembrane Helix 2 – Extracellular Loop 1

The shape of TM2 at the extracellular part, which bends towards TM1 and leans away from TM3, is similar in all structures (Fig. 2.3); despite the amino acid sequence is strongly divergent with, for instance, Pro residues at either position 2.58 (CXCR4, opioid receptors, PAR1), 2.59 (biogenic amine receptors, NTSR1) or 2.60 (sRho). The only exception is TM2 of $A_{2A}R$, which contains Pro at position 2.59 but kinks towards TM3 due to the Cys-bridge between ECL1 and ECL2 exclusive of this family (not shown); and TM2 of $S1P_1R$ that lacks Pro in the helix (see below). Contrarily to $S1P_1R$, the also Pro-lacking bRho and muscarinic receptors possess TM2 structurally similar to the other Pro-containing receptors due to the presence of the GGxTT motif in bRho and $N^{2.59}$ in muscarinic receptors that hydrogen bonds the backbone carbonyl at position 2.55 (Gonzalez et al. 2012). Interestingly, the superimposition of structures reveals that the highly $D^{2.50}$ and the Pro residue located at position 2.58, 2.59 or 2.60 are perfectly overlaid (Gonzalez et al. 2012). Thus, the backbone helical conformation of the amino acids located between these two residues must differ. In this region, TM2 of CXCR4, opioid receptors, and PAR1 adopts a 3_{10} or tight turn (~3.0 residues per turn), TM2 of biogenic amine receptors and NTSR1 adopts a π-bulge or wide turn (~4.8 residues per turn), and TM2 of sRho presents an extreme distortion (~9 residues per turn) characterized by a *cis* $P^{2.60}$ backbone conformation, which is stabilized by two water molecules (Gonzalez et al. 2012). In contrast to these receptors, $S1P_1R$ contains a canonical α-helix at the extracellular part (~3.6 residues per turn). This conformation of TM2 moves its extracellular part away from the TM bundle, relative to the other structures, and modifies the orientation of the side chains at the extracellular side. In order to translate these structural observations into the sequence space, a two-residue gap in the sequences of $S1P_1R$, CXCR4, opioid receptors and PAR1, or one-residue gap in the sequences of bRho, biogenic amine receptors and NTSR1, relative to sRho, must be inserted (Gonzalez et al. 2012) (Fig. 2.2).

Importantly, the conserved Trp residue in ECL1, part of the (W/F) × (F/L)G motif previously identified (Klco et al. 2006), points toward the helical bundle, between TMs 2 and 3, in the crystal structures with the exception of $S1P_1R$ (not shown).

2.4.3 Transmembrane Helix 3

TM3 is the longest and most tilted helix in the receptor structures. No major deviations among structures are observed with the exception of $A_{2A}R$, due to the Cys-bridge between ECL1 and ECL2 exclusive of this family (see above). The highly conserved $C^{3.25}$ forms a disulphide bridge with a Cys residue located at various positions in ECL2. The cytoplasmic side of TM3 contains the highly conserved $(D/E)R^{3.50}(Y/W)$ motif involved in receptor activation (see below). Importantly, the central location of TM3 within the TM bundle allows the helix to interact with the ligand at the extracellular part and with the G protein at the intracellular part (Venkatakrishnan et al. 2013).

2.4.4 Transmembrane Helix 4

TM4, the shortest helix, is almost perpendicular to the membrane. However, significant structural divergences at the extracellular part of TM 4 are found among structures, which may be related to the structural requirements necessary to accommodate the diverse ECL2 architectures (see below). For instance, in contrast to TM4 of other biogenic amine receptors, TM4 of muscarinic receptors bends towards outside of the bundle, away from TM3, due to the hydrogen bond interactions between the side chain of $Q^{4.65}$ and the backbone carbonyl oxygen at position 4.62. Significantly, the shape of TM4 at the extracellular part, in peptide receptors (in which ECL2 is formed by two β-strands, see below), bends towards TM3. In CXCR4, TM4 is longer and substantially deviate from the conformation observed in other peptide receptors.

2.4.5 Transmembrane Helix 5

$P^{5.50}$ (conserved in 76 % of the rhodopsin-like sequences) induces a local opening of TM5, at the 5.43–5.48 turn (Pro-unwinding), in all crystal structures except $S1P_1R$ (see below), which has been proposed to be involved in the mechanism of ligand-induced receptor activation (Sansuk et al. 2011; Rasmussen et al. 2011a). Thus, $P^{5.50}$ triggers a π-bulge or wide turn conformation (~5 residues per turn). However, $A_{2A}R$ displays an extended opening of TM5 from positions 5.35–5.48, in contrast to other $P^{5.50}$-containing structures in which the opening of the helix is restricted to the 5.43–5.48 range of amino acids.

Moreover, $P^{5.50}$ is absent in melanocortin, glycoprotein hormone, lysosphingolipid, prostanoid, and cannabinoid receptors. In these cases, the similarly conserved $Y^{5.58}$ (73 % of the sequences), functionally involved in the stabilization of the active state of the receptor by interacting with $R^{3.50}$ of the (D/E)RY motif in TM3, as revealed by the crystal structures of β_2AR in complex with Gs (Rasmussen et al. 2011b) and the ligand-free opsin (Park et al. 2008), is used as reference for sequence alignment of TM5 (Fig. 2.2). The absence of Pro in TM5 of $S1P_1R$ leads to a regular a-helical conformation (~3.6 residues per turn). Thus, the alignment of the $S1P_1R$ sequence to the other receptors requires two-residue gap relative to $A_{2A}R$ and one-residue gap relative to all other structures, which overlays $Y^{5.37}$ (*i-13* relative to $P^{5.50}$) of $A_{2A}R$ with $F/Y^{5.38}$ of the $P^{5.50}$-containing structures (*i-12* relative to $P^{5.50}$) and $F/Y/W^{5.39}$ of the $P^{5.50}$-lacking structures (Fig. 2.2).

2.4.6 Transmembrane Helix 6

TM6 presents the most pronounced kink in the TM bundle. This severe distortion is energetically stabilized through two structural and functional elements. First, $P^{6.50}$ of the highly conserved $CWxP^{6.50}(Y/F)$ motif introduces a flexible point in TM6 facilitating this extreme conformation. Second, a structural water molecule located in a small cavity between TMs 6 and 7 help to maintain the Pro induced distortion. This water acts as a hydrogen-bond acceptor in the interaction with the backbone N-H amide at position 6.51, and as a hydrogen bond donor in the interactions with the backbone carbonyl at position 6.47 and 7.38. Thus, in addition to stabilizing the kink of TM6, this water molecule links TMs 6 and 7.

2.4.7 Transmembrane Helix 7

TM7 start at different position among receptors. TM7 in CXCR4 is two helical turns longer than in other GPCRs. In this case, the longer TM7 allows $C^{7.25}$ to be placed at the tip of the helix in a favorable position to form a disulphide bond with Cys28 in the N-terminal region. TM7 is kinked at $P^{7.50}$ of the highly conserved NPxxY motif. This region of TM7, involved in key conformational changes associated with GPCR activation (Rosenbaum et al. 2009), is highly irregular. A network of water molecules stabilizes the helical deformation of TM7 and provides hydrogen-bonding partners to polar side chains. For instance, the unusual $P^{7.50}$ deformation removes the intrahelical hydrogen bond between the carbonyl group and the N-H amide at positions 7.45 and 7.49, respectively. A conserved water molecule is located between the backbone carbonyl at position 7.45 and the backbone N-H amide at position 7.49.

2.5 The Extracellular Surface in Class A G Protein-Coupled Receptors

The extracellular surface of GPCRs is defined by the conformation of the N-terminus region and ECLs1-3. Notably, the N-terminus and ECL2 in particular are highly variable in sequence, length, and structure (Peeters et al. 2011) (Fig. 2.4). In rhodopsin, the N-terminus (formed by two β-strands) and ECL2 (two β-strands) block the access of the extracellular ligand to the core of the receptor (Palczewski et al. 2000). Similarly, in $S1P_1R$, the N-terminus (contains a short α-helix) covers half the binding pocket and ECL2 (formed by a family-specific disulphide bridge within ECL 2, but lacking the conserved disulphide bridge between TM3 and ECL 2) covers the other half (Hanson et al. 2012). In these cases, retinal (Hildebrand et al. 2009; Park et al. 2008) and sphingosine-1-phosphate (Hanson et al. 2012) may gain access to the binding pocket from the lipid bilayer (Martin-Couce et al. 2012). In contrast, ECL2 in biogenic amine receptors, adenosine and peptide receptors adopt different spatial conformations that maintain the binding site rather accessible from the extracellular environment (Fig. 2.4). ECL2 of peptide receptors are formed by two β-strands, whereas a helical segment forms ECL2 of adrenergic receptors. This α-helix between TM4 and the disulphide bridge is not conserved in the other members of the biogenic amine receptor family. Thus, each receptor subfamily has probably developed, during evolution, a specific N-terminus/ECL2 to adjust the structural characteristics of its cognate ligands, and to

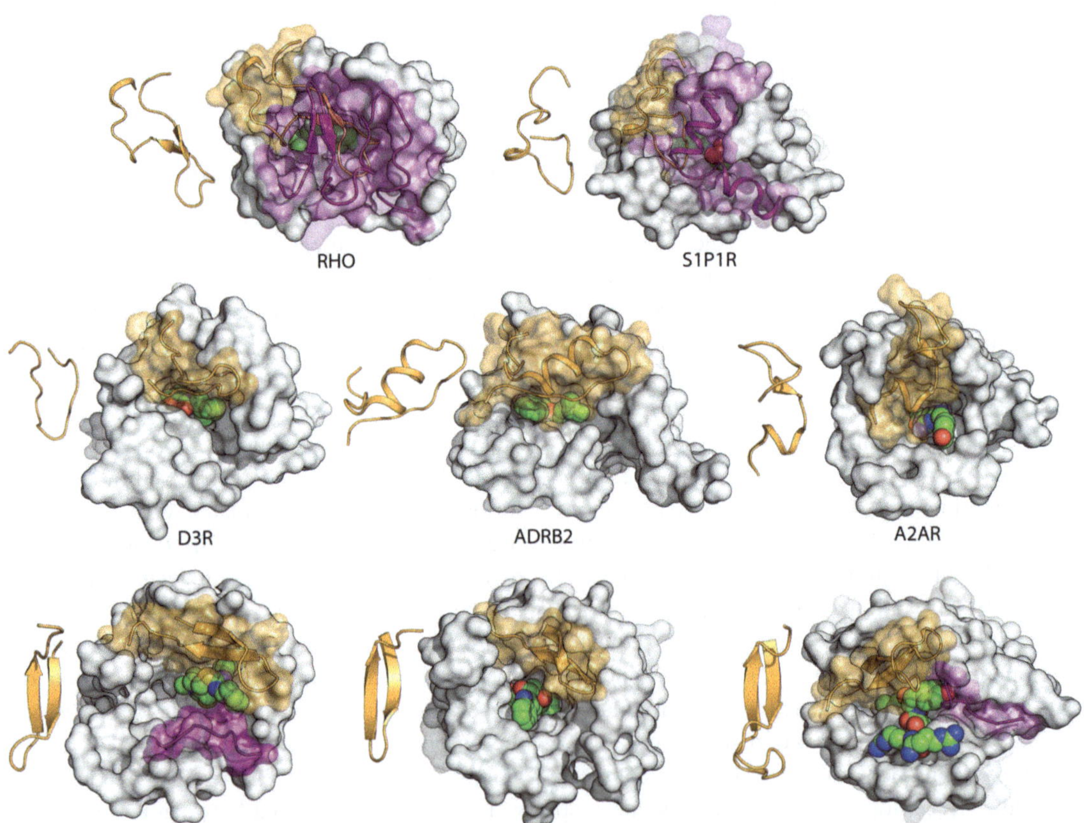

Fig. 2.4 Molecular surface of the extracellular domain in known crystal structures of GPCRs. The N-terminus domain is shown in *red*, ECL2 is shown in *yellow*, and the ligand in the binding site is shown as *spheres*

modulate the ligand binding/unbinding events (Hurst et al. 2010; Dror et al. 2011; Gonzalez et al. 2011).

2.6 Ligand Binding to G Protein-Coupled Receptors

Analysis of the known crystal structures of GPCRs shows that ligand binding mostly occurs in a main cavity located between the extracellular segments of TMs 3, 5, 6, and 7 or in a minor binding cavity located between the extracellular segments of TMs 1, 2, 3, and 7 (Rosenkilde et al. 2010) (Fig. 2.5a). Despite these common pockets, different ligands penetrate to different depths within the TM bundle (Venkatakrishnan et al. 2013) (Fig. 2.5b–f). A major issue in these common binding modes is the specificity of ligands among subfamilies of receptors.

2.7 Intracellular Structural Changes Associated with Activation of G Protein-Coupled Receptors

The publication of the crystal structure of the ligand-free opsin (Park et al. 2008), which contains several distinctive features of the active state as it has been confirmed in the recent structure of the β_2-adrenergic receptor bound to Gs (Rasmussen et al. 2011b), showed that during the process of receptor activation the intracellular part of TM6 tilts outwards, TM5 nears TM6, and $R^{3.50}$ within the (D/E)RY motif in TM3 adopts an extended conformation pointing towards the protein core, to interact with the highly conserved $Y^{5.58}$ in TM5 and $Y^{7.53}$ of the (N/D)PxxY motif in TM7 (Fig. 2.6). As shown in the original publication of the opsin structure, these conformational

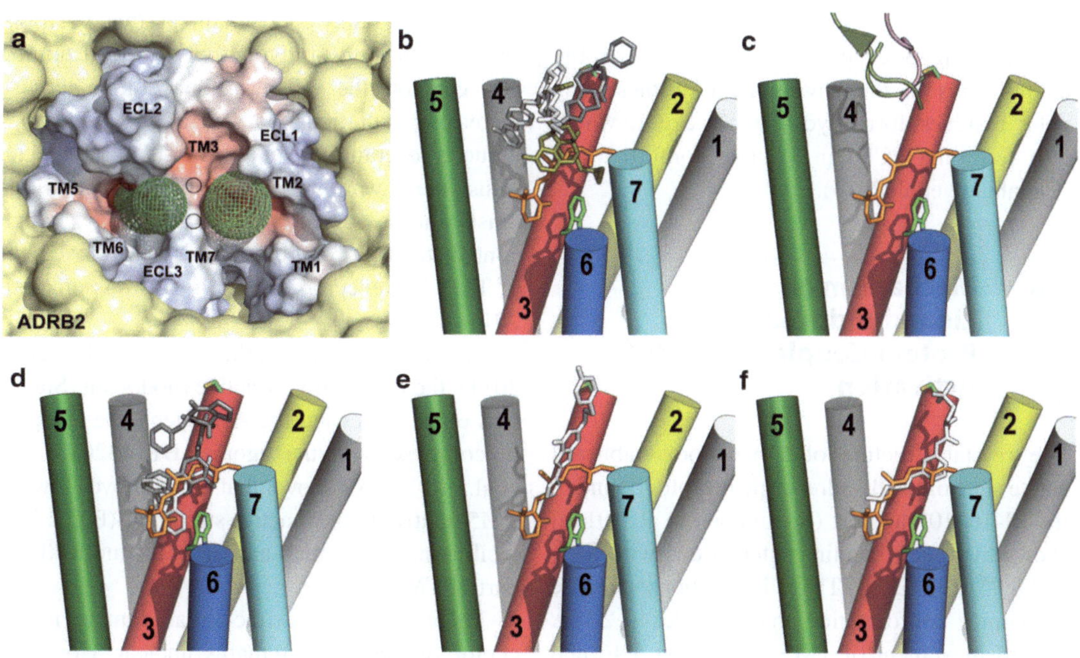

Fig. 2.5 Ligand binding to GPCRs. (**a**) Binding cavities in β_2AR. (**b**) The binding of vorapaxar (*white*) to PAR1 (Zhang et al. 2012), IT1t (*gray*) to CXCR4 (Wu et al. 2010), and morphinan (*olive*) to μ-OR (Manglik et al. 2012). (**c**) The binding of the CVX15 cyclic peptide (*olive*) to CXCR4 (Wu et al. 2010) and aminoacids 8–13 of neurotensin (*pink*) to NTSR1 (White et al. 2012). (**d**) The binding of doxepin (*white*) to H_1R (Shimamura et al. 2011) and ergotamine (*gray*) to $5HT_{1b}$R (Wang et al. 2013a). (**e**) The binding of ZM241385 (*white*) to A_{2A}R (Jaakola et al. 2008). (**f**) The binding of ML056 to $S1P_1$R (Hanson et al. 2012). The structures of retinal (*orange sticks*) and $C^{3.25}$ and $W^{6.48}$ (*green sticks*) are shown in panels B-F for comparison purposes

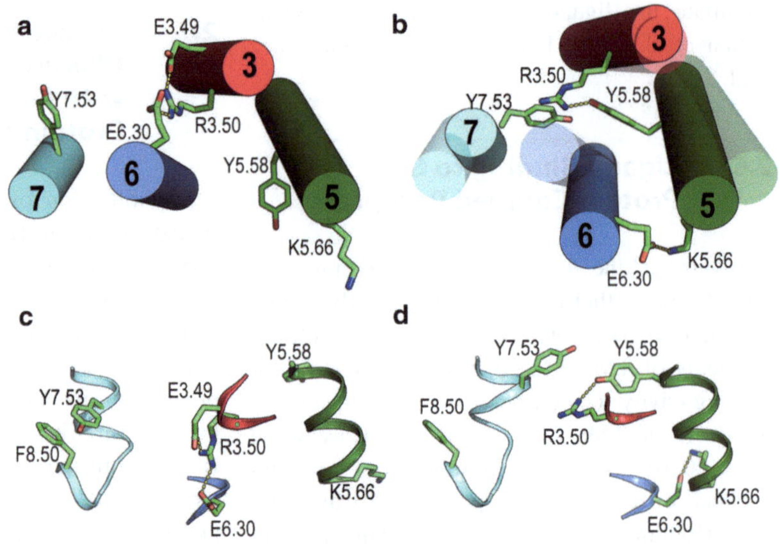

Fig. 2.6 Intracellular structural changes associated with receptor activation. Comparison of (**a, c**) the crystal structure of inactive rhodopsin (1GZM) with (**b, d**) the crystal structure of the ligand-free opsin (3CAP), which contains several distinctive features of the active state, in views parallel (**c, d**) and perpendicular (**a, b**) to the membrane. Panel *B* shows the positions of TMs 3, 5–7 in rhodopsin (transparent cylinders) and opsin (opaque cylinders) for comparison purposes

changes disrupt the ionic interaction between $R^{3.50}$ with negatively charged side chains at positions 3.49 in TM3 and 6.30 in TM6 (Fig. 2.6a, c) and facilitates the interaction between $K^{5.66}$ in TM 5 and $E^{6.30}$ in TM 6 (Fig. 2.6b, d). It has been suggested that conserved hydrophobic amino acids in the environment of these key polar residues form hydrophobic cages, which also restrain GPCRs in inactive conformations (Caltabiano et al. 2013).

2.8 Mechanism of Ligand-Induced G Protein-Coupled Receptor Activation

The crystal structure of a nanobody-stabilized active state of the β_2-adrenergic receptor bound to the BI-167107 agonist (Rasmussen et al. 2011a) shows hydrogen bonding interactions with $S^{5.42}$ and $S^{5.46}$ (Fig. 2.7a). These interactions stabilize a receptor conformation that includes a 2.1 Å inward movement of TM5 at position 5.46 and 1.4 Å inward movement of the conserved $P^{5.50}$ relative to the inactive, carazolol-bound structure (Rosenbaum et al. 2007). This key distortion is stabilized in the known crystal structures by a bulky hydrophobic side chain at position 3.40 (Fig. 2.7a), highly conserved in the whole Class A GPCR family (I:40 %, V:25 %, L:11 %). Mutation of $I^{3.40}$ to either Ala or Gly, i.e. removing the bulky side chain at this position, abolishes the constitutive activity of the histamine H_1 receptor, the effect of constitutive-activity increasing mutations, as well as the histamine-induced receptor activation (Sansuk et al. 2011). Thus, the inward movement of $P^{5.50}$ upon agonist binding repositions $I^{3.40}$ and $F^{6.44}$, which contributes to a rotation and outward movement of TM6 for receptor activation (Rasmussen et al. 2011a).

The structures of metarhodopsin II (Choe et al. 2011), the constitutively active rhodopsin (Standfuss et al. 2011) and the A_{2A} adenosine receptor in complex with the agonist UK-432097 (Xu et al. 2011) have shown that $W^{6.48}$ moves toward TM5 relative to the inactive structures (Fig. 2.7b), facilitating the rotation and tilt of the intracellular part of TM6.

The role of the extracellular domain in receptor function still remains unclear. However, NMR studies on the β_2-adrenergic receptor have shown ligand-specific conformational changes on the extracellular domain (Bokoch et al. 2010). Similarly, it has recently been reported that a

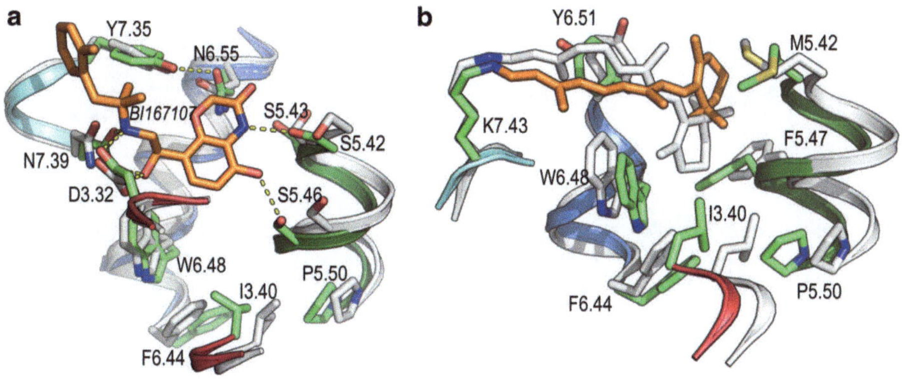

Fig. 2.7 Mechanisms of ligand-induced receptor activation. (**a**) Detailed view of the β_2-adrenergic receptor bound to the full agonist BI-167107 (in *orange*). The hydrogen bond interaction between full agonists and $S^{5.46}$ stabilizes a receptor conformation that includes an inward movement of TM5 relative to the inactive (shown in *white* for comparison purposes), carazolol-boundstructure. (**b**) The conformational change of inactive 11-*cis* retinal (in *white*) to the active 11-*trans* retinal (in *orange*) stabilizes a receptor conformation that includes an inward movement of TM5 together with a movement of $W^{6.48}$ toward TM5 relative to the inactive structures (shown in *white* for comparison purposes)

small cavity (vestibule) present at the entrance of the ligand-binding cavity controls the extent of receptor movement to govern a hierarchical order of G-protein coupling (Bock et al. 2012). Finally, the N-terminal domain of melanocortin receptors plays a significant role in their constitutive, ligand-independent, activity (Ersoy et al. 2012).

2.9 G Protein-Coupled Receptor Oligomerization

GPCRs have been classically described as monomeric TM receptors that form a ternary complex: a ligand, the GPCR, and its associated G protein. This is compatible with observations that monomeric rhodopsin and β_2-adrenergic receptor are capable of activating G proteins (Ernst et al. 2007; Whorton et al. 2007). Nevertheless, it is now well accepted that many GPCRs have been observed to oligomerize in cells (Pin et al. 2007; Ferre et al. 2009). It has been shown that receptor activation is modulated by allosteric communication between protomers of dopamine class A GPCR dimers (Han et al. 2009). The minimal signaling unit, two receptors and a single G protein, is maximally activated by agonist binding to a single protomer. Inverse agonist binding to the second protomer enhances signaling, whereas agonist binding to the second protomer blunts signaling. Moreover, binding of agonists or the G protein to β_2- regulates receptor oligomerization (Fung et al. 2009). Cysteine cross-linking experiments have suggested that receptor oligomerization involves hydrophobic interactions via the surfaces of TMs1, 4, and/or 5 (Klco et al. 2003; Guo et al. 2005, 2008). Nevertheless, electrostatic interactions of the intracellular domains are key in the formation of receptor heteromers (Navarro et al. 2010).

The recent release of the high-resolution crystal structures of μOR (Manglik et al. 2012) and β_1-AR (Huang et al. 2013) in the form of homo-oligomers (Fig. 2.8) facilitates the task of modeling GPCR dimers and higher order oligomers. The structure of μOR shows receptor molecules associated into pairs through two different interfaces (Fig. 2.8a). The first interface is via TMs1 and 2 and Hx8, and the second interface comprises TMs 5 and 6. The structure of β_1-AR contains a similar TMs1 and 2 and Hx8 interface but the other interface engages residues from TMs4 and 5 (Fig. 2.8b).

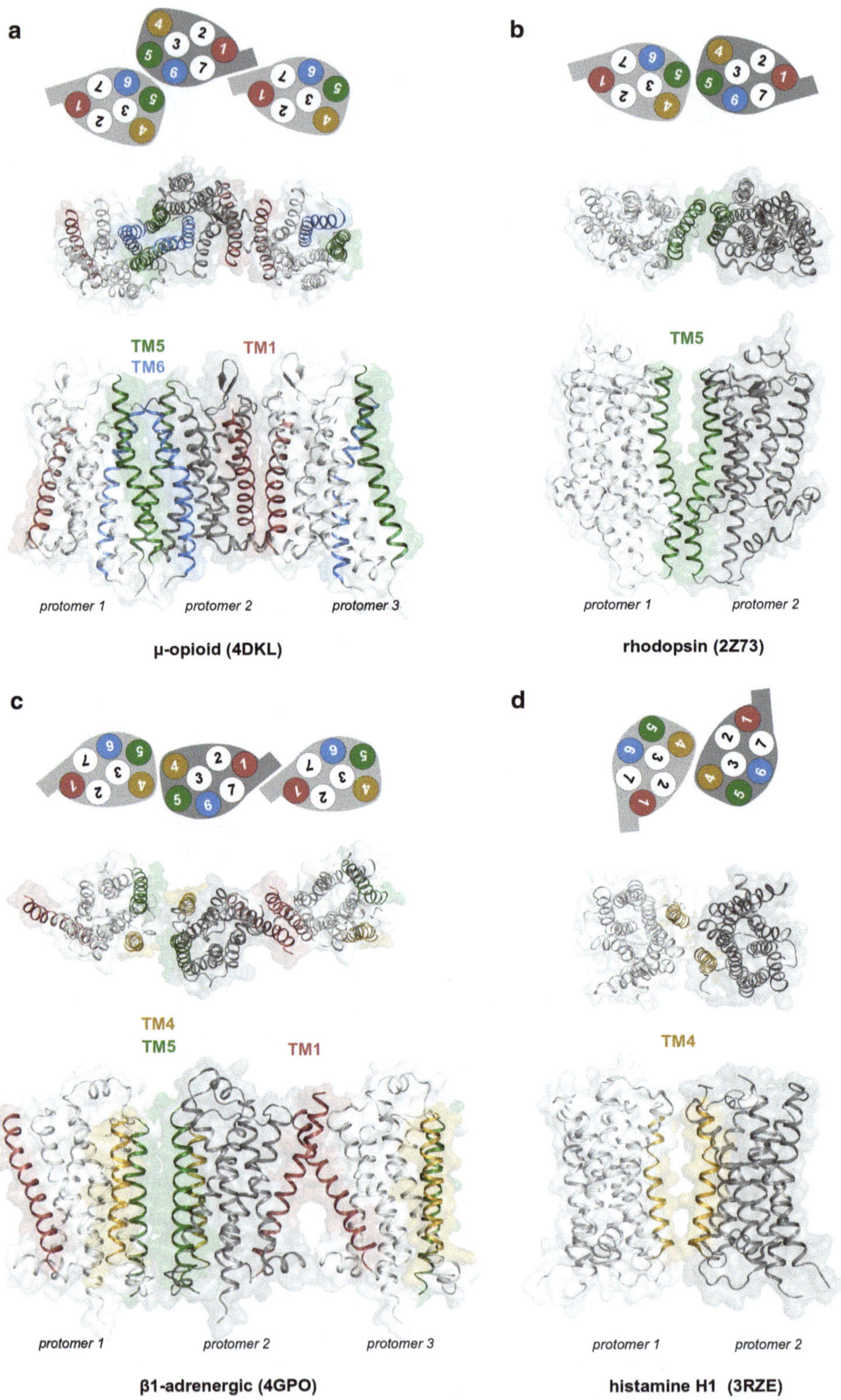

Fig. 2.8 **GPCR oligomerization**. The recent high-resolution crystal structures of (**a**) μOR (Manglik et al. 2012) and (**b**) β_1-AR (Huang et al. 2013) in the form of homo-oligomers, and sRho (Murakami and Kouyama 2008) and H_1R (Shimamura et al. 2011) in the form of homo-dimers

Additional crystal structures with GPCR dimers have been published. Interestingly, a TM1 interface, similar to the one observed for μOR and β_1-AR, is present in the structures of the κOR (Wu et al. 2012), opsin (Scheerer et al. 2008), and metarhodopsin II (Choe et al. 2011). Moreover, the TM4/5 interface of β_1-AR resembles the interface previously obtained for rhodopsin using atomic force microscopy (Fotiadis et al. 2003). The crystal structure of the histamine H_1 receptor (Shimamura et al. 2011) contains a TM4 interface (Fig. 2.8d), which is different from the TM4/5 interface of β_1-AR due to the absence of TM5 contacts. Similarly, the structures of CXCR4 (Wu et al. 2010) and squid rhodopsin (Murakami and Kouyama 2008) contain a TM5 interface (Fig. 2.8c), which are different from the TM4/5 interface of β_1-AR or the TM5/6 interface of μOR.

2.10 The Binding of G Protein-Coupled Receptors to the G Protein

The formation of the complex between the active conformation of the receptor and the heterotrimeric G protein triggers GDP release from the Gα-subunit, GTP binding to the Gα-subunit and dissociation of the Gβγ-subunits (Chung et al. 2011), which finally leads to a cascade of signals depending on the G-protein type. Noteworthy, more than 800 known GPCRs can bind 17 different Gα subunits, which have been grouped into four different classes ($G\alpha_s$, $G\alpha_i$, $G\alpha_q$ and $G\alpha_{12}$) (Simon et al. 1991). To date, the crystal structures of the ligand-free opsin (Scheerer et al. 2008), metarhodopsin II (Choe et al. 2011) and the constitutively active rhodopsin mutant $E^{3.28}Q$ (Standfuss et al. 2011) in complex with a peptide derived from the carboxy terminus of the α-subunit of the G protein transducin, together with the structure of the β_2-adrenergic receptor bound to Gs (Rasmussen et al. 2011b) have been released. These structures have shown that the C-terminal α5 helix of Gα binds to the intracellular cavity that is opened by the movement of the cytoplasmic end of TM6 away from TM3 and towards TM5 (see above). The C-terminal α5 helix of the α-subunit interacts with the extended conformation of $R^{3.50}$, the short loop connecting TM7 and Hx8, and the inner side of the cytoplasmic TMs 5 and 6 (Fig. 2.9).

2.11 The Binding of the C-Tail of G Protein-Coupled Receptors to Arrestin

Phosphorylation of several residues of the C-tail of GPCRs, by Ser/Thr kinases called G protein-coupled receptor kinases (GRKs), promotes the interactions between the receptor and arrestin, leading to receptor desensitization (Lefkowitz and Shenoy 2005). GPCRs can bind four different arrestin proteins: arrestin-1 and arrestin-4 (known as visual arrestins) bind to the phosphorylated form of active rhodopsin, whereas arrestin-2 and arrestin-3 interact and regulate the activity of non-visual GPCRs (Gurevich and Gurevich 2006).

Arrestin comprises two domains (N- and C- domains) of antiparallel β-sheets connected through a hinge region (Granzin et al. 1998) (Fig. 2.10). The binding region for phosphorylated ligand-activated receptor is located at the N-terminal domain, which is occupied by the long C-terminal tail in the basal state (blue peptide in Fig. 2.10a). The crystal structure of arrestin-2 in complex with a phosphorylated 29-aminoacid carboxy-terminal peptide derived from the human V2 vasopressin receptor (V2Rpp) (Shukla et al. 2013) has recently released. This structure shows that the phosphorylated C-tail region of GPCRs (yellow peptide in Fig. 2.10a) displaces the C-tail of arrestin. Moreover, an active conformation of arrestin-1, mimicked by C-tail truncation, has also been published (Kim et al. 2013). Both structures show significant conformational changes relative to inactive, basal, arrestin. These include rotation of the N- and C-terminal

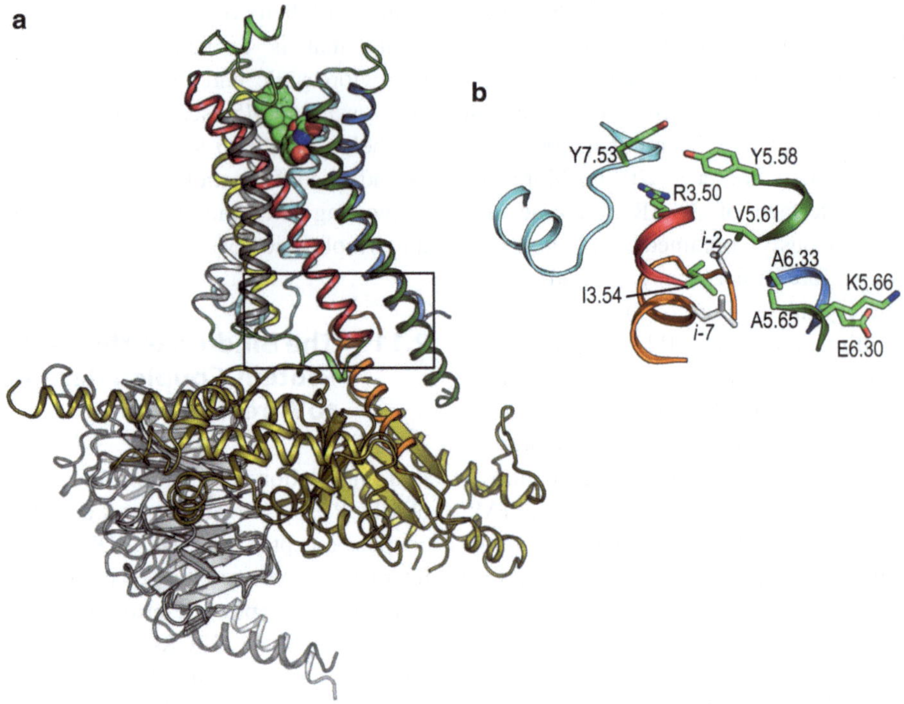

Fig. 2.9 G-protein binding. (**a**) Crystal structure of the β_2-adrenergic receptor in complex with the Gs heterotrimer (α-subunit in *olive*, β-subunit in *white*, and γ-subunit in *gray*). The C-terminal $\alpha 5$ helix of the α-subunit is shown in *orange*. The *rectangle* shows the part of the complex depicted in *panel B*. (**b**) Detailed view of the interaction between the C-terminal $\alpha 5$ helix of the α-subunit (in *orange*) with the short loop connecting TM7 and Hx8 (*light blue*), TM3 (*red*), and the inner side of the cytoplasmic TMs 5 (*green*) and 6 (*blue*)

domains relative to each other, and major reorientations of the lariat, middle, and finger loops (Fig. 2.10b).

2.12 Conclusions

GPCRs are disordered allosteric proteins that exhibit modulator behavior with a number of guests in both the extracellular (ligand) and intracellular (G protein, arrestin) spaces (Kenakin and Miller 2010). This considers GPCRs as monomeric TM receptors. Nevertheless, it is now well accepted that many GPCRs form homo- and hetero-oligomers (Khelashvili et al. 2010). Since 2007, innovative crystallographic techniques (Venkatakrishnan et al. 2013) have resulted in an exponential growth in the number of solved structures that include several members of the GPCR family (bound to either agonists, antagonists, or inverse agonists), in the form of monomers or homo-oligomers, in complex with the G protein, or the C-tail bound to arrestin. Thus, the used of these structures as templates allows molecular modelers to simulate the process of signal transduction through the cell membrane. These tailor-made models can study ligand binding, receptor specificity, receptor activation, G protein coupling, allosteric communication among protomers, among others. However, we want to emphasize that homology modeling of GPCRs is far from being a routine technique. Clearly, the inclusion of experimental results can improve the reliability of the models, and their predictive character.

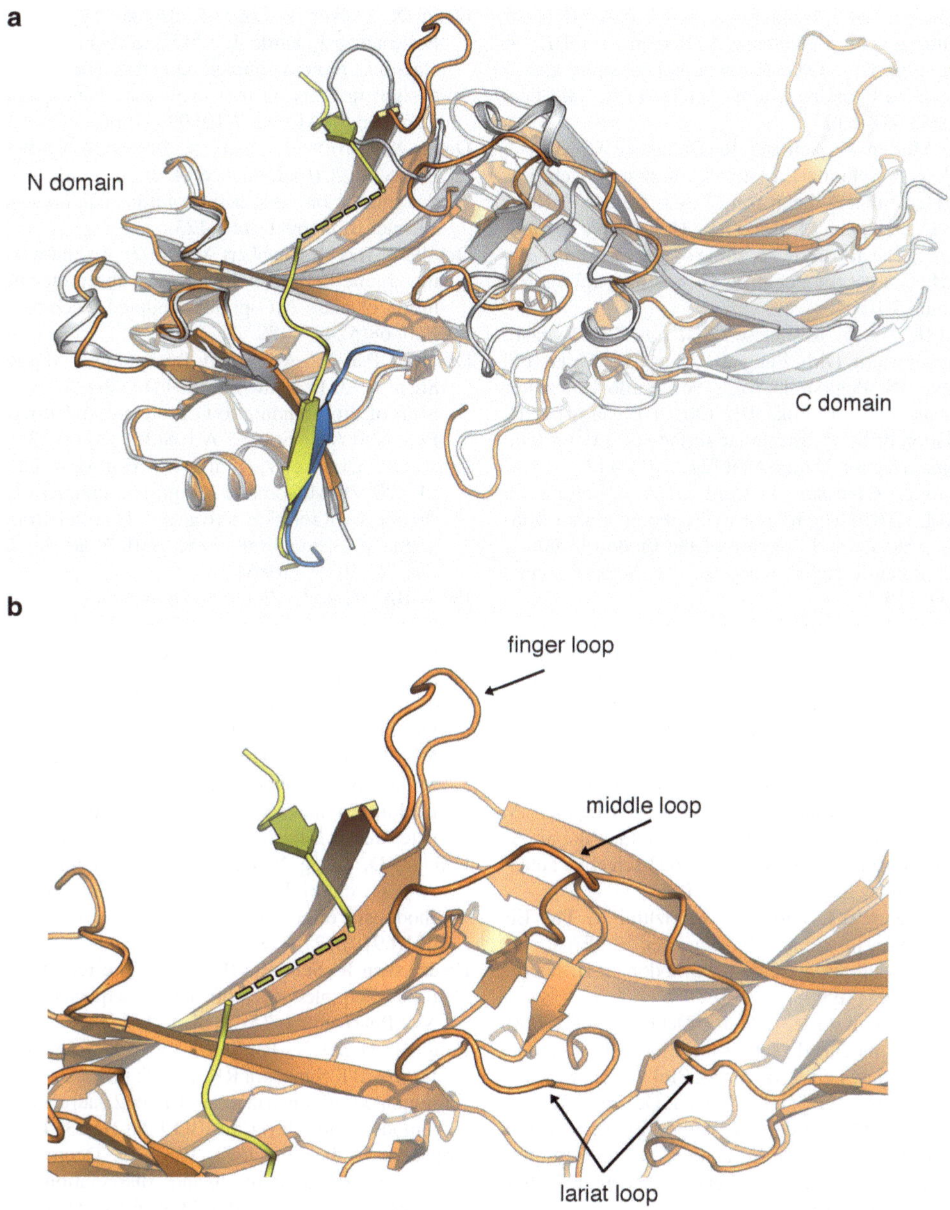

Fig. 2.10 The binding of the C-tail of GPCRs to arrestin. (**a**) The active conformation of arrestin-2 (PDB id 4JQI, shown in *orange*) is superimposed to inactive arrestin-2 (1G4M, in *gray*). The phosphorylated C-tail region of GPCRs (*yellow peptide*) displaces the C-tail of inactive arrestin (*blue peptide*). (**b**) Detailed view of the finger, middle and lariat loops, in the presumably active conformation of arrestin-2, which interact with the phosphorylated C-tail of GPCRs

References

Arinaminpathy Y, Khurana E, Engelman DM, Gerstein MB (2009) Computational analysis of membrane proteins: the largest class of drug targets. Drug Discov Today 14(23–24):1130–1135

Ballesteros JA, Weinstein H (1995) Integrated methods for the construction of three dimensional models and computational probing of structure-function relations in G-protein coupled receptors. Methods Neurosci 25:366–428

Blattermann S, Peters L, Ottersbach PA, Bock A, Konya V, Weaver CD, Gonzalez A, Schroder R, Tyagi R,

Luschnig P, Gab J, Hennen S, Ulven T, Pardo L, Mohr K, Gutschow M, Heinemann A, Kostenis E (2012) A biased ligand for OXE-R uncouples G alpha and G beta gamma signaling within a heterotrimer. Nat Chem Biol 8(7):631–638

Bock A, Merten N, Schrage R, Dallanoce C, Batz J, Klockner J, Schmitz J, Matera C, Simon K, Kebig A, Peters L, Muller A, Schrobang-Ley J, Trankle C, Hoffmann C, De Amici M, Holzgrabe U, Kostenis E, Mohr K (2012) The allosteric vestibule of a seven transmembrane helical receptor controls G-protein coupling. Nat Commun 3:1044

Bokoch MP, Zou Y, Rasmussen SG, Liu CW, Nygaard R, Rosenbaum DM, Fung JJ, Choi HJ, Thian FS, Kobilka TS, Puglisi JD, Weis WI, Pardo L, Prosser RS, Mueller L, Kobilka BK (2010) Ligand-specific regulation of the extracellular surface of a G-protein-coupled receptor. Nature 463(7277):108–112

Caltabiano G, Gonzalez A, Cordomi A, Campillo M, Pardo L (2013) The role of hydrophobic amino acids in the structure and function of the rhodopsin family of G protein-coupled receptors. Methods Enzymol 520:99–115

Cherezov V, Rosenbaum DM, Hanson MA, Rasmussen SG, Thian FS, Kobilka TS, Choi HJ, Kuhn P, Weis WI, Kobilka BK, Stevens RC (2007) High-resolution crystal structure of an engineered human beta2-adrenergic G protein-coupled receptor. Science 318(5854):1258–1265

Chien EY, Liu W, Zhao Q, Katritch V, Han GW, Hanson MA, Shi L, Newman AH, Javitch JA, Cherezov V, Stevens RC (2010) Structure of the human dopamine D3 receptor in complex with a D2/D3 selective antagonist. Science 330(6007):1091–1095

Choe HW, Kim YJ, Park JH, Morizumi T, Pai EF, Krauss N, Hofmann KP, Scheerer P, Ernst OP (2011) Crystal structure of metarhodopsin II. Nature 471(7340):651–655

Choma C, Gratkowski H, Lear JD, DeGrado WF (2000) Asparagine-mediated self-association of a model transmembrane helix. Nat Struct Biol 7(2):161–166

Chung KY, Rasmussen SG, Liu T, Li S, DeVree BT, Chae PS, Calinski D, Kobilka BK, Woods VL Jr, Sunahara RK (2011) Conformational changes in the G protein Gs induced by the beta2 adrenergic receptor. Nature 477(7366):611–615

Day PW, Rasmussen SG, Parnot C, Fung JJ, Masood A, Kobilka TS, Yao XJ, Choi HJ, Weis WI, Rohrer DK, Kobilka BK (2007) A monoclonal antibody for G protein-coupled receptor crystallography. Nat Methods 4(11):927–929

de la Fuente T, Martin-Fontecha M, Sallander J, Benhamu B, Campillo M, Medina RA, Pellissier LP, Claeysen S, Dumuis A, Pardo L, Lopez-Rodriguez ML (2010) Benzimidazole derivatives as new serotonin 5-HT(6) receptor antagonists. Molecular mechanisms of receptor inactivation. J Med Chem 53(3):1357–1369

Deupi X, Olivella M, Govaerts C, Ballesteros JA, Campillo M, Pardo L (2004) Ser and Thr residues modulate the conformation of pro-kinked transmembrane alpha-helices. Biophys J 86(1):105–115

Deupi X, Dolker N, Lopez-Rodriguez M, Campillo M, Ballesteros J, Pardo L (2007) Structural models of class a G protein-coupled receptors as a tool for drug design: insights on transmembrane bundle plasticity. Curr Top Med Chem 7(10):999–1006

Deupi X, Olivella M, Sanz A, Dolker N, Campillo M, Pardo L (2010) Influence of the g- conformation of Ser and Thr on the structure of transmembrane helices. J Struct Biol 169(1):116–123

Deville J, Rey J, Chabbert M (2009) An indel in transmembrane helix 2 helps to trace the molecular evolution of class A G-protein-coupled receptors. J Mol Evol 68(5):475–489

Dror RO, Pan AC, Arlow DH, Borhani DW, Maragakis P, Shan Y, Xu H, Shaw DE (2011) Pathway and mechanism of drug binding to G-protein-coupled receptors. Proc Natl Acad Sci U S A 108(32):13118–13123

Ernst OP, Gramse V, Kolbe M, Hofmann KP, Heck M (2007) Monomeric G protein-coupled receptor rhodopsin in solution activates its G protein transducin at the diffusion limit. Proc Natl Acad Sci U S A 104(26):10859–10864

Ersoy BA, Pardo L, Zhang S, Thompson DA, Millhauser G, Govaerts C, Vaisse C (2012) Mechanism of N-terminal modulation of activity at the melanocortin-4 receptor GPCR. Nat Chem Biol 8(8):725–730

Ferre S, Baler R, Bouvier M, Caron MG, Devi LA, Durroux T, Fuxe K, George SR, Javitch JA, Lohse MJ, Mackie K, Milligan G, Pfleger KD, Pin JP, Volkow ND, Waldhoer M, Woods AS, Franco R (2009) Building a new conceptual framework for receptor heteromers. Nat Chem Biol 5(3):131–134

Fotiadis D, Liang Y, Filipek S, Saperstein DA, Engel A, Palczewski K (2003) Atomic-force microscopy: rhodopsin dimers in native disc membranes. Nature 421(6919):127–128

Fredriksson R, Schioth HB (2005) The repertoire of G-protein-coupled receptors in fully sequenced genomes. Mol Pharmacol 67(5):1414–1425

Fung JJ, Deupi X, Pardo L, Yao XJ, Velez-Ruiz GA, Devree BT, Sunahara RK, Kobilka BK (2009) Ligand-regulated oligomerization of beta(2)-adrenoceptors in a model lipid bilayer. EMBO J 28(21):2384–2392

Gonzalez A, Perez-Acle T, Pardo L, Deupi X (2011) Molecular basis of ligand dissociation in beta-adrenergic receptors. PLoS One 6(9):e23815

Gonzalez A, Cordomí A, Caltabiano G, Campillo M, Pardo L (2012) Impact of helix irregularities on sequence alignment and homology modelling of G protein-coupled receptors. Chembiochem 13(10):1393–1399

Granier S, Manglik A, Kruse AC, Kobilka TS, Thian FS, Weis WI, Kobilka BK (2012) Structure of the delta-opioid receptor bound to naltrindole. Nature 485(7398):400–404

Granzin J, Wilden U, Choe HW, Labahn J, Krafft B, Buldt G (1998) X-ray crystal structure of arrestin from bovine rod outer segments. Nature 391(6670):918–921

Guo W, Shi L, Filizola M, Weinstein H, Javitch JA (2005) Crosstalk in G protein-coupled receptors:

changes at the transmembrane homodimer interface determine activation. Proc Natl Acad Sci U S A 102:17495–17500
Guo W, Urizar E, Kralikova M, Mobarec JC, Shi L, Filizola M, Javitch JA (2008) Dopamine D2 receptors form higher order oligomers at physiological expression levels. EMBO J 27(17):2293–2304
Gurevich VV, Gurevich EV (2006) The structural basis of arrestin-mediated regulation of G-protein-coupled receptors. Pharmacol Ther 110(3):465–502
Haga K, Kruse AC, Asada H, Yurugi-Kobayashi T, Shiroishi M, Zhang C, Weis WI, Okada T, Kobilka BK, Haga T, Kobayashi T (2012) Structure of the human M2 muscarinic acetylcholine receptor bound to an antagonist. Nature 482(7386):547–551
Han Y, Moreira IS, Urizar E, Weinstein H, Javitch JA (2009) Allosteric communication between protomers of dopamine class A GPCR dimers modulates activation. Nat Chem Biol 5(9):688–695
Hanson MA, Roth CB, Jo E, Griffith MT, Scott FL, Reinhart G, Desale H, Clemons B, Cahalan SM, Schuerer SC, Sanna MG, Han GW, Kuhn P, Rosen H, Stevens RC (2012) Crystal structure of a lipid G protein-coupled receptor. Science 335(6070):851–855
Hildebrand PW, Scheerer P, Park JH, Choe HW, Piechnick R, Ernst OP, Hofmann KP, Heck M (2009) A ligand channel through the G protein coupled receptor opsin. PLoS One 4(2):e4382
Hopf TA, Colwell LJ, Sheridan R, Rost B, Sander C, Marks DS (2012) Three-dimensional structures of membrane proteins from genomic sequencing. Cell 149(7):1607–1621
Huang J, Chen S, Zhang JJ, Huang XY (2013) Crystal structure of oligomeric beta1-adrenergic G protein-coupled receptors in ligand-free basal state. Nat Struct Mol Biol 20(4):419–425
Hurst DP, Grossfield A, Lynch DL, Feller S, Romo TD, Gawrisch K, Pitman MC, Reggio PH (2010) A lipid pathway for ligand binding is necessary for a cannabinoid G protein-coupled receptor. J Biol Chem 285(23):17954–17964
Imming P, Sinning C, Meyer A (2006) Drugs, their targets and the nature and number of drug targets. Nat Rev Drug Discov 5(10):821–834
Jaakola VP, Griffith MT, Hanson MA, Cherezov V, Chien EY, Lane JR, Ijzerman AP, Stevens RC (2008) The 2.6 angstrom crystal structure of a human A2A adenosine receptor bound to an antagonist. Science 322(5905):1211–1217
Katritch V, Cherezov V, Stevens RC (2012) Diversity and modularity of G protein-coupled receptor structures. Trends Pharmacol Sci 33(1):17–27
Katritch V, Cherezov V, Stevens RC (2013) Structure-function of the G protein-coupled receptor superfamily. Annu Rev Pharmacol Toxicol 53:531–556
Kenakin T, Miller LJ (2010) Seven transmembrane receptors as shapeshifting proteins: the impact of allosteric modulation and functional selectivity on new drug discovery. Pharmacol Rev 62(2):265–304
Khelashvili G, Dorff K, Shan J, Camacho-Artacho M, Skrabanek L, Vroling B, Bouvier M, Devi LA, George SR, Javitch JA, Lohse MJ, Milligan G, Neubig RR, Palczewski K, Parmentier M, Pin JP, Vriend G, Campagne F, Filizola M (2010) GPCR-OKB: the G protein coupled receptor oligomer knowledge base. Bioinformatics 26(14):1804–1805
Kim YJ, Hofmann KP, Ernst OP, Scheerer P, Choe HW, Sommer ME (2013) Crystal structure of pre-activated arrestin p44. Nature 497(7447):142–146
Klco JM, Lassere TB, Baranski TJ (2003) C5a Receptor oligomerization. I. Disulfide trapping reveals oligomers and potential contact surfaces in a G protein-coupled receptor. J Biol Chem 278(37):35345–35353
Klco JM, Nikiforovich GV, Baranski TJ (2006) Genetic analysis of the first and third extracellular loops of the C5a receptor reveals an essential WXFG motif in the first loop. J Biol Chem 281(17):12010–12019
Krissinel E, Henrick K (2004) Secondary-structure matching (SSM), a new tool for fast protein structure alignment in three dimensions. Acta Crystallogr D Biol Crystallogr 60(Pt 12 Pt 1):2256–2268
Kristiansen K (2004) Molecular mechanisms of ligand binding, signaling, and regulation within the superfamily of G-protein-coupled receptors: molecular modeling and mutagenesis approaches to receptor structure and function. Pharmacol Ther 103(1):21–80
Kruse AC, Hu J, Pan AC, Arlow DH, Rosenbaum DM, Rosemond E, Green HF, Liu T, Chae PS, Dror RO, Shaw DE, Weis WI, Wess J, Kobilka BK (2012) Structure and dynamics of the M3 muscarinic acetylcholine receptor. Nature 482(7386):552–556
Lefkowitz RJ, Shenoy SK (2005) Transduction of receptor signals by beta-arrestins. Science 308(5721):512–517
Li J, Edwards PC, Burghammer M, Villa C, Schertler GF (2004) Structure of bovine rhodopsin in a trigonal crystal form. J Mol Biol 343(5):1409–1438
Liu W, Chun E, Thompson AA, Chubukov P, Xu F, Katritch V, Han GW, Roth CB, Heitman LH, IJzerman AP, Cherezov V, Stevens RC (2012) Structural basis for allosteric regulation of GPCRs by sodium ions. Science 337(6091):232–236
Manglik A, Kruse AC, Kobilka TS, Thian FS, Mathiesen JM, Sunahara RK, Pardo L, Weis WI, Kobilka BK, Granier S (2012) Crystal structure of the micro-opioid receptor bound to a morphinan antagonist. Nature 485(7398):321–326
Martin-Couce L, Martin-Fontecha M, Palomares O, Mestre L, Cordomi A, Hernangomez M, Palma S, Pardo L, Guaza C, Lopez-Rodriguez ML, Ortega-Gutierrez S (2012) Chemical probes for the recognition of cannabinoid receptors in native systems. Angewandte Chemie 51(28):6896–6899
Mirzadegan T, Benko G, Filipek S, Palczewski K (2003) Sequence analyses of G-protein-coupled receptors: similarities to rhodopsin. Biochemistry 42(10):2759–2767
Mobarec JC, Sanchez R, Filizola M (2009) Modern homology modeling of G-protein coupled receptors: which structural template to use? J Med Chem 52(16):5207–5216

Murakami M, Kouyama T (2008) Crystal structure of squid rhodopsin. Nature 453(7193):363–367
Navarro G, Ferre S, Cordomi A, Moreno E, Mallol J, Casado V, Cortes A, Hoffmann H, Ortiz J, Canela EI, Lluis C, Pardo L, Franco R, Woods AS (2010) Interactions between intracellular domains as key determinants of the quaternary structure and function of receptor heteromers. J Biol Chem 285(35):27346–27359
Oldham WM, Hamm HE (2008) Heterotrimeric G protein activation by G-protein-coupled receptors. Nat Rev Mol Cell Biol 9(1):60–71
Olivella M, Gonzalez A, Pardo L, Deupi X (2013) Relation between sequence and structure in membrane proteins. Bioinformatics 29(13):1589–1592. doi:10.1093/bioinformatics/btt249
Palczewski K, Kumasaka T, Hori T, Behnke CA, Motoshima H, Fox BA, Trong IL, Teller DC, Okada T, Stenkamp RE, Yamamoto M, Miyano M (2000) Crystal structure of rhodopsin: a G protein-coupled receptor. Science 289(5480):739–745
Pardo L, Deupi X, Dolker N, Lopez-Rodriguez ML, Campillo M (2007) The role of internal water molecules in the structure and function of the rhodopsin family of G protein-coupled receptors. Chembiochem 8(1):19–24
Park JH, Scheerer P, Hofmann KP, Choe HW, Ernst OP (2008) Crystal structure of the ligand-free G-protein-coupled receptor opsin. Nature 454(7201):183–187
Peeters MC, van Westen GJ, Li Q, Ijzerman AP (2011) Importance of the extracellular loops in G protein-coupled receptors for ligand recognition and receptor activation. Trends Pharmacol Sci 32(1):35–42
Pieper U, Schlessinger A, Kloppmann E, Chang GA, Chou JJ, Dumont ME, Fox BG, Fromme P, Hendrickson WA, Malkowski MG, Rees DC, Stokes DL, Stowell MH, Wiener MC, Rost B, Stroud RM, Stevens RC, Sali A (2013) Coordinating the impact of structural genomics on the human alpha-helical transmembrane proteome. Nat Struct Mol Biol 20(2):135–138
Pin JP, Neubig R, Bouvier M, Devi L, Filizola M, Javitch JA, Lohse MJ, Milligan G, Palczewski K, Parmentier M, Spedding M (2007) International union of basic and clinical pharmacology. LXVII. Recommendations for the recognition and nomenclature of G protein-coupled receptor heteromultimers. Pharmacol Rev 59(1):5–13
Rasmussen SG, Choi HJ, Fung JJ, Pardon E, Casarosa P, Chae PS, Devree BT, Rosenbaum DM, Thian FS, Kobilka TS, Schnapp A, Konetzki I, Sunahara RK, Gellman SH, Pautsch A, Steyaert J, Weis WI, Kobilka BK (2011a) Structure of a nanobody-stabilized active state of the beta(2) adrenoceptor. Nature 469(7329):175–180
Rasmussen SG, DeVree BT, Zou Y, Kruse AC, Chung KY, Kobilka TS, Thian FS, Chae PS, Pardon E, Calinski D, Mathiesen JM, Shah ST, Lyons JA, Caffrey M, Gellman SH, Steyaert J, Skiniotis G, Weis WI, Sunahara RK, Kobilka BK (2011b) Crystal structure of the beta2 adrenergic receptor-Gs protein complex. Nature 477(7366):549–555
Riek RP, Rigoutsos I, Novotny J, Graham RM (2001) Non-alpha-helical elements modulate polytopic membrane protein architecture. J Mol Biol 306(2):349–362
Rosenbaum DM, Cherezov V, Hanson MA, Rasmussen SG, Thian FS, Kobilka TS, Choi HJ, Yao XJ, Weis WI, Stevens RC, Kobilka BK (2007) GPCR engineering yields high-resolution structural insights into beta2-adrenergic receptor function. Science 318(5854):1266–1273
Rosenbaum DM, Rasmussen SG, Kobilka BK (2009) The structure and function of G-protein-coupled receptors. Nature 459(7245):356–363
Rosenkilde MM, Benned-Jensen T, Frimurer TM, Schwartz TW (2010) The minor binding pocket: a major player in 7TM receptor activation. Trends Pharmacol Sci 31(12):567–574
Sansuk K, Deupi X, Torrecillas IR, Jongejan A, Nijmeijer S, Bakker RA, Pardo L, Leurs R (2011) A structural insight into the reorientation of transmembrane domains 3 and 5 during family a G protein-coupled receptor activation. Mol Pharmacol 79(2):262–269
Scheerer P, Park JH, Hildebrand PW, Kim YJ, Krauss N, Choe HW, Hofmann KP, Ernst OP (2008) Crystal structure of opsin in its G-protein-interacting conformation. Nature 455(7212):497–502
Senes A, Gerstein M, Engelman DM (2000) Statistical analysis of amino acid patterns in transmembrane helices: the GxxxG motif occurs frequently and in association with β-branched residues at neighboring positions. J Mol Biol 296:921–936
Serrano-Vega MJ, Magnani F, Shibata Y, Tate CG (2008) Conformational thermostabilization of the beta1-adrenergic receptor in a detergent-resistant form. Proc Natl Acad Sci U S A 105(3):877–882
Shimamura T, Shiroishi M, Weyand S, Tsujimoto H, Winter G, Katritch V, Abagyan R, Cherezov V, Liu W, Han GW, Kobayashi T, Stevens RC, Iwata S (2011) Structure of the human histamine H1 receptor complex with doxepin. Nature 475(7354):65–70
Shukla AK, Manglik A, Kruse AC, Xiao K, Reis RI, Tseng WC, Staus DP, Hilger D, Uysal S, Huang LY, Paduch M, Tripathi-Shukla P, Koide A, Koide S, Weis WI, Kossiakoff AA, Kobilka BK, Lefkowitz RJ (2013) Structure of active beta-arrestin-1 bound to a G-protein-coupled receptor phosphopeptide. Nature 497(7447):137–141
Simon MI, Strathmann MP, Gautam N (1991) Diversity of G proteins in signal transduction. Science 252(5007):802–808
Smit MJ, Vischer HF, Bakker RA, Jongejan A, Timmerman H, Pardo L, Leurs R (2007) Pharmacogenomic and structural analysis of constitutive G protein-coupled receptor activity. Annu Rev Pharmacol Toxicol 47:53–87
Standfuss J, Edwards PC, D'Antona A, Fransen M, Xie G, Oprian DD, Schertler GF (2011) The structural basis of agonist-induced activation in constitutively active rhodopsin. Nature 471(7340):656–660
Thompson AA, Liu W, Chun E, Katritch V, Wu H, Vardy E, Huang XP, Trapella C, Guerrini R, Calo G, Roth

BL, Cherezov V, Stevens RC (2012) Structure of the nociceptin/orphanin FQ receptor in complex with a peptide mimetic. Nature 485(7398):395–399

Venkatakrishnan AJ, Deupi X, Lebon G, Tate CG, Schertler GF, Babu MM (2013) Molecular signatures of G-protein-coupled receptors. Nature 494(7436):185–194

Von Heijne G (1991) Proline kinks in transmembrane alpha-helices. J Mol Biol 218(3):499–503

Wacker D, Wang C, Katritch V, Han GW, Huang XP, Vardy E, McCorvy JD, Jiang Y, Chu M, Siu FY, Liu W, Xu HE, Cherezov V, Roth BL, Stevens RC (2013) Structural features for functional selectivity at serotonin receptors. Science 340(6132):615–619

Wang C, Jiang Y, Ma J, Wu H, Wacker D, Katritch V, Han GW, Liu W, Huang XP, Vardy E, McCorvy JD, Gao X, Zhou EX, Melcher K, Zhang C, Bai F, Yang H, Yang L, Jiang H, Roth BL, Cherezov V, Stevens RC, Xu HE (2013a) Structural basis for molecular recognition at serotonin receptors. Science 340(6132):610–614

Wang C, Wu H, Katritch V, Han GW, Huang XP, Liu W, Siu FY, Roth BL, Cherezov V, Stevens RC (2013b) Structure of the human smoothened receptor bound to an antitumour agent. Nature 497(7449):338–343

Warne T, Serrano-Vega MJ, Baker JG, Moukhametzianov R, Edwards PC, Henderson R, Leslie AG, Tate CG, Schertler GF (2008) Structure of a beta(1)-adrenergic G-protein-coupled receptor. Nature 454:486–491

Warne T, Moukhametzianov R, Baker JG, Nehme R, Edwards PC, Leslie AG, Schertler GF, Tate CG (2011) The structural basis for agonist and partial agonist action on a beta(1)-adrenergic receptor. Nature 469(7329):241–244

White JF, Noinaj N, Shibata Y, Love J, Kloss B, Xu F, Gvozdenovic-Jeremic J, Shah P, Shiloach J, Tate CG, Grisshammer R (2012) Structure of the agonist-bound neurotensin receptor. Nature 490(7421):508–513

Whorton MR, Bokoch MP, Rasmussen SG, Huang B, Zare RN, Kobilka B, Sunahara RK (2007) A monomeric G protein-coupled receptor isolated in a high-density lipoprotein particle efficiently activates its G protein. Proc Natl Acad Sci U S A 104(18):7682–7687

Wu B, Chien EY, Mol CD, Fenalti G, Liu W, Katritch V, Abagyan R, Brooun A, Wells P, Bi FC, Hamel DJ, Kuhn P, Handel TM, Cherezov V, Stevens RC (2010) Structures of the CXCR4 chemokine GPCR with small-molecule and cyclic peptide antagonists. Science 330(6007):1066–1071

Wu H, Wacker D, Mileni M, Katritch V, Han GW, Vardy E, Liu W, Thompson AA, Huang XP, Carroll FI, Mascarella SW, Westkaemper RB, Mosier PD, Roth BL, Cherezov V, Stevens RC (2012) Structure of the human kappa-opioid receptor in complex with JDTic. Nature 485(7398):327–332

Xu F, Wu H, Katritch V, Han GW, Jacobson KA, Gao ZG, Cherezov V, Stevens RC (2011) Structure of an agonist-bound human A2A adenosine receptor. Science 332(6027):322–327

Zhang C, Srinivasan Y, Arlow DH, Fung JJ, Palmer D, Zheng Y, Green HF, Pandey A, Dror RO, Shaw DE, Weis WI, Coughlin SR, Kobilka BK (2012) High-resolution crystal structure of human protease-activated receptor 1. Nature 492(7429):387–392

Part II

GPCRs in Motion: Insights from Simulations

Structure and Dynamics of G-Protein Coupled Receptors

3

Nagarajan Vaidehi, Supriyo Bhattacharya, and Adrien B. Larsen

Abstract

G-protein coupled receptors (GPCRs) are seven helical transmembrane proteins that mediate cell-to-cell communication. They also form the largest superfamily of drug targets. Hence detailed studies of the three dimensional structure and dynamics are critical to understanding the functional role of GPCRs in signal transduction pathways, and for drug design. In this chapter we compare the features of the crystal structures of various biogenic amine receptors, such as β_1 and β_2 adrenergic receptors, dopamine D_3 receptor, M2 and M3 muscarinic acetylcholine receptors. This analysis revealed that conserved residues are located facing the inside of the transmembrane domain in these GPCRs improving the efficiency of packing of these structures. The NMR structure of the chemokine receptor CXCR1 without any ligand bound, shows significant dynamics of the transmembrane domain, especially the helical kink angle on the transmembrane helix6. The activation mechanism of the β_2-adrenergic receptor has been studied using multiscale computational methods. The results of these studies showed that the receptor without any ligand bound, samples conformations that resemble some of the structural characteristics of the active state of the receptor. Ligand binding stabilizes some of the conformations already sampled by the apo receptor. This was later observed in the NMR study of the dynamics of human β_2-adrenergic receptor. The dynamic nature of GPCRs leads to a challenge in obtaining purified receptors for biophysical studies. Deriving thermostable mutants of GPCRs has been a successful strategy to reduce the conformational heterogeneity and stabilize the receptors. This has lead to several crystal

N. Vaidehi (✉) • S. Bhattacharya • A.B. Larsen
Division of Immunology, Beckman Research Institute of the City of Hope, 1500, E. Duarte Road, Duarte, CA 91010, USA
e-mail: NVaidehi@coh.org

M. Filizola (ed.), *G Protein-Coupled Receptors - Modeling and Simulation*, Advances in Experimental Medicine and Biology 796, DOI 10.1007/978-94-007-7423-0_3,

structures of GPCRs. However, the cause of how these mutations lead to thermostability is not clear. Computational studies are beginning to shed some insight into the possible structural basis for the thermostability. Molecular Dynamics simulations studying the conformational ensemble of thermostable mutants have shown that the stability could arise from both enthalpic and entropic factors. There are regions of high stress in the wild type GPCR that gets relieved upon mutation conferring thermostability.

Keywords
Thermostable mutants • Predictions • Conformational states • Drug design • Biogenic amines • Multiscale method • LItiCon

3.1 Introduction

3.1.1 Introduction to Class A GPCRs

G-protein coupled receptors (GPCRs), also known as the seven transmembrane (TM) receptors form a superfamily of membrane bound proteins that mediate cell signaling and are responsible for many physiological functions of the cell. GPCRs belong to one of the largest superfamily of receptors that serve as important drug targets. GPCRs have been broadly classified into three classes based on the lengths of their amino termini. In this chapter we discuss only the class A GPCRs (denoted as just GPCRs hereafter) that have short, about 10–75 amino acids in their amino terminus. GPCRs have a structural topology consisting of seven transmembrane (TM) helices connected by intracellular (ICL) and extracellular (ECL) loops.

Ligands ranging from small molecules to peptides and proteins activate GPCRs by binding either in the TM domain or in the extracellular loop region. Ligand binding leads to conformational changes in the receptor that results in coupling with cytosolic proteins such as trimeric G-protein or β-arrestin to elicit cellular response (Reiter et al. 2012). Understanding the mechanism of how ligand binding in the extracellular region of the receptor, causing an allosteric effect of coupling to cytosolic proteins, triggering the signal transduction cascade is an important step in modulating various cellular processes initiated by GPCRs.

Although the class A GPCRs have a common seven helical transmembrane topology, their cellular functions are divergent. Further, the same GPCR can bind to different ligands and thereby elicit different responses depending on the nature of the ligand that binds to the GPCR (Audet and Bouvier 2012). They do so by varying the three dimensional structural conformation of the receptor. Different ligands also elicit different types of response at a receptor level (Vaidehi and Kenakin 2010). A full agonist elicits a maximal response while a partial agonist elicits only a fraction of the maximal response at saturating concentrations of the ligand. A very low efficacy ligand that may not produce any measurable response but could prevent further response from a full or a partial agonist is known as neutral antagonist. Ligands that reduce the basal response exhibited in a ligand-independent fashion in many GPCRs are known as an inverse agonist. GPCRs exist in multiple "active" and "inactive" conformational states that are in equilibrium even in the absence of any ligand (Niesen et al. 2011; Vaidehi and Bhattacharya 2011). This leads to the ligand independent basal activation of the receptor and the level of the basal activation is different for each GPCR. In the ligand independent state, the population of the receptor in the inactive state of the receptor is still higher than the active state of the receptor (Niesen et al. 2011). The "inactive state" is defined as a conformation of the receptor that does not activate the G-protein. Upon binding of an agonist (full or partial) and the G-protein, the GPCR undergoes conformational changes that lead to the "active state" of the receptor (Rasmussen et al. 2011). Upon binding of an agonist or inverse agonist the relative population of the active and inactive states shift depending on the efficacy of the ligand (Yao et al. 2006, 2009; Vaidehi and Kenakin 2010). However binding of the agonist alone may not be sufficient to activate the receptor, which is evident from a few agonist bound crystal structures of GPCRs (e.g. β_2-adrenergic receptor bound to BI-167107, β_1-adrenergic receptor bound to isoproterenol). These structures are very close to the inactive conformations of various GPCRs. It has been suggested that binding of both the agonist and G-protein are possibly required to shift the conformational equilibrium of the receptor to the fully active state. However, the agonist bound adenosine receptor A2A shows active-like state movement (Lebon et al. 2011b). Exactly how do the receptor conformations change upon binding of various ligands of different efficacies and how the relative population of the various conformational states are modulated by ligand binding is a current topic in GPCR research (Granier and Kobilka 2012). An understanding of the structural and dynamic features of the various active and inactive conformations of GPCRs, and the molecular mechanism of the conformational changes from the inactive to the active states is important in designing functionally specific drugs (Galandrin et al. 2007; Mailman and Murthy 2010; Kenakin et al. 2012). In fact this dynamic flexibility of GPCRs poses one of the major bottlenecks in obtaining pure protein for biophysical structural studies of GPCRs (Tate and Schertler 2009; Tate 2012). These bottlenecks have been overcome partially, resulting in many recently published three dimensional crystal structures of GPCRs. There are two major strategies that have been used to stabilize GPCRs for crystallization: (1) Attachment of an antibody (Steyaert and Kobilka 2011) or a protein such as T4Lysozyme (denoted as T4L hereafter) to the ICL or ECL to facilitate crystallization (Cherezov et al. 2007; Katritch et al. 2012) and (2) screening for thermostable mutations that stabilize a particular receptor conformation to enable crystallization. In fact many GPCRs such as β_1-adrenergic (Warne et al. 2008) and β_2-adrenergic receptors (Cherezov et al. 2007; Rasmussen et al. 2011), muscarinic acetyl choline receptors M2 and M3 (Haga et al. 2012; Kruse et al. 2012), dopamine D_3DR receptor (Chien et al. 2010), histamine H1HR receptor (Shimamura et al. 2011), adenosine A2A receptors (Lebon et al. 2011b, Xu et al. 2011) chemokine receptor CXCR4 (Wu et al. 2010), peptide receptor neurotensin receptor 1 (White et al. 2012) and protease activated receptor 1 (Zhang et al. 2012), and opioid receptors (Granier and Kobilka 2012; Manglik et al. 2012; Thompson et al. 2012) have been crystallized using these two strategies for stabilizing specific conformations. With the publication of all these GPCR structures, one can understand the structural differences between closely related receptors as well as those that are far removed in sequence space. Additionally β_2-adrenergic receptor has been crystallized both in the "inactive" and "fully active" conformational states. Adenosine A_{2A} receptor has been crystallized in the inactive and partially active state (Lebon et al. 2011b; Xu et al. 2011). Many of these GPCRs have also been crystallized with only the agonist bound. In the next section we compare various GPCR crystal structures and analyze the similarities and differences.

3.1.2 Comparison of Various GPCR Structures

GPCRs have a seven helical topology and the structural differences between two structures arise from the differences in the helical tilts and kinks, helical rotational and translational orientations, packing of the helices, and the loop structures. Two GPCR structures are compared by aligning the C_α atoms, and the root mean squared deviation (RMSD) in coordinates of a set of chosen atoms that is common between the two compared structures is reported. RMSD is a crude measure of the differences between structures and does not describe the nuances in structural similarities and differences especially for two GPCR structures with divergent sequences. Therefore we developed a computational method (*GPCRCompare*) that provides structural

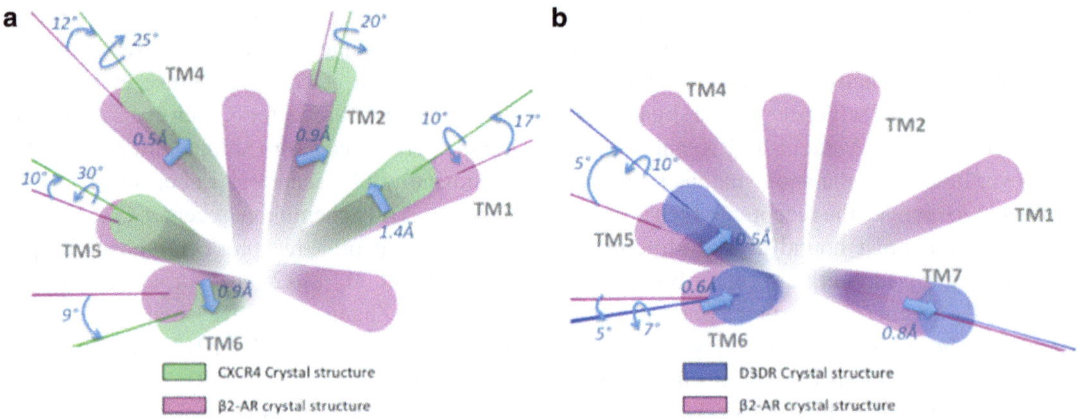

Fig. 3.1 Comparison of rigid body structural differences (helical translation, tilt and axial rotation) of CXCR4 (pdb ID: 3ODU) and D_3DR (pdb ID: 3PBL) crystal structures to that of β_2AR (pdb ID: 2RH1); (**a**) CXCR4; (**b**) D3DR; Helices are represented by cylinders; *block arrows* represent center of mass translations, *curved arrows* represent tilts and rotations. Figures are schematic, and not to actual scale

differences in terms of the relative helical tilt, rotational and translational orientations (Bhattacharya et al. 2013). Such a comparison provides a more intuitive understanding of the structural similarities and differences. Figure 3.1 shows the comparison of two crystal structures of similar receptors such as the dopamine D_3DR receptor and β_2 adrenergic receptor and also comparison of two different receptors such as the chemokine receptor, CXCR4 and the dopamine receptor D_3DR that have very different amino acid sequences. The structural comparisons here have been made based on relative translational, rotational and tilt orientations of the seven helices. Such an analysis provides insight into the type of helical movements that occur when comparing GPCR structures. The structural differences in the helical movements between CXCR4 and β_2 adrenergic receptor are more predominant as anticipated, compared to the differences between D_3DR and β_2 adrenergic receptor. Significant structural differences between CXCR4 and β_2 adrenergic receptor are in the rotational and tilt orientations of TM2, TM4 and TM5 as shown in Fig. 3.1a. The difference in the tilt angle of TM1 is also significant, but this difference could stem from the truncated amino terminus of each of these receptors. In D_3DR, TM4 shows a small rotation (10°), while TM6 shows a minor translation (0.6 Å), tilt (5°), and rotation (8°) compared to the β_2-adrenergic receptor.

The crystal structures of several class A GPCRs have been compared in previous publications (Katritch et al. 2012; Deupi 2012; Bhattacharya et al. 2013; Filizola and Devi 2013). In the next section we demonstrate the advantages of comparing GPCR structures using helical structural parameters with *GPCRCompare* program to compare the crystal structures of the inactive states of the biogenic amine receptors (β_1 and β_2 adrenergic receptors, muscarinic acetyl choline receptors M2 and M3, dopamine receptor D_3DR and histamine receptor H1HR) crystallized so far.

3.1.3 Comparison of the Inactive States of the Biogenic Amine Receptors

Here we have compared the adrenergic receptor structures to the dopamine, histamine and muscarinic acetylcholine receptors. Since the structures of β_1 and β_2-adrenergic receptors are similar, we report a comparison of the adrenergic receptor structures with those of the other biogenic amine family receptors. The major differences between the dopamine D_3

receptor and the β_1 and β_2–adrenergic receptors are tilts of TM4, TM5 and TM6 (10°, 9°, 8° respectively) and rotations of TM4 and TM5 (9° and 10°). The histamine H1 receptor shows a 7° tilt of TM4 and major rotation of TM6, and TM7 (13°, 8°). The muscarinic acetylcholine M2 receptor shows a 1.2 Å translation of TM5, tilts of TM4 and TM5 (7°, 6°), and rotations of TM1, TM5, TM6 (11°, 6°, 13°). The M3 muscarinic receptor shows significant differences in the rotational orientations of TM5 and TM6 (9°, 13° respectively). Thus in comparison to the β_1 and β_2–adrenergic receptors, TM4, TM5 and TM6 are the most diverse in conformation among the biogenic amine family of receptors. In these receptors, TM3, TM5 and TM6 contain a large proportion of the residues that line the ligand binding pocket of the orthosteric ligands. While TM3 consists of a conserved aspartate residue that forms a salt bridge with the amine group of the endogenous biogenic amines, the conformations of TM5 and TM6 give selectivity to the specific ligands each of these receptors binds. The diverse conformation of TM4 could be related to selectivity of dimerization with other GPCRs, since TM4 has been shown to form the dimer interface in several GPCRs (Johnston et al. 2011).

With the availability of representative structures of the biogenic amine receptors an in depth analysis of the location of the conserved residues and their role in structure packing would be useful for structure prediction protocols (Bhattacharya et al. 2008a, b, 2013; Bhattacharya and Vaidehi 2010; Hall et al. 2009; Nedjai et al. 2011). We have analyzed the extent to which the residues are conserved in each one of these receptors within their subfamilies and compared the location of the conserved residues in the respective receptor structures as shown in Fig. 3.2. The conserved residues are shown in blue and the non-conserved ones are shown in red. It can be seen from Fig. 3.2, that the residues in the intracellular half of the transmembrane domains of the receptors are more conserved than the extracellular half of the receptors. As shown in Fig. 3.2c, d, the residues that point inwards into the transmembrane barrel are more conserved than the residues that point to the membrane in all these receptors. TM3 has more conserved residues than any other TM helix. The extracellular loops show less conserved residues than the intracellular loops. ICL1 is highly conserved in all the receptors. We have omitted histamine H1HR in this analysis since it had only 20 % sequence identity to other subtypes of the histamine receptors and skewed this analysis.

3.1.4 Analysis of the NMR Structures of the Chemokine Receptor CXCR1

We have also compared the extent of variability and the dynamics observed in the recently published three dimensional NMR structures of the chemokine receptor CXCR1 (Park et al. 2012). This structure is the first, class A GPCR structure to be solved without any ligand bound to it. The NMR structures of CXCR1 in phospholipids show significant dynamics. We have analyzed the extent of the dynamics of these structures using *GPCRCompare* program in this section. The RMSD in coordinates among these ten NMR models ranges from 1.8 to 2.2 Å. The ensemble of NMR conformations is shown in Fig. 3.3. The RMSD in coordinates among these ten NMR models range from 1.8 to 2.2 Å. The kink angle on TM6 ranges from 21° to 34° showing significant dynamics in the ten NMR structures as shown in Figs. 3.3 and 3.4a. To quantify the dynamic fluctuations of different regions of the receptor, we computed the average root mean square fluctuations (RMSF) of the Cα coordinate of each residue from the average NMR structure. Figure 3.4b shows a representative NMR structure with the residues colored according to their mean RMSF, with red being large fluctuations and blue being less than 1 Å change in RMSD. TM6 and helix 8 are the most dynamic segments in the CXCR1 structure, followed by TM3 and TM5. Among the loops, ECL2 is packed within the TM bundle and shows very little fluctuations, whereas ICL1 and ICL3 are the most dynamic. In order to further "dissect" the nature of helical motions observed in CXCR1, we compared the rigid body orientations of the seven TM helices of

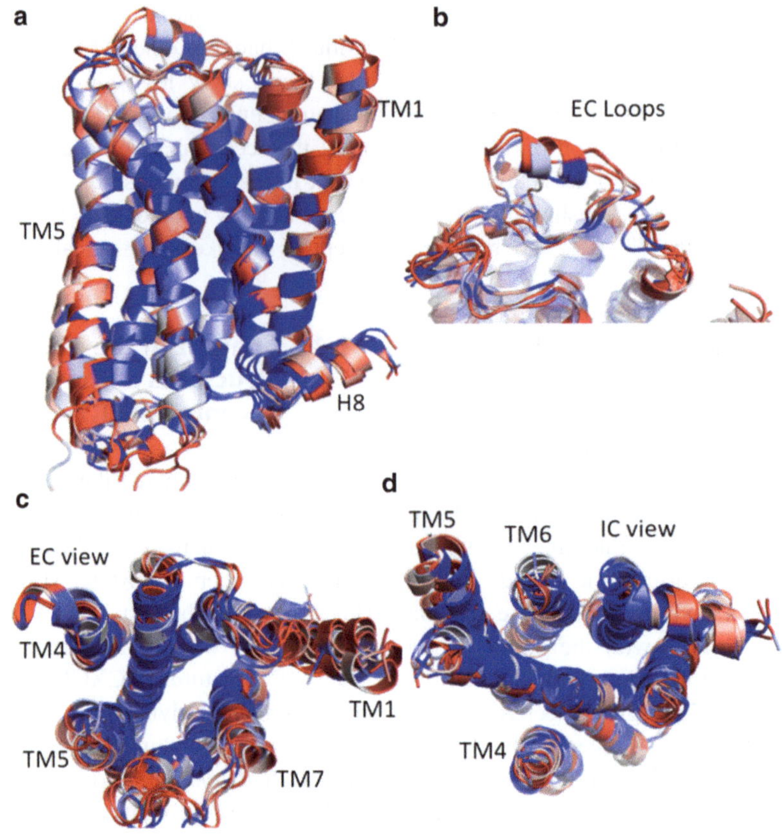

Fig. 3.2 Comparison of the location of conserved residues in the crystal structures of β_1 and β_2 adrenergic receptors, M2 and M3 muscarinic acetylcholine receptors, D_3 dopamine receptor. The residues that are highly conserved within their respective subtypes are shown in *blue* and the non-conserved residues are shown in *red*. (**a**) The side view of the seven TM domains showing that more conserved residues are in the intracellular region of the receptors. (**b**) The extracellular loops shows a lot of variability that could confer specificity to ligand binding; (**c**) The view from the extracellular side of the receptor showing that the conserved residues mostly face inside the TM bundle, and (**d**) The view from the intracellular side of the receptors showing more conserved residues pointing inwards towards the TM bundle

the ten NMR conformations against one another using the program *GPCRCompare*. In the course of this analysis, we measured the Cα RMSD of individual helices, first by aligning the overall structures (referred to as "in place" in Fig. 3.4), and then by aligning the TM helix of one structure to that of the other (referred to as "aligned" in Fig. 3.4). Figure 3.4c compares the mean RMSD of the TM domains both before and after alignment. Since the alignment removes the rigid body orientational differences between the helices, the "aligned" RMSD gives a measure of the internal deformations (kinks) of the helices. For CXCR1, the aligned RMSDs of the TM helices range from 0.4 to 0.8 Å with TM7 showing the highest aligned RMSD. Next we compared the rigid body orientations of the TM helices. TM5, TM6, and TM7 show significant variations in the tilt and rotational motion, whereas TM3 shows less tilt and larger rotation. A schematic of the dynamic motions of CXCR1 is shown in Fig. 3.4d. The table in Fig. 3.4e lists the magnitudes of rigid body movements of the different helices, as calculated by *GPCRCompare*.

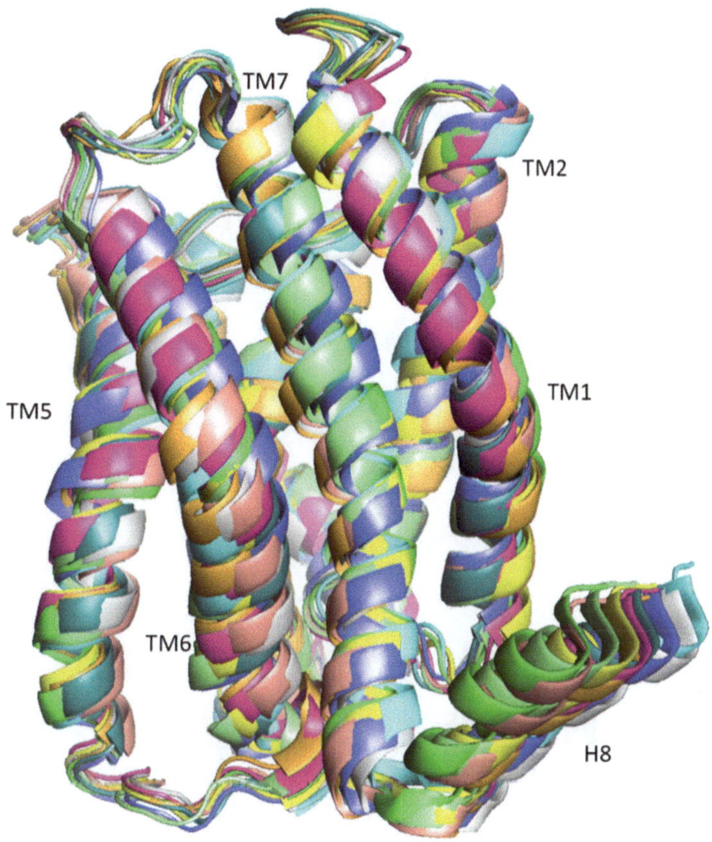

Fig. 3.3 Superposition of the 10 NMR derived structures from the pdb file 2LNL. The backbone of each structure is shown in a *different color*. The kink on helix6 is dynamic in the NMR time scale

3.2 Flexibility of GPCR Structures and Relevance to Drug Design

GPCRs exhibit multiple functions in the cell, that is achieved by adopting various receptor conformations influenced by binding of the ligand, G-protein, other proteins coupling to the receptor, and by the lipid bilayer environment. Hence an understanding of the receptor flexibility and conformational dynamics would lead to more selective design of drugs. The fact that the GPCR conformations are dynamic has been substantiated by many experimental and computational techniques (Swaminath et al. 2004, 2005; Bhattacharya et al. 2008a; Ahuja and Smith 2009; Dror et al. 2009; Khelashvili et al. 2009; Bhattacharya and Vaidehi 2010; Vaidehi and Kenakin 2010; Vaidehi 2010; Niesen et al. 2011; Provasi et al. 2011; Park et al. 2012; Nygaard et al. 2013). Further the effect of ligands of varied efficacies on the dynamics of the receptor has also been well studied using several biophysical techniques.

3.2.1 NMR and Fluorescence Spectroscopy Shows Flexibility in GPCR Conformations

The NMR structure of the chemokine receptor CXCR1 in phospholipid bilayers has been published recently (Park et al. 2012). As discussed in the previous section, comparison of the ten NMR models shows significant level of dynamics in the system. The structural variations in the ten NMR structures are in the intracellular part of TM3, kink modulation on TM6 and helix8 as shown in Figs. 3.3 and 3.4. Interestingly, GPCR crystal structures show the TM helices to be less flexible (as evident by low b-factors),

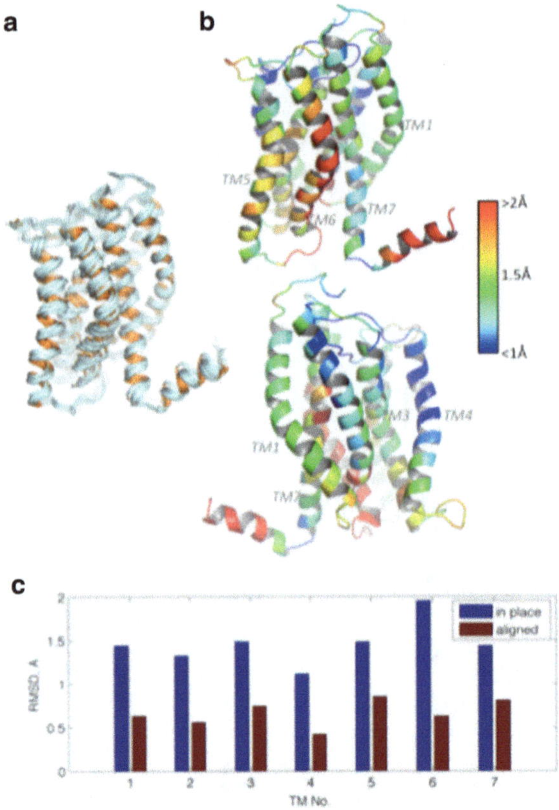

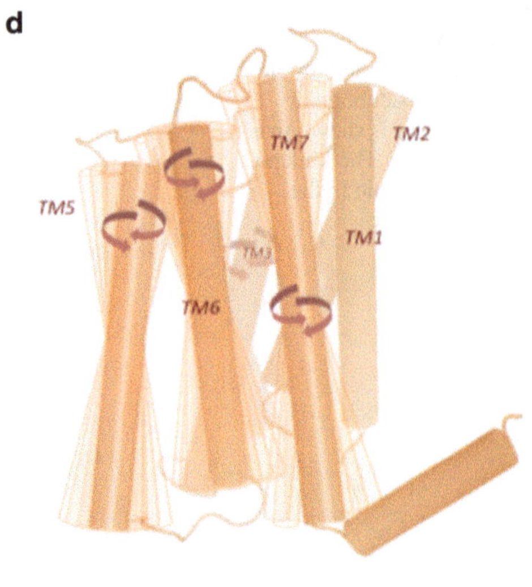

e

TM no.	Mean translation	Mean tilt	Mean rotation
1	0.33 Å	3.2°	3.2°
2	0.39 Å	4.2°	3.8°
3	0.39 Å	5.4°	9.6°
4	0.41 Å	5.2°	5.3°
5	0.58 Å	7.0°	9.3°
6	0.86 Å	5.5°	15.5°
7	0.66 Å	5.7°	19.5°

Fig. 3.4 Variability among the 10 NMR derived conformations of CXCR1 (pdb ID: 2LNL); (**a**) Cartoon representation of the 10 NMR conformations; (**b**) Two representative NMR structures with residues colored according to the mean RMSD fluctuations in coordinates; (**c**) RMSDs of each transmembrane helix before and after alignment; (**d**) Schematic representation showing movements exhibited in TM5, TM6 and TM7 deduced from the ensemble of NMR structures; (**e**) Table showing the average value of the translation, tilt and rotational orientations of seven helices for the ten NMR derived structures calculated by the program *GPCRCompare* (Bhattacharya et al. 2013)

whereas the loops are more mobile with higher b-factors. However in the apoprotein structure of CXCR1 the TM helices show high mobility. This is a marked distinction from crystal structures, which are cryogenic conformations, whereas the NMR structures represent the dynamic nature of GPCRs at the physiological temperature. Also the use of phospholipid bilayers in the NMR measurements could contribute to the increased mobility of the TM regions. However the high mobility of the TM helices is in direct agreement with the established notion on GPCR activation, which suggests that the helical domains (most importantly TM6) undergo conformational change to trigger activation. It is noteworthy that in the NMR conformations of CXCR1, TM6 is the most dynamic of all the TMs. The kink angle on transmembrane helix 6 is also dynamic and varies from 21° to 34°. Based on earlier studies on activation of GPCRs, the side chain conformation of the $Trp^{6.48}$ of the highly conserved motif WxP in TM6, was expected to toggle upon activation (Yao et al. 2006). This toggle was observed in the NMR studies of the activation of rhodopsin (Ahuja and Smith 2009) but was not seen in the crystal structure of the active state of β_2-adrenergic receptor (Rasmussen et al. 2011), or the partial active state of adenosine A_{2A} receptor (Lebon et al. 2011a, b; Xu et al. 2011). The side chain conformation of $Trp255^{6.48}$ in CXCR1 shows significant variation among the ten NMR structures, denoting the dynamic nature of this residue in the absence of any ligand. Here we have used the Ballesteros and Weinstein numbering for residues that were defined for class A GPCRs (Ballesteros and Weinstein 1995). The first number denotes the TM helix in which the residue is located and the second number is the position of the residue from the most conserved residue position denoted as x.50 in that helix. In the CXCR1 NMR structures, the distance between the intracellular regions of TM3 and TM6 (for example, the distance between Arg135 and Met241) varies between 9.1 and 12.6 Å in the ten NMR structures. This analysis shows that the GPCR without any ligand samples a wide range of conformations. Future studies on NMR structures with ligand on CXCR1 would reveal if conformational selection from the subset of conformations sampled by the apoprotein could be one of the predominant mechanisms of ligand binding in GPCRs.

Biophysical studies on the visual receptor rhodopsin, using spin-labeling techniques, solid-state NMR, fluorescent spectroscopy, computational methods, and crystal structure of the partially active state opsin show that the intracellular region of the TM3 and TM6 move apart upon activation (Park et al. 2008). Additionally, there are several inter-residue contacts also known as "conformational switches" made or broken upon activation of rhodopsin (Krishna et al. 2002; Hubbell et al. 2003; Schertler 2005; Park et al. 2008; Ahuja and Smith 2009; Zaitseva et al. 2010). Briefly, comparison of the rhodopsin and opsin crystal structures showed considerable conformational changes in the intracellular region of TM5 and TM6. TM5 also shows elongation of the helix in this region. Comparatively smaller changes were observed in the extracellular loop 2 (ECL2) (Ahuja and Smith 2009). There are substantial rearrangements in all the three intracellular loops in the opsin structure. Thus it is evident that activation is associated with considerable conformational changes especially in the intracellular region of the receptor.

Fluorescence spectroscopic life time measurements with purified receptors, have shown that binding of ligands of varied efficacies to the human β_2-adrenergic receptor lead to stabilization of ligand specific conformations of the receptor (Swaminath et al. 2004, 2005; Yao et al. 2006). Bioluminescence Resonance Energy Transfer (BRET) studies pioneered by Bouvier and coworkers, on the kinetics of G protein coupling to β_2-adrenergic receptor and β_1-adrenergic receptor in living cells, showed differential activation kinetics by full, partial and inverse agonists (Vilardaga et al. 2003; Galés et al. 2005; Galandrin et al. 2007). Fluorescence experiments of reconstituted human β_2-adrenergic receptor in lipid nanoparticles coupled with the G-protein, showed that ligands modulate the binding affinity of the receptor to the G-protein and similarly binding of the G-protein also modulates the efficacy of the receptor to the ligand (Yao et al. 2009). Since the experimental results on activation of GPCRs are fragmented, computational methods play an important role in combining these observations to provide atomic level insight into the mechanism of the conformational transitions and activation in GPCRs.

3.2.2 Multiscale Computational Methods for Understanding the Mechanism of Conformational Transitions and Activation of GPCRs

Computational methods have been used to delineate the atomic level mechanisms of activation of GPCRs. All-atom molecular dynamics (MD) simulations combined with inter-residue distances measured by NMR as restraints provided insight into the movement of ECL2 in the activation of rhodopsin (Ahuja and Smith 2009). Starting from the inactive state crystal structure of rhodopsin, several MD simulation studies have provided insight into the early events of activation of rhodopsin (Saam et al. 2002; Crozier et al. 2007; Grossfield et al. 2008; Khelashvili et al. 2009). The activation mechanism of rhodopsin from MD trajectories has been described in terms of structural changes in inter-residue contacts also known

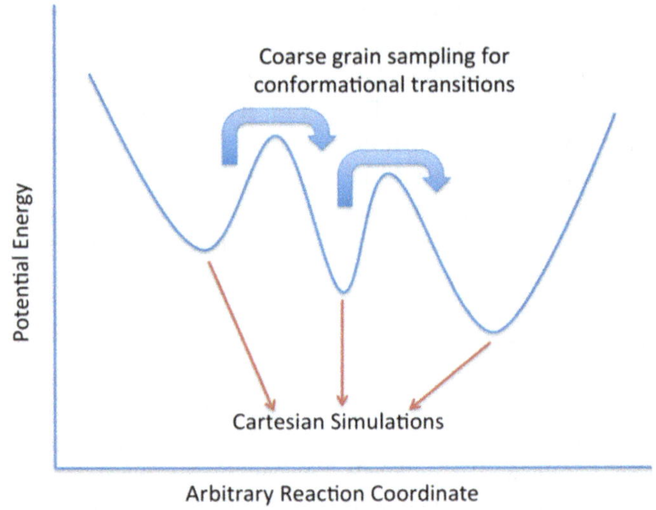

Fig. 3.5 A scheme of the conformational sampling in multiscale simulation method. The coarse grained simulation methods in this figure represents methods that sample wider range of conformations with accurate all-atom forcefield. The coarse grained methods capture conformational transitions. All-atom Cartesian simulations are then performed starting from the low energy and well populated structures from the coarse grained simulations. This would lead to more accurate sampling of the low energy states

as microdomain switches. These microdomain switches are also modulated by interaction with cholesterol (Khelashvili et al. 2009).

Dror et al. performed microseconds of MD simulations on the inactive state of β2AR crystal structure with and without T4L lysozyme (T4L) bound to show that there are two inactive states of the receptor with and without the ionic lock formed between Arg$^{3.50}$ and Glu$^{6.30}$ (Dror et al. 2009). More recently, long time scale MD simulations on the pathway of ligand entry into β_2-adrenergic receptor showed that the ligand initially binds to the extracellular loop region where they shed the water molecules and hence show a finite lifetime in this region. Following this event the ligand trickles down into the orthosteric binding site. The pathway is the same for agonists and antagonists tested in these simulations (Dror et al. 2011). Although all-atom MD simulations provide valuable insights into atomic level mechanism of the early events of GPCR activation, there still exists a bottleneck in simulating long time scale biological processes such as GPCR activation using all atom MD simulations. Starting from the inactive state of the receptor, it is not feasible to sample the active state using all-atom MD simulations. This is because the system gets trapped in the potential energy minimum of the inactive state with high free energy barriers to transition to the active state, and the dynamic movement to the active state has to occur through a series of rare events as the system moves from one potential energy basin to another. Thus advanced multiscale dynamics methods are required to study the activation mechanism of GPCRs.

3.2.3 Multiscale Method Simulations of Activation of GPCRs

Multiscale simulation methods comprise of using (1) a coarse grained MD technique that samples multiple conformational states by overcoming energy barriers, (2) followed by finer scale all-atom MD simulations starting from conformations extracted from the coarse grained simulations (see Fig. 3.5). In here, by "coarse grained" we mean a wider and enhanced sampling dynamics method than all-atom MD simulations and not necessarily the coarse grained forcefield.

Steered MD (Isralewitz et al. 2001), Accelerated MD (Hamelberg et al. 2007), metadynamics MD (Provasi et al. 2011), methods are examples of methods that sample multiple conformational states of proteins. Both Steered MD (Gouldson et al. 2004) and metadynamics simulations have been used to study the ligand specific conformational states of β_2-adrenergic receptor. This method provides insight into how ligands of varied efficacies modulate the free energy surface of the β_2-adrenergic receptor. The method however requires prior knowledge of the active and inactive states of the receptor as well as predetermined activation pathways (Provasi et al. 2011).

For cases where the active state of the receptor is not known, we developed a multiscale dynamics approach by combining a coarse-grained discrete conformational sampling method, called LITICon (Bhattacharya et al. 2008a, b) with fine-grained molecular dynamics investigating the effect of various ligand binding on the ensemble of conformations sampled by human β_2-adrenergic receptor (Bhattacharya and Vaidehi 2010; Niesen et al. 2011). The discrete conformational sampling method allows traversing over energy barriers and samples multiple conformations of the receptor (Fig. 3.5). We showed that the receptor, in the absence of any ligand samples an extensive conformational space that includes a shear motion of the intracellular regions of TM5 and TM6 and a breathing motion of the ligand binding site. This shear motion is similar to the reorganization of the TM helices observed in the crystal structure of the active state of β_2-adrenergic receptor (Rasmussen et al. 2011). However the extent of the movement in the intracellular regions of TM5 and TM6 with respect to TM3, is considerably less when no ligand is present compared to when the agonist and G-protein are bound to the receptor. Thus, the computational data prior to experiments lead to the prediction that the receptor samples "active-like" conformational states even in the absence of any ligand, and the population of these "active-like" states is much less compared to the population of the inactive state (Niesen et al. 2011). Subsequent NMR experiments studying the ensemble of conformations without the ligand present have actually shown that the β_2-adrenergic receptor without any ligand is lot more flexible and samples multiple states, than when a ligand is bound (Nygaard et al. 2013). This is an example, where the computational methods provided predictions and insights into the conformational sampling of the receptor without any ligand prior to the experiments. Binding of agonist norepinephrine or partial agonist salbutamol leads to selection of a subset of these conformations sampled by the receptor, while inverse agonist carazolol selects only inactive state conformations. This demonstrates the usefulness of the computational methods to provide insights that are difficult to probe experimentally. Thus we conclude that ligand binding in GPCRs lead to conformational selection to a greater extent than the induced conformational changes. However the role of ligand induced receptor movements cannot be eliminated completely.

3.3 Insights into the Structural Basis of Thermostability in GPCRs

3.3.1 Thermostabilization of GPCR Conformations

The dynamic nature of GPCRs confers its adaptability and versatility to get activated by multiple ligands and trigger multiple signaling pathways in the cell. At the same time the dynamic nature has been a great impediment and challenge to purify GPCRs in detergent solution and amenable to crystallization. Although membrane proteins are extremely stable in bilayer environment they are often unstable when solubilized in detergents (Bowie 2001). Two major strategies have been used to circumvent this challenge. One is to attach T4L or antibody to the loop regions (Hanson and Stevens 2009; Steyaert and Kobilka 2011) and the other is to derive thermostable mutants that remain stable in detergents (Tate 2012).

These two strategies have proved successful in obtaining GPCR crystal structures. The experimental procedure for deriving the thermostable mutant GPCRs is tedious and therefore an understanding of how do these thermostable mutations affect the stability of GPCRs is unknown. In this section we will focus on the thermostable mutants and demonstrate how computational methods have been used to understand the effect of thermostable mutations on the stability of the receptor structures.

Membrane proteins show mutations that enhance the protein stability occurring with a high frequency, thus showing that these wild type proteins are probably not optimized for high stability (Bowie 2001). Thermostable mutants have been engineered for GPCRs by time consuming processes as described in detail in the next section. Various types of assays have been used to measure thermostability for membrane proteins (Kawate and Gouaux 2006; Alexandrov et al. 2008). Among these techniques, the radiolabelled ligand binding is the most used technique for GPCRs (Tate 2012). Thermostable mutants have lead to crystal structures for turkey β_1-adrenergic receptor (Warne et al. 2008), human adenosine A_{2A} receptor (Doré et al. 2011, Lebon et al. 2011a, b) and neurotensin receptor 1 (White et al. 2012). Analysis of the amino acid positions that lead to thermostability shows that there is no sequence consensus in the positions among various receptors, nor there is clustering of the thermostable mutant positions. Thus the process of engineering thermostable GPCR mutants is tedious and time consuming. Therefore a molecular level understanding of the effect of thermostable mutations would throw light into rational engineering of mutants for other GPCRs.

3.3.2 Structural Basis of Thermostability from Computational Methods

The fact that GPCRs are dynamic and exhibit multiple active and inactive states have been demonstrated amply with experimental and computational data. This poses a great challenge in purifying these proteins. There are multiple challenges in obtaining purified GPCRs for structural studies. Some of these challenges are: (1) Low level of expression of mammalian GPCRs, (2) optimization of detergents for GPCR solubilization. When solubilized in detergents for purification, GPCRs tend to aggregate (Tate 2012; Grisshammer 2009; Chiu et al. 2008) and, (3) The flexibility of GPCR conformations poses the greatest challenge for the receptor purification due to the conformational heterogeneity. Rational single point mutations to engineer stabilizing disulfide bridges (Standfuss et al. 2007), or mutating certain non-conserved residues that would strengthen helical interfaces were tested (Roth et al. 2008) for producing thermostable mutants of rhodopsin and β_2-adrenergic receptor. Although these strategies were successful, they could not be generalized and applied to any GPCR. Tate and coworkers developed a systematic procedure for deriving thermostable GPCR mutants (Serrano-Vega et al. 2008; Magnani et al. 2008; Shibata et al. 2009; Tate 2012; Cooke et al. 2013). This procedure consists of several steps:

- Ala/Leu scanning mutagenesis where every residue was mutated to Ala (except for Ala residues that were mutated to Leu), and each mutant was expressed and its thermostability was determined relative to the wild type receptor. The thermostability was determined by measuring the antagonist binding affinity at high temperatures.
- Once thermostabilizing single point mutations were identified, these positions were mutated to other amino acid residues to identify improvements in thermostability.
- The best thermostabilizing mutations were then combined to give the optimally stable mutant.

This is a systematic but time-consuming process, and most of the mutation positions are not transferable from one GPCR to another and this fact makes it more tedious (Serrano-Vega and Tate 2009). Insights into why certain mutations stabilize the receptor, as well as why certain mutations specifically stabilize agonist binding versus antagonist binding (Lebon et al.

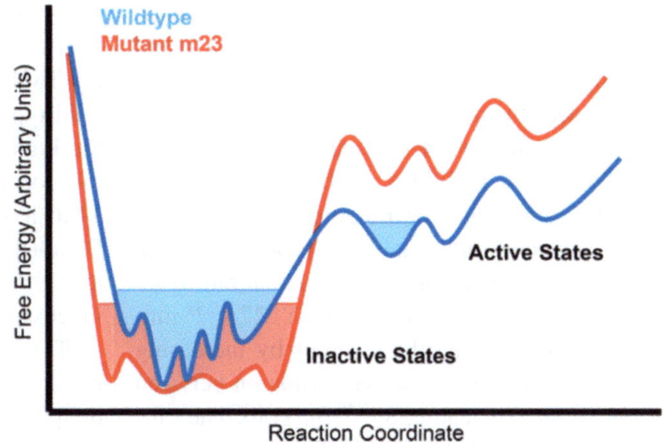

Fig. 3.6 Schematic representation of the receptor population distribution in wt-β_1AR (*blue*) and thermostable mutant m23-β_1AR (*red*). Note that in the mutant m23-β_1AR the receptor conformational population is shifted towards the inactive state of the receptor

2011b) states would greatly aid the design process (Balaraman et al. 2010).

The structural basis of thermostability of soluble proteins has been studied extensively from the perspectives of structure, biophysics and thermodynamics (Sterpone and Melchionna 2012). In contrast, studies on membrane protein stability are less common, which is probably because they are far less tractable experimentally (White and Wimley 1999; Bowie 2011). Measurements of reversible folding and unfolding processes to determine the thermodynamic and kinetic properties of folding have been done only for a few membrane proteins (Booth and Curnow 2009). These measurements have not yet been possible to study the role of thermostabilizing mutations in GPCRs. Therefore computational methods offer a complementary and feasible solution to study the effect of thermostable mutations on the structure and dynamics of GPCRs.

3.3.3 Insights into the Structural Basis of Thermostability from Computational Studies

Serrano-Vega et al. derived a thermally stable mutant of the inactive state of turkey β_1-adrenergic receptor (β_1AR) and showed that a combination of six single point mutations resulted in a 20 °C increase in thermal stability of the inactive state of β_1AR (Serrano-Vega et al. 2008). This inactive state mutant of β_1AR known as m23-β_1AR has six point mutations namely R68S$^{1.59}$, M90V$^{2.53}$, Y227A$^{5.58}$, A282L^{ICL3}, F327A$^{7.37}$ and F338M$^{7.48}$. Using long time scale all-atom MD simulations in explicit lipid bilayer and water, we have studied the effect of the mutations on the stability and dynamics of the m23-β_1AR as compared to the wild type β_1AR referred to as wt-β_1AR hereafter (Niesen et al. 2013).

We found that thermostabilization results in an increase in the number of accessible microscopic conformational substates arising from side chain fluctuations within the inactive state ensemble, effectively increasing the entropy of the inactive state. Figure 3.6 shows the schematic representation of the energy landscape of the inactive state of the receptor for both wild type and m23 mutant. The number of states shown in the figure comes from computational data (Niesen et al. 2013). It is seen that the number of microscopic states in the mutant m23-β_1AR has increased at room temperature. The thermostable mutant m23-β_1AR showed an increase in side chain movement within the inactive state (multiple microscopic states within the inactive state). This increase in side chain flexibility in the m23-β_1AR plays a critical role in absorbing the extra thermal energy at higher temperatures, thus retaining the native fold even at higher temperatures thus preventing unfolding and aggregation. Such entropic stabilization has been observed in soluble thermophilic proteins (Colombo and Merz 1999; Hernandez et al. 2000; Wintrode et al. 2003; Meinhold et al. 2008). The dynamics of the mutant m23-β_1AR also shows loss of correlated motion between residues that are in the ligand binding site to the G-protein coupling site and this reduces the ability of the receptor to go into the active state. The thermostabilizing mutations, Y227A$^{5.58}$ and F338M$^{7.48}$ stabilize the inactive state by removing long range correlated movement between residues in the G-protein coupling site to the extracellular region thus inhibiting activation (Niesen et al. 2013). The Y227A$^{5.58}$ mutation also stabilized the receptor by increasing the propensity for salt-bridge formation between R139$^{3.50}$ on TM3 and E285$^{6.30}$ on TM6. The salt-bridge between TM3 and TM6, a putative signature of the inactive state (Hubbell et al. 2003; Yao et al. 2006; Palczewski et al. 2000), is stabilized by these mutations and hence contribute to increase in stability. The mutation F338M$^{7.48}$ on TM7 alters the interaction in the conserved motif NPxxY(x)$_{5,6}$F and hence modulates the interaction of TM2-TM7-helix8 micro-domain (Balaraman et al. 2010). The R68S mutation that is present in the intracellular loop 1, leads to formation of a hydrogen bond with R355 on helix 8 to gain stabilization. We observed the formation of the hydrogen bond in the simulations and this was also found subsequently in the crystal structure of the mutant (Warne et al. 2012).

The distribution of the net intra and intermolecular forces, intramolecular repulsive and attractive forces within a protein structure governs how the protein reacts to thermal stress, since non-optimal distribution of internal stress can lead to structural destabilization at elevated temperatures (Stadler et al. 2012). However, internal stress also leads to movement of those regions of the receptor that lead to the active state. Thus a balance of these internal forces is required to maintain receptor stability and activity. The distribution of the net force on each residue calculated from the snapshots from the MD simulations of both m23-β_1AR and wt-β_1AR showed that Y227$^{5.58}$, F327$^{7.37}$ and F338$^{7.48}$ are all in positions of high stress in wt-β_1AR that get relieved upon mutation, leading to thermostabilization. Y227$^{5.58}$ is also a position that shows long range torsional correlation to the extracellular regions of the receptor. Therefore mutating this residue leads not only stress relieving stabilization but also loss of activation of the receptor. P305$^{6.50}$ on TM6 showed a high level of stress in the wild type receptor. P305$^{6.50}$ forms the hinge for the kink in TM6 and modulation of the proline kink could be crucial for the outward motion of TM6 leading to activation (Hornak et al. 2010; Bhattacharya et al. 2008a). Thus reduced stress at P305$^{6.50}$ could lead to increased stability of the inactive state in m23-β_1AR.

Thus the basis of thermostabilization of the mutants are multifold. They stem from both enthalpic and entropic contributions. They not only affect the neighborhood residues of the location of the mutation sites, but also show long range allosteric effects upon mutation. Thus studying the effect of the mutations on the overall structure and the conformational ensemble is important rather than looking for specific changes in microdomains of protein structures.

3.4 Conclusions

GPCRs are seven helical transmembrane proteins that play a critical physiological role and are implicated in many serious diseases. The dynamics of GPCR structure plays an important role in their activation and signal transduction function. Here we have described a method, *GPCRCompare* to compare the three dimensional structures of GPCRs in terms of the relative rotational, tilt and translational orientations of the seven helices. This software is freely available for users by contacting the corresponding author. Comparison of the ten NMR structures of the chemokine receptor shows a significantly dynamic range of rotation and tilt angles for transmembrane helices 5, 6 and 7. The kink angle on TM6 ranges from 21° to 34° showing significant dynamics in the ten NMR structures. Comparison of the crystal structures of the biogenic amine receptors available to date, showed that the conserved amino acids face the interior of the transmembrane barrel. We have also elaborated the structural dynamics and activation mechanism of β_2-adrenergic receptor as derived from multiscale MD simulation methods. We observed that the β_2-adrenergic receptor samples conformations with active state like characteristics and the population of such states are much lower than those of the inactive state. Ligand binding leads to conformational selection of substates (Niesen et al. 2011). Finally, study of the dynamics of the conformational ensemble of thermostable mutants showed that the stability could arise from both enthalpic and entropic factors. There are regions of high stress in the wild type GPCR that get relieved upon mutation conferring thermostability.

Acknowledgements We acknowledge the helpful discussions with Dr. Christopher Tate and Dr. Reinhard Grisshammer. We thank NIH-R01097261 for providing funding for our study on GPCR thermostable mutants.

References

Ahuja S, Smith SO (2009) Multiple switches in G protein-coupled receptor activation. Trends Pharmacol Sci 30:494–502

Alexandrov AI, Mileni M, Chien EY, Hanson MA, Stevens RC (2008) Microscale fluorescent thermal stability assay for membrane proteins. Structure 16:351–359

Audet M, Bouvier M (2012) Restructuring G-protein coupled receptor activation. Cell 151:14–23

Balaraman GS, Bhattacharya S, Vaidehi N (2010) Structural insights into conformational stability of wild type and mutant beta1-adrenergic receptor. Biophys J 99:568–577

Ballesteros J, Weinstein J (1995) Integrated methods for modeling G-protein coupled receptors. Methods Neurosci 25:366–428

Bhattacharya S, Vaidehi N (2010) Computational mapping of the conformational transitions in agonist selective pathways of a G-protein coupled receptor. J Am Chem Soc 132(14):5205–5214

Bhattacharya S, Hall SE, Vaidehi N (2008a) Agonist-induced conformational changes in bovine rhodopsin: insight into activation of G-protein-coupled receptors. J Mol Biol 382:539–555

Bhattacharya S, Hall SE, Li H, Vaidehi N (2008b) Ligand-stabilized conformational states of human beta(2) adrenergic receptor: insight into G-protein-coupled receptor activation. Biophys J 94(6):2027–2042

Bhattacharya S, Lam AR, Li H, Balaraman G, Niesen MJ, Vaidehi N (2013) Critical analysis of the successes and failures of homology models of G protein-coupled receptors. Proteins 81:729–739. doi:10.1002/prot.24195

Booth PJ, Curnow P (2009) Folding scene investigation: membrane proteins. Curr Opin Struct Biol 19(1):8–13

Bowie JU (2001) Stabilizing membrane proteins. Curr Opin Struct Biol 11:397–402

Bowie JU (2011) Membrane protein folding: how important are hydrogen bonds? Curr Opin Struct Biol 21:42–49

Cherezov V, Rosenbaum DM, Hanson MA, Rasmussen SG, Thian FS, Kobilka TS, Choi HJ, Kuhn P, Weis WI, Kobilka BK, Stevens RC (2007) High-resolution crystal structure of an engineered human beta2-adrenergic G-protein coupled receptor. Science 318: 1258–1265

Chien EY, Liu W, Zhao Q, Katritch V, Han GW, Hanson MA, Shi L, Newman AH, Javitch JA, Cherezov V, Stevens RC (2010) Structure of the human dopamine D3 receptor in complex with a D2/D3 selective antagonist. Science 330:1091–1095

Chiu ML, Tsang C, Grihalde N, MacWilliams MP (2008) Over-expression, solubilization and purification of G-protein coupled receptors for structural biology. Comb Chem High Throughput Screen 11:439–462

Colombo G, Merz KM (1999) Stability and activity of mesophilic subtilisin E and its thermophilic homolog: insights from molecular dynamics simulations. J Am Chem Soc 121:6895–6903

Cooke RM, Koglin M, Errey JC, Marshall FH (2013) Preparation of purified GPCRs for structural studies. Biochem Soc Trans 41(1):185–190

Crozier PS, Stevens MJ, Woolf TB (2007) How a small change in retinal leads to G-protein activation: initial events suggested by molecular dynamics calculations. Proteins 66:559–574

Deupi X (2012) Quantification of structural distortions in the transmembrane helices of GPCRs. Methods Mol Biol 914:219–235

Doré AS, Robertson N, Errey JC, Ng I, Hollenstein K, Tehan B, Hurrell E, Bennett K, Congreve M, Magnani F, Tate CG, Weir M, Marshall FH (2011) Structure of the adenosine A2A receptor in complex with ZM241385 and xanthines XAC and caffeine. Structure 19(9):1283–1293

Dror RO, Arlow DH, Borhani DW, Jensen MØ, Piana S, Shaw DE (2009) Identification of two distinct inactive conformations of the beta2-adrenergic receptor reconciles structural and biochemical observations. Proc Natl Acad Sci U S A 106:4689–4694

Dror RO, Pan AC, Arlow DH, Borhani DW, Maragakis P, Shan Y, Xu H, Shaw DE (2011) Pathway and mechanism of drug binding to G-protein coupled receptors. Proc Natl Acad Sci U S A 108:13118–13123

Filizola M, Devi LA (2013) Grand opening of structure-guided design for novel opioids. Trends Pharmacol Sci 34(1):6–12

Galandrin S, Oligny-Longpré G, Bouvier M (2007) The evasive nature of drug efficacy: implications for drug discovery. Trends Pharmacol Sci 2007(28):423–430

Galés C, Rebols RV, Hogue M, Trieu P, Brelt A, Hébert TE, Bouvier M (2005) Real-time monitoring of receptor and G-protein interactions in living cells. Nat Methods 2:177–184

Gouldson PR, Kidley NJ, Bywater RP, Psaroudakis G, Brooks HD, Diaz C, Shire D, Reynolds CA (2004) Toward the active conformations of rhodopsin and the beta2-adrenergic receptor. Proteins 56(1):67–84

Granier S, Kobilka B (2012) A new era of GPCR structural and chemical biology. Nat Chem Biol 8:670–673

Granier S, Manglik A, Kruse AC, Kobilka TS, Thian FS, Weis WI, Kobilka BK (2012) Structure of the δ-opioid receptor bound to naltrindole. Nature 485(7398):400–404

Grisshammer R (2009) Purification of recombinant G-protein coupled receptors. Methods Enzymol 463:631–645

Grossfield A, Pittman MC, Feller SE, Soubias O, Gawrisch K (2008) Internal hydration increases during activation of the G-protein-coupled receptor rhodopsin. J Mol Biol 381:478–486

Haga K, Kruse AC, Asada H, Yurugi-Kobayashi T, Shiroishi M, Zhang C, Weis WI, Okada T, Kobilka BK, Haga T, Kobayashi T (2012) Structure of the human M2 muscarinic acetylcholine receptor bound to an antagonist. Nature 482:547–551

Hall SE, Mao A, Nicolaidou V, Finelli M, Wise EL, Nedjai B, Kanjanapangka J, Harirchian P, Chen D, Selchau V, Ribeiro S, Schyler S, Pease JE, Horuk R, Vaidehi N (2009) Elucidation of binding sites of dual antagonists in the human chemokine receptors CCR2 and CCR5. Mol Pharm 75:1325–1336

Hamelberg D, de Oliveira CA, McCammon JA (2007) Sampling of slow diffusive conformational transitions with accelerated molecular dynamics. J Chem Phys 127(15):155102–155110

Hanson MA, Stevens RC (2009) Discovery of new GPCR biology: one receptor structure at a time. Structure 17(1):8–14

Hernandez G, Jenney FE, Adams MWW, LeMaster DM (2000) Millisecond time scale conformational flexibility in a hyperthermophile protein at ambient temperature. Proc Natl Acad Sci U S A 97:3166–3170

Hornak V, Ahuja S, Eilers M, Goncalves JA, Sheves M et al (2010) Light activation of rhodopsin: insights from molecular dynamics simulations guided by solid-state NMR distance restraints. J Mol Biol 396: 510–527

Hubbell WL, Altenbach C, Hubbell CM, Khorana HG (2003) Rhodopsin structure, dynamics, and activation: a perspective from crystallography, site directed spin labeling, sulfhydryl reactivity, and disulfide cross-linking. Adv Protein Chem 63:243–290

Isralewitz B, Gao M, Schulten K (2001) Steered molecular dynamics and mechanical functions of proteins. Curr Opin Struct Biol 11:224–230

Johnston JM, Aburi M, Provasi D, Bortolato A, Urizar E, Lambert NA, Javitch JA, Filizola M (2011) Making structural sense of dimerization interfaces of delta opioid receptor homodimers. Biochemistry 50(10):1682–1690

Katritch V, Cherezov V, Stevens RC (2012) Diversity and modularity of G-protein coupled receptor structures. Trends Pharmacol Sci 33:17–27

Kawate T, Gouaux E (2006) Fluorescence-detection size exclusion chromatography for precrystallization screening of integral membrane proteins. Structure 14:673–681

Kenakin T, Watson C, Muniz-Medina V, Christopoulos A, Novick S (2012) A simple method for quantifying functional selectivity and agonist bias. ACS Chem Neurosci 3:193–203

Khelashvili G, Grossfield A, Feller SE, Pitman MC, Weinstein H (2009) Structural and dynamic effects of cholesterol at preferred sites of interaction with rhodopsin identified from microsecond length molecular dynamics simulations. Proteins 76:403–417

Krishna A, Menon ST, Terry TJ, Sakmar TP (2002) Evidence that helix 8 of rhodopsin acts as a membrane-dependent conformational switch. Biochemistry 41:8298–8309

Kruse AC, Hu J, Pan AC, Arlow DH, Rosenbaum DM, Rosemond E, Green HF, Liu T, Chae PS, Dror RO, Shaw DE, Weis WI, Wess J, Kobilka BK (2012) Structure and dynamics of the M3 muscarinic acetylcholine receptor. Nature 482:552–556

Lebon G, Bennett K, Jazayeri A, Tate CG (2011a) Thermostabilization of an agonist bound conformation of the human adenosine A2A receptor. J Mol Biol 409:298–310

Lebon G, Warne T, Edwards PC, Bennett K, Langmead CJ, Leslie AG, Tate CG (2011b) Agonist bound adenosine A2A receptor structures reveal common features of GPCR activation. Nature 474:521–525

Magnani F, Shibata Y, Serrano-Vega MJ, Tate CG (2008) Co-evolving stability and conformational homogeneity of the human adenosine A2a receptor. Proc Natl Acad Sci U S A 105(31):10744–10749

Mailman RB, Murthy V (2010) Ligand functional selectivity advances our understanding of drug mechanisms and drug discovery. Neuropsychopharmacology 35:345–346

Manglik A, Kruse AC, Kobilka TS, Thian FS, Mathiesen JM, Sunahara RK, Pardo L, Weis WI, Kobilka BK, Granier S (2012) Crystal structure of the μ-opioid receptor bound to a morphinan antagonist. Nature 485(7398):321–326

Meinhold L, Clement D, Tehei M, Daniel R, Finney JL et al (2008) Protein dynamics and stability: the distribution of atomic fluctuations in thermophilic and mesophilic dihydrofolate reductase derived using elastic incoherent neutron scattering. Biophys J 94:4812–4818

Nedjai B, Li H, Stroke IL, Wise EL, Webb ML, Merritt JR, Henderson I, Klon AE, Cole AG, Horuk R, Vaidehi N, Pease JE (2011) Small-molecule chemokine mimetics suggest a molecular basis for the observation that CXCL10 and CXCL11 are allosteric ligands of CXCR3. Br J Pharmacol 166(3):912–923

Niesen M, Bhattacharya S, Vaidehi N (2011) Conformational selection upon ligand binding in G-protein coupled receptors. J Am Chem Soc 133(33):13197–13204

Niesen M, Bhattacharya S, Tate CG, Grisshammer R, Vaidehi N (2013) Thermostabilization of the β_1-

adrenergic receptor Correlates with Increased Entropy of the Inactive State. J Phys Chem 117(24):7283–7291

Nygaard R, Zou Y, Dror RO, Mildorf TJ, Arlow DH, Manglik A, Pan AC, Liu CW, Fung JJ, Bokoch MP, Thian FS, Kobilka TS, Shaw DE, Mueller L, Prosser RS, Kobilka BK (2013) The dynamic process of β2-adrenergic receptor activation. Cell 152(3):532–542

Palczewski K, Kumasaka T, Hori T, Behnke CA, Motoshima H, Fox BA, Le Trong I, Teller DC, Okada T, Stenkamp RE, Yamamoto M, Miyano M (2000) Crystal structure of rhodopsin: a G protein-coupled receptor. Science 289:739–745

Park JH, Scheerer P, Hoffman KP, Choe H-W, Ernst OP (2008) Crystal structure of the ligand-free G-protein coupled receptor opsin. Nature 454:183–187

Park SH, Das BB, Casagrande F, Tian Y, Nothnagel HJ, Chu M, Kiefer H, Maier K, De Angelis AA, Marassi FM, Opella SJ (2012) Structure of the chemokine receptor CXCR1 in phospholipid bilayers. Nature 491(7426):779–783

Provasi D, Artacho MC, Negri A, Mobarec JC, Filizola M (2011) Ligand induced modulation of the free energy landscape of the G-protein coupled receptors explored by adaptive biasing techniques. PLoS Comput Biol 7(10):e1002193

Rasmussen SG, DeVree BT, Zou Y, Kruse AC, Chung KY, Kobilka TS, Thian FS, Chae PS, Pardon E, Calinski D, Mathiesen JM, Shah ST, Lyons JA, Caffrey M, Gellman SH, Steyaert J, Skiniotis G, Weis WI, Sunahara RK, Kobilka BK (2011) Crystal structure of the β2-adrenergic receptor-Gs protein complex. Nature 477:549–555

Reiter E, Ahn S, Shukla AK, Lefkowitz RJ (2012) Molecular mechanism of β-arrestin biased agonism at seven transmembrane receptors. Ann Rev Pharmacol Toxicol 52:179–197

Roth CB, Hanson MA, Stevens RC (2008) Stabilization of the human beta2-adrenergic receptor TM4-TM3-TM5 helix interface by mutagenesis of Glu122(3.41), a critical residue in GPCR structure. J Mol Biol 376(5):1305–1319

Saam J, Tajkhorshid E, Havashi S, Schulten K (2002) Molecular dynamics investigation of primary photoinduced events in the activation of rhodopsin. Biophys J 83:3097–3112

Schertler GF (2005) Structure of rhodopsin and the metarhodopsin I photointermediate. Curr Opin Struct Biol 15:408–15

Serrano-Vega MJ, Tate CG (2009) Transferability of thermostabilizing mutations between beta adrenergic receptors. Mol Membr Biol 26:385–396

Serrano-Vega MJ, Magnani F, Shibata Y, Tate CG (2008) Conformational thermostabilization of the beta1-adrenergic receptor in a detergent resistant form. Proc Natl Acad Sci U S A 105(3):877–882

Shibata Y, White JF, Serrano-Vega MJ, Magnani F, Aloia AL, Grisshammer R, Tate CG (2009) Thermostabilization of the neurotensin receptor NTS1. J Mol Biol 2390(2):262–277

Shimamura T, Shiroishi M, Weyand S, Tsujimoto H, Winter G, Katritch V, Abagyan R, Cherezov V, Liu W, Han GW, Kobayashi T, Stevens RC, Iwata S (2011) Structure of the human histamine H1 receptor complex with doxepin. Nature 475:65–70

Stadler AM, Garvey CJ, Bocahut A, Sacquin-Mora S, Digel I et al (2012) Thermal fluctuations of haemoglobin from different species: adaptation to temperature via conformational dynamics. J R Soc Interface 9(76):2845–2855

Standfuss J, Xie G, Edwards PC, Burghammer M, Oprian DD, Schertler GF (2007) Crystal structure of a thermally stable rhodopsin mutant. J Mol Biol 372(5):1179–1188

Sterpone F, Melchionna S (2012) Thermophilic proteins: insight and perspective from in silico experiments. Chem Soc Rev 41(5):1665–1676

Steyaert J, Kobilka BK (2011) Nanobody stabilization of G-protein coupled receptor conformational states. Curr Opin Struct Biol 21:567–572

Swaminath G, Xiang Y, Lee TW, Steenhuis J, Parnot C, Kobilka BK (2004) Sequential binding of agonists to the beta2 adrenoceptor-kinetic evidence for intermediate conformational states. J Biol Chem 279:686–691

Swaminath G, Deupi X, Lee TW, Zhu W, Thian FS, Kobilka TS, Kobilka BK (2005) Probing the beta2 adrenoceptor binding site with catechol reveals differences in binding and activation by agonists and partial agonists. J Biol Chem 280:22165–22171

Tate CG (2012) A crystal clear solution for determining G-protein coupled receptor structures. Trends Biochem Sci 37:343–352

Tate CG, Schertler GF (2009) Engineering G protein-coupled receptors to facilitate their structure determination. Curr Opin Struct Biol 19:386–395

Thompson AA, Liu W, Chun E, Katritch V, Wu H, Vardy E, Huang XP, Trapella C, Guerrini R, Calo G, Roth BL, Cherezov V, Stevens RC (2012) Structure of the nociceptin/orphanin FQ receptor in complex with a peptide mimetic. Nature 485(7398):395–399

Vaidehi N (2010) Dynamics and flexibility of G-protein coupled receptor conformations and their relevance to drug design. Drug Discov Today 15:951–957

Vaidehi N, Bhattacharya S (2011) Multiscale computational methods for mapping conformational ensembles of G-protein coupled receptors. Adv Protein Chem Struct Biol 85:253–280

Vaidehi N, Kenakin T (2010) The role of conformational ensembles of seven transmembrane receptors in functional selectivity. Curr Opin Pharmacol 10:775–781

Vilardaga JP, Bunemann M, Krasel C, Castro M, Lohse M (2003) Measurement of the millisecond activation switch of G protein-coupled receptors in living cells. Nat Biotechnol 21:807–812

Warne T, Serrano-Vega MJ, Baker JG, Moukhametzianov R, Edwards PC, Henderson R, Leslie AG, Tate CG, Schertler GF (2008) Structure of a beta1-adrenergic G-protein-coupled receptor. Nature 454(7203):486–491

Warne T, Edwards PC, Leslie AGW, Tate CG (2012) Crystal structures of a stabilized β_1-adrenoceptor bound to the biased agonists bucindolol and carvedilol. Structure 20:841–849

White SH, Wimley WC (1999) Membrane protein folding and stability: physical principles. Annu Rev Biophys Biomol Struct 28:319–365

White JF, Noinaj N, Shibata Y, Love J, Kloss B, Xu F, Gvozdenovic-Jeremic J, Shah P, Shiloach J, Tate CG, Grisshammer R (2012) Structure of the agonist bound neurotensin receptor. Nature 490:508–513

Wintrode PL, Zhang DQ, Vaidehi N, Arnold FH, Goddard WA (2003) Protein dynamics in a family of laboratory evolved thermophilic enzymes. J Mol Biol 327:745–757

Wu B, Chien EY, Mol CD, Fenalti G, Liu W, Katritch V, Abagyan R, Brooun A, Wells P, Bi FC, Hamel DJ, Kuhn P, Handel TM, Cherezov V, Stevens RC (2010) Structures of the CXCR4 chemokine GPCR with small-molecule and cyclic peptide antagonists. Science 330(6007):1066–1071

Xu F, Wu H, Katritch V, Han GW, Jacobson KA, Gao ZG, Cherezov V, Stevens RC (2011) Structure of an agonist bound human A2A adenosine receptor. Science 332(6027):322–327

Yao X, Parnot C, Deupi X, Ratnala VRP, Swaminath G, Farrens D, Kobilka BK (2006) Coupling ligand structure to specific conformational switches in the beta2-adrenoceptor. Nat Chem Biol 2006(2): 417–422

Yao XJ, Ruiz G, Whorton MR, Rasmussen SG, De Vree BT, Deupi X, Sunahara RK, Kobilka B (2009) The effect of ligand efficacy on the formation and stability of a GPCR-G-protein complex. Proc Natl Acad Sci U S A 106:9501–9506

Zaitseva E, Brown MF, Vogel R (2010) Sequential rearrangement of interhelical networks upon rhodopsin activation in membranes: the Meta II(a) conformational substate. J Am Chem Soc 132:4815–4821

Zhang C, Srinivasan Y, Arlow DH, Fung JJ, Palmer D, Zheng Y, Green HF, Pandey A, Dror RO, Shaw DE, Weis WI, Coughlin SR, Kobilka BK (2012) High-resolution crystal structure of human protease-activated receptor 1. Nature 492(7429):387–392

How the Dynamic Properties and Functional Mechanisms of GPCRs Are Modulated by Their Coupling to the Membrane Environment

4

Sayan Mondal, George Khelashvili, Niklaus Johner, and Harel Weinstein

Abstract

Experimental observations of the dependence of function and organization of G protein-coupled receptors (GPCRs) on their lipid environment have stimulated new quantitative studies of the coupling between the proteins and the membrane. It is important to develop such a quantitative understanding at the molecular level because the effects of the coupling are seen to be physiologically and clinically significant. Here we review findings that offer insight into how membrane-GPCR coupling is connected to the structural characteristics of the GPCR, from sequence to 3D structural detail, and how this coupling is involved in the actions of ligands on the receptor. The application of a recently developed computational approach designed for quantitative evaluation of membrane remodeling and the energetics of membrane-protein interactions brings to light the importance of the radial asymmetry of the membrane-facing surface of GPCRs in their interaction with the surrounding membrane. As the radial asymmetry creates adjacencies of hydrophobic and polar residues at specific sites of the GPCR, the ability of membrane remodeling to achieve complete hydrophobic matching is limited, and the residual mismatch carries a significant energy cost. The adjacencies are shown to be affected by ligand-induced conformational changes. Thus, functionally important organization of GPCRs in the cell membrane can depend both on ligand-determined properties

S. Mondal • G. Khelashvili • N. Johner
Department of Physiology and Biophysics, Weill Cornell Medical College, Cornell University, Room E-509, 1300 York Avenue, 10065 New York City, NY, USA

H. Weinstein (✉)
Department of Physiology and Biophysics, Weill Cornell Medical College, Cornell University, Room E-509, 1300 York Avenue, 10065 New York City, NY, USA

The HRH Prince Alwaleed Bin Talal Bin Abdulaziz Alsaud Institute for Computational Biomedicine, Weill Cornell Medical College, Cornell University, New York, NY 10065, USA
e-mail: haw2002@med.cornell.edu

M. Filizola (ed.), *G Protein-Coupled Receptors - Modeling and Simulation*, Advances in Experimental Medicine and Biology 796, DOI 10.1007/978-94-007-7423-0_4,

and on the lipid composition of various membrane regions with different remodeling capacities. That this functionally important reorganization can be driven by oligomerization patterns that reduce the energy cost of the residual mismatch, suggests a new perspective on GPCR dimerization and ligand-GPCR interactions. The relation between the modulatory effects on GPCRs from the binding of specific cell-membrane components, e.g., cholesterol, and those produced by the non-local energetics of hydrophobic mismatch are discussed in this context.

Keywords
GPCR • Transmembrane proteins • Structure and remodeling of lipid membranes • Hydrophobic mismatch • Lipid/protein interactions • CTMD • Continuum elastic theory • Molecular dynamics simulations • Residual exposure • Residual mismatch • Membrane deformation energy • Oligomerization • Serotonin receptor • Dopamine receptor • Rhodopsin • Cholesterol/GPCR binding • DHA • Structure-function relations in GPCRs • Functional microdomains in GPCRs

Abbreviations

5HT	5-hydroxytryptamine
CTMD	Continuum-Molecular Dynamics hybrid approach
DHA	ω-3 Docosahexaenoic Acid
GPCR	G protein-coupled receptor
MD	Molecular Dynamics
SASA	Solvent Accessible Surface Area
SDPC	1-stearoyl-2-docosahexaenoyl-*sn*-Glycero-3-phosphocholine
SDPE	1-stearoyl-2-docosahexaenoyl-*sn*-Glycero-3-phosphoethanolamine
SM/FM	Sequence Motif/ Functional Microdomain
TM	Transmembrane segment

4.1 Introduction

A rich body of knowledge and information concerning the molecular pharmacology of the receptors, and their coupling to signal transduction cascades in the cell, is offering a functional context for the recently determined structures of GPCRs (Palczewski et al. 2000; Ruprecht et al. 2004; Cherezov et al. 2007; Rasmussen et al. 2007, 2011; Jaakola et al. 2008; Park et al. 2008; Scheerer et al. 2008; Warne et al. 2008; Chien et al. 2010; Wu et al. 2010; Choe et al. 2011; Lebon et al. 2011; Rosenbaum et al. 2011; Xu et al. 2011; Granier et al. 2012; Haga et al. 2012; Hanson et al. 2012; Liu et al. 2012; Wu et al. 2012; Manglik et al. 2012). The literature abounds in such information, obtained from decades of detailed examination of the molecular structural determinants for the pharmacologically and physiologically measured ligand-recognition properties and mechanisms of receptor activation. Such structure-based details have emerged from experiments in which the receptors were probed with structurally defined ligands and mutations to reveal mechanistic details in activation and dimerization (Ballesteros et al. 2001; Han et al. 2009; Strader et al. 1987; Suryanarayana et al. 1992; Luttrell et al. 1999; Liapakis et al. 2000; Barak et al. 1995; Chelikani et al. 2007; Fritze et al. 2003; Green and Liggett 1994; Guo et al. 2005; Javitch et al. 1995a, b, 1998; Kahsai et al. 2011; Kofuku et al. 2012; O'Dowd et al. 1988; Prioleau et al. 2002; Shi and Javitch 2004; Shi et al. 2002). Together, computational modeling, mutagenesis, and crystallography have shown that each of the five GPCR families share highly conserved sequence motifs (SMs) that constitute functional microdomains (FMs) mediating receptor activation (Ballesteros et al. 1998;

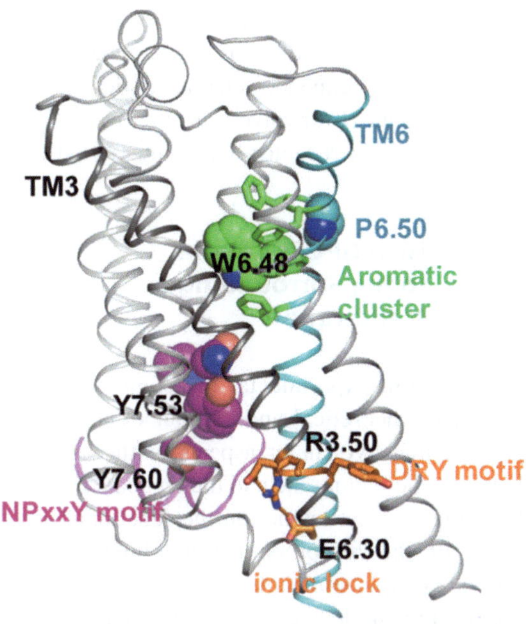

Fig. 4.1 The position of Structural Motifs recognized as Functional Microdomains (SM/FMs) in GPCRs, illustrated in the molecular model of 5HT$_{2A}$R (Reprinted with permission from Shan et al. 2012)

Visiers et al. 2002; Prioleau et al. 2002; Barak et al. 1995; Fritze et al. 2003; Lagerstrom and Schioth 2008; Rosenbaum et al. 2009; Palczewski et al. 2000; Warne et al. 2008; Cherezov et al. 2007; Weinstein 2005). Prominent examples of these SM/FMs from Class-A GPCRs include the "arginine cage" around the E/DRY motif in TM3 (Ballesteros et al. 1998), and the NPxxY motif in TM7 (Ballesteros et al. 1998; Barak et al. 1995; Fritze et al. 2003; Prioleau et al. 2002). Figure 4.1 illustrates some of these SM/FMs in a structural model of the serotonin receptor 5HT$_{2A}$R.

With the focus on the conformational changes of structural elements of the GPCR proteins, the mechanistic models have largely ignored the involvement of the membrane environment in the discriminant properties of the GPCRs, and the differential ligand-determined effects. This situation is now changing rapidly, because the membrane environment has emerged as an essential part of functional mechanisms including the activation, coupling, and cell-surface organization of the GPCRs. In particular, the conformations of the GPCRs in the various states of activation – both ligand-determined and constitutive – couple the response of specific membrane regions to the transition among such conformations. This coupling is becoming recognized as a determinant factor in (i)-the specificity of ligand binding, and (ii)-the responses elicited from the various activated states. Importantly, the effects of the membrane environment are strongly dependent on the lipid composition and hence can vary between different cell types, and also within different regions of the plasma membrane of the same cell. For example, cholesterol, a component of biological membranes with a relative abundance that varies among cell types and even within different regions of the same plasma membrane, has been shown to affect GPCR signaling. Functionally important receptor properties such as thermal stability, ligand binding, and localization in the cell membrane have been shown to be cholesterol-dependent for beta2-adrenergic, 5HT1A, oxytocin, and metabotropic glutamate receptors (Xiang et al. 2002; Pucadyil and Chattopadhyay 2004; Gimpl and Fahrenholz 2002; Eroglu et al. 2003). Rhodopsin has served as a prototype for elucidating the role of the lipid environment. In systematic studies, the activation and the photochemical function of rhodopsin were found to be affected by the cholesterol concentration, as well as by molecular-level characteristics of the lipids such as their acyl chain length, unsaturation of the acyl chain, headgroup charge, and headgroup size (Brown 1994; Brown et al. 2002; Botelho et al. 2002, 2006; Gibson and Brown 1991, 1993). To take the example of acyl chain unsaturation, the equilibrium between the inactive Metarhodopsin I and the active Metarhodopsin II was shown to shift in favor of Metarhodopsin II with increasing molar fraction of lipids containing the polyunsaturated ω-3 docosahexaenoic (DHA) tail, as compared to monounsaturated lipid tails (Brown 1994). Such modulation is physiologically and clinically significant, as dietary deficiency in ω-3 fatty acids is well-known to be implicated in a wide variety of diseases (Innis 2008; Lavie et al. 2009; Neuringer et al. 1984; Stahl et al. 2008), at least some of which involve the modulation of GPCR function by the lipid environment. In rat models, for example, deficiency of dietary ω-3 fatty acids impaired photochemical function, with the Metarhodopsin II formation being both lower and slower in response to dim light stimulus (Niu et al. 2002).

As the structures of GPCRs become known from crystallography at increasingly high resolution, and various biophysical methods of investigation offer increasingly detailed quantitative information about mechanistic elements of GPCR function, it becomes possible to focus on the structural context of the mechanisms. In parallel, this development requires a corresponding evaluation of the demonstrated role of the lipid environment using appropriate quantitative methods to assess the membrane-protein coupling components. Achieving this goal requires the development and application of quantitative approaches, capable of relating the lipid-protein interactions to the structural characteristics of the receptor in its various states, and especially in the functional context of ligand-induced conformational rearrangements. Significant progress in this direction has been achieved with biophysical methods anchored in fundamental experimental data, and formulated in the framework of physics-based computational modeling approaches ranging from molecular dynamics simulations to continuum-based and mean field methods. We review, in Sect. 4.2, the new types of insights attained from the application of such approaches. In Sect. 4.3, we present the essential methodological advances that enabled quantitative investigation of the relation between lipid-protein interactions and the GPCR structure. To enable clear references to structural elements in different GPCRs, residues and motifs are identified throughout the chapter with the Ballesteros-Weinstein notation (Ballesteros and Weinstein 1995).

4.2 Relation Between Receptor Structure and Its Interaction with the Membrane in a Functional Context

This section reviews findings that demonstrate how the membrane interacts with the embedded GPCR, i.e., the nature of the membrane-GPCR coupling. The focus is on the manner in which such coupling is (a)-connected to the structural/sequence characteristics of the receptor, and (b)-involved in the modes and consequences of receptor-ligand interactions.

4.2.1 How GPCRs Couple to the Surrounding Membrane

Results from experimental investigations and a variety of theoretical and computational studies suggest that two major aspects of the coupling between the membrane and the embedded GPCR have the greatest impact on the functional properties of the receptors: (1) the interaction of individual lipid molecules with specific sites of the receptor molecule, and (2) the mismatch between the hydrophobic/hydrophilic character of the membrane and the membrane-facing surface of the receptor. As detailed below, a significant part of this mismatch persists even after the membrane is remodeled around the embedded protein, giving rise to an energy cost that affects the functional properties and spatial organization of the GPCRs. This is due to the inherent asymmetry of the membrane-facing surface of the embedded protein, caused by the differences among the TM segments. Consequently, it is an important common characteristic of the energetics of membrane coupling for multi-TM segment proteins.

4.2.1.1 Specific Interactions of Lipid Molecules with Membrane-Embedded GPCRs

With the details obtained from atomistic Molecular Dynamics (MD) simulations it has become possible to discern the dynamics of membrane components surrounding the receptor molecules, such as cholesterol, DHA, and various other lipids. The simulations revealed both the nature of the interactions of the various lipid molecules with specific sites of the GPCR, and the effect on the membrane and the proteins as a whole. For example, in a 1.6 μs

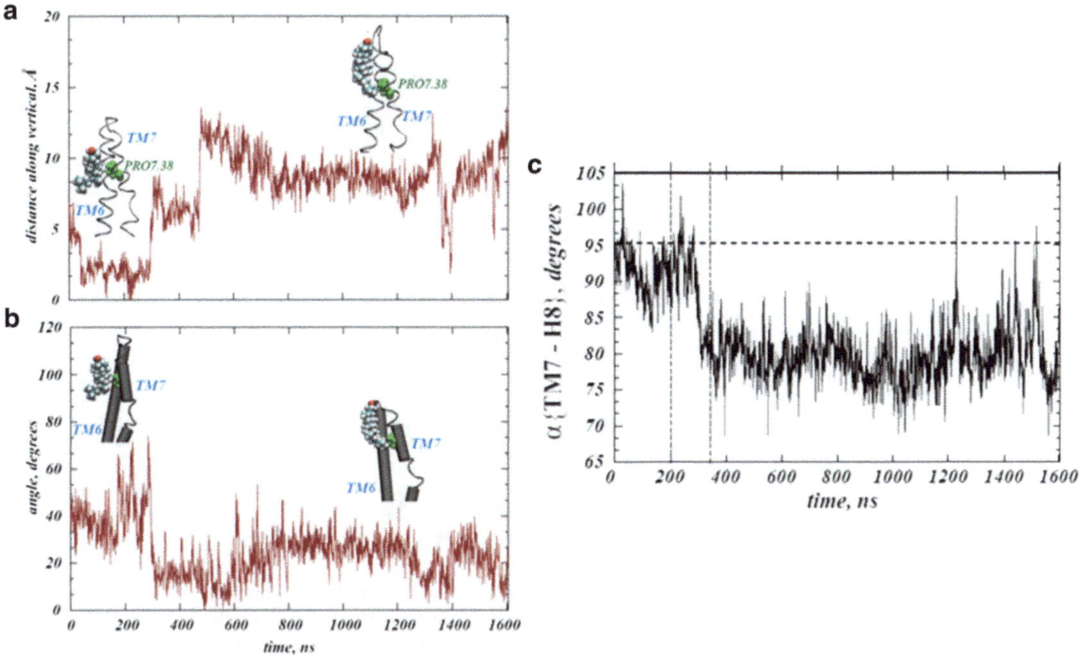

Fig. 4.2 Cholesterol dynamics around rhodopsin from atomistic MD simulation. Cholesterol interacting with Pro7.38 in TM7. The time-sequence in panel (**a**) is for the z-directional distance between the cholesterol oxygen and the C_{α} atom of Pro7.38. Panel (**b**) shows the angle between the cholesterol ring and the axis of the extracellular segment of TM7 as a function of trajectory time. Cholesterol is shown in space-fill, and Pro7.38 in *green*. The TM6-TM7 bundle of rhodopsin is shown as *ribbons* (**a**), or as cylinders (**b**). (**c**) The angle between TM7 and H8 over the course of the trajectory. (Reprinted with permission from Khelashvili et al. 2009. Copyright by Wiley Periodicals Inc.)

all-atom MD simulation of the prototypical GPCR rhodopsin, three molecules of cholesterol were found to be localized preferentially at three distinct sites of the protein (Khelashvili et al. 2009). Similar preferential localization of cholesterol molecules to specific sites was observed in μs-scale MD simulations of ligand-bound serotonin receptor 5-$HT_{2A}R$ (Shan et al. 2012). Such observations of specific cholesterol-receptor interactions at identified sites in MD simulations are consistent with structural data for several GPCRs (Khelashvili et al. 2009). Indeed, electron microscopy of the inactive photo-intermediate Metarhodopsin I (Ruprecht et al. 2004), and several X-ray structures of GPCRs, show individual cholesterol molecules associated with specific receptor sites (Hanson et al. 2012; Liu et al. 2012; Cherezov et al. 2007). A proposed cholesterol binding motif identified by groups of residues [4.39–4.43 (R,K)] – [4.50 (W,Y)] – [4.46 (I,V,L)] – [2.41 (F,Y)] was calculated to occur in 21 % of human class A GPCRs (Hanson et al. 2008), suggesting that such specific interactions are a general feature of GPCR-membrane coupling.

Results from the MD simulations suggested as well a mechanistic role for the specific binding of cholesterol. This was based on observed correlations between the dynamics of cholesterol molecules and the conformations adopted by the SM/FM regions of the receptor structure that had been identified for their involvement in activation. For example, the binding mode of a cholesterol molecule in the TM6-TM7 region of rhodopsin was found to correlate with the angle between TM7 and H8, which involves the NPxxY SM/FM (see Fig. 4.2) (Khelashvili et al. 2009). Similarly, in simulations of the agonist-bound $5HT_{2A}R$ (Shan et al. 2012), the movement of cholesterol away from

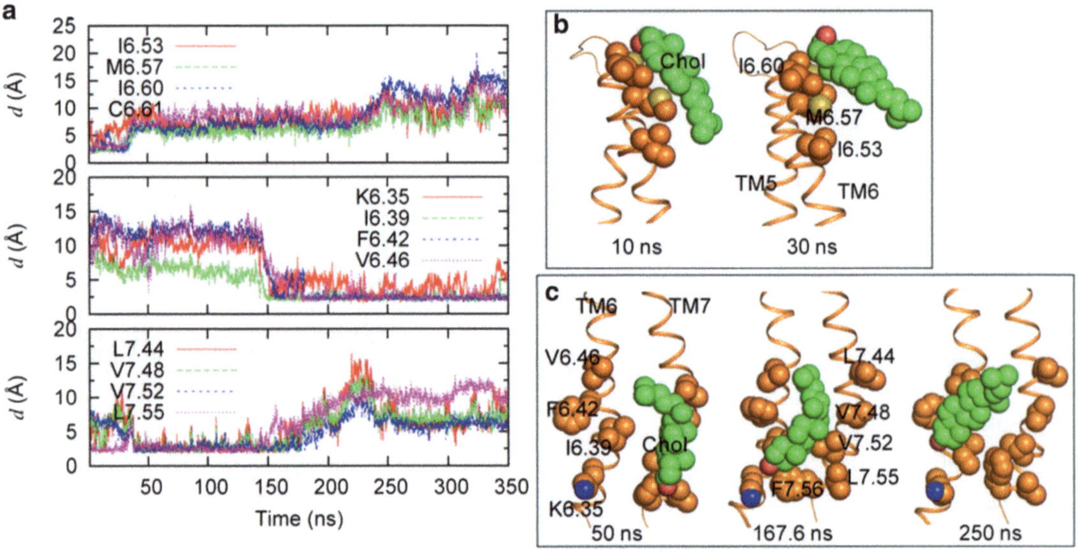

Fig. 4.3 Cholesterol dynamics correlates with the structural transitions in agonist-bound 5-HT$_{2A}$R. (**a**) Evolution of the minimum distances between the cholesterol at the extracellular (*EC*) end of TM6 and selected TM6 residues in the 5-HT simulation (*top panel*). Time traces of the minimum distances between the cholesterol at the intracellular (*IC*) ends of TM6–7 and selected residues on TM6 and 7 (*middle* and *bottom panels*). The cholesterol initially in contact with the L7.44, V7.48, V7.52, and L7.55 residues on TM7 moves towards TM6 and engages in interactions with the residues K6.35, I6.39, F6.42, and V6.46 on TM6. (**b**) Snapshots at 10 and 30 ns showing the cholesterol from the top panel of (**a**) interacting with EC end of TM6. (**c**) Snapshots at 50, 167.6 and 250 ns showing the cholesterol from the bottom panels of (**a**) interacting with either IC end of TM6 or of TM7 (Reprinted with permission from Shan et al. 2012)

its association with TM7 and towards TM6 (see Fig. 4.3), coincided with the intracellular end of TM6 beginning to bend away from TM3. Overall, the cholesterol-TM6 distance was found to be correlated with agonist-induced conformational changes in 5-HT$_{2A}$R (Shan et al. 2012).

Similarly, DHA lipid chains were found to localize preferentially at a small number of specific sites of the GPCR from a statistical analysis of 26 independent 100 ns atomistic MD simulations of rhodopsin embedded in a membrane composed of a 2:2:1 molar mixture of SDPC (1-stearoyl-2-docosahexaenoyl-*sn*-Glycero-3-phosphocholine)/SDPE (1-stearoyl-2-docosahexaenoyl-*sn*-Glycero-3-phosphoethanolamine)/cholesterol (Grossfield et al. 2006b). The suggestion of specific sites for DHA, an important component of the cell membrane environment, in binding to rhodopsin (Grossfield et al. 2006b), is consistent with spectroscopic data from NMR experiments measuring magnetization transfer from rhodopsin to lipid components in the first shell (Soubias et al. 2006). Statistical analyses of the MD trajectories further suggested that DHA preferentially solvates rhodopsin due to the lower entropic cost for its association with rhodopsin, compared to that incurred by the stearoyl chain (Grossfield et al. 2006a).

4.2.1.2 Hydrophobic Mismatch and Membrane Remodeling

Alongside the specific sites of interactions between lipid molecules and the GPCR molecules, the remodeling of the membrane surrounding the protein constitutes another important form of coupling between membranes and integral proteins. This remodeling is driven by the mismatch of the hydrophobic-hydrophilic interfaces between the membrane and the protein. Deformation of the membrane to match the hydrophobic length of an embedded peptide has been demonstrated in studies employing single-helical peptides as models to

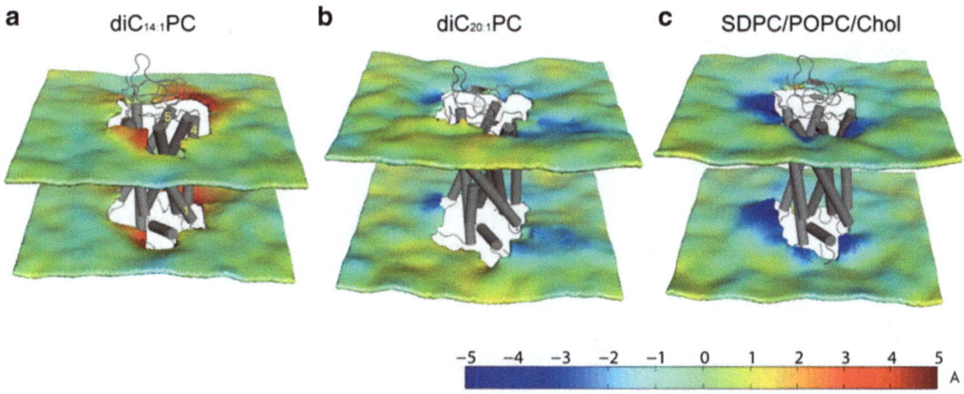

Fig. 4.4 Membrane-deformation profiles $u(x,y)$ **for rhodopsin immersed in lipid bilayers of different bulk thicknesses.** $u(x,y)$ was calculated directly from MD trajectories for rhodopsin in bilayers composed of (**a**) diC$_{14:1}$PC, (**b**) diC$_{20:1}$PC, and (**c**) 7:7:6 SDPC/POPC/cholesterol membranes (Reprinted with permission from Mondal et al. 2011. Copyright by Elsevier)

investigate membrane-protein coupling (Harroun et al. 1999a). Furthermore, the treatment of a membrane-inserted peptide as a cylinder enabled the assumption of radially symmetric bilayer deformations and perfect hydrophobic matching between the peptide and the membrane so that the energy cost of bilayer remodeling could be evaluated using the continuum theory of membrane elasticity (Huang 1986; Andersen and Koeppe 2007; Lundbæk and Andersen 1999; Lundbaek et al. 2003; Nielsen et al. 1998; Goforth et al. 2003). This energy cost has been shown to affect the function of membrane-embedded molecules (Andersen and Koeppe 2007; Lundbaek et al. 2010).

However, the membrane remodeling pattern is much more complex for GPCRs. This was shown by results from calculations with a novel method that did not require the assumption of radially symmetric membrane deformations and cylindrical proteins. We developed the theoretical framework of the hybrid Continuum-MD (CTMD) method, described in detail in Sect. 4.3, to make possible the quantitative evaluation of radially asymmetric bilayer deformations around membrane-inserted proteins (Mondal et al. 2011). The CTMD calculations confirmed the expectation that the average remodeling of a thick membrane will make it thinner around an embedded GPCR, and a thin membrane will become thicker on average. But, freed of the symmetry assumptions, the calculations also revealed that the membrane deforms differently near different parts of the 7-TM GPCR, in a specific pattern dependent on the properties of the embedded GPCR. This is illustrated in Fig. 4.4 by results obtained for the systems of rhodopsin embedded in a thinner di(C14:1)PC membrane and a thicker di(C20:1)PC bilayer that had been examined earlier by the Gawrisch lab (Soubias et al. 2008). Indeed, the di(C14:1)PC membrane is seen to become thicker on average around rhodopsin, and the di(C20:1)PC membrane to become thinner. The calculated extents of average membrane deformation around the GPCR agreed with the findings from solid state NMR measurements on these systems (Soubias et al. 2008). But in addition, our CTMD calculations brought to light the radial asymmetry of the membrane deformations that exhibit both local thickening and local thinning near different regions of the protein (Mondal et al. 2011). For example, the di(C14:1)PC lipid bilayer thickened by 2 Å on average near the protein, similar to results from NMR experiments, but it thickened locally by ~5 Å near TM4 (Mondal et al. 2011). The theoretical framework allowed the conclusion that the radial asymmetry of the membrane deformation is due to the radially asymmetric hydrophobic surface of the GPCR, which has TMs of different hydrophobic lengths (Mondal et al. 2011).

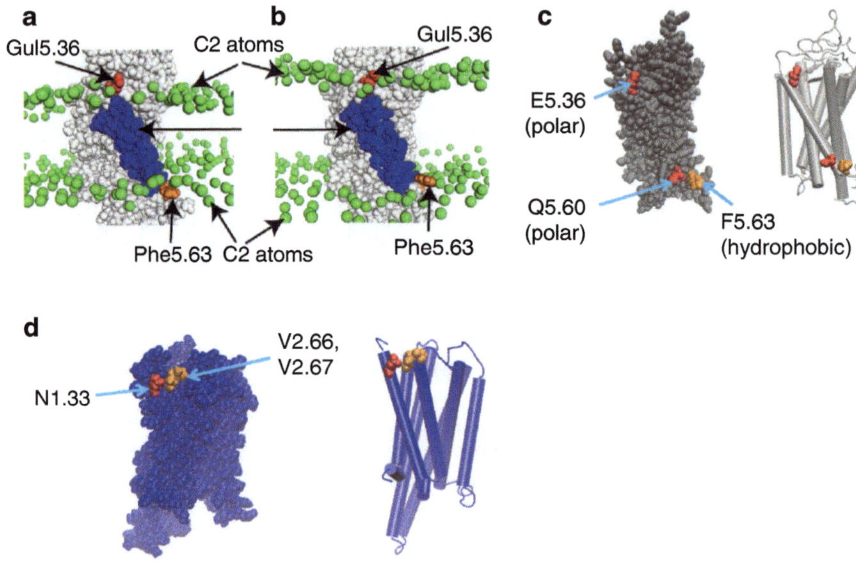

Fig. 4.5 Illustration of hydrophobic adaptation and residual exposure at the molecular level. (**a**) A snapshot from the MD trajectory of rhodopsin in the thin diC$_{14:1}$PC bilayer. TM5 is shown in *blue*, and two residues (Glu5.36 and Phe5.63) are highlighted. (**b**) A snapshot of rhodopsin in the thick diC$_{20:1}$PC bilayer. The diC$_{20:1}$PC bilayer thins near Glu5.36, which substantially reduces its exposure to the hydrophobic core of the bilayer, thus showing hydrophobic adaptation. Phe5.63 remains unfavorably exposed to the polar environment in the thin diC$_{14:1}$PC bilayer but not in the thick diC$_{20:1}$PC bilayer. Thus, Phe5.63 contributes to residual exposure energy penalty at TM5 for rhodopsin in diC$_{14:1}$PC (*Panels A* and *B* are reprinted with permission from Mondal et al. 2011. Copyright by Elsevier). (**c**) In Rhodopsin, adjacency of the hydrophobic Phe5.63 to the polar Gln5.60. The *left panel* shows rhodopsin in VdW representation in order to depict the irregular surface with which the membrane interacts. The *right panel* shows the protein in the cartoon representation to show the helices. (**d**) In the dopamine D2 receptor model, adjacency of polar N1.33 and hydrophobic V.266, V2.67 in the membrane-facing surface, with the protein shown in both, VdW (*left panel*) and cartoon representations (*right panel*)

4.2.1.3 The Residual Hydrophobic Mismatch and the Energy Cost of Protein-Membrane Coupling

Although hydrophobic mismatch is energetically costly, membrane deformations do not achieve a complete hydrophobic matching at all TM segments (Mondal et al. 2011), and residues in the TM-bundle can remain exposed to unfavorable hydrophobic-polar interactions. For example, in MD simulations of rhodopsin in di(C14:1)PC, TM5 maintains a residual hydrophobic mismatch on its extracellular side where a hydrophobic phenylalanine is exposed to the solvent in spite of the overall membrane remodeling around the protein (see Fig. 4.5a). Such incomplete adaptation of the membrane to the hydrophobic surface has also been inferred from EPR experiments measuring lipid-protein associations for a number of transmembrane proteins (Marsh 2008). Furthermore, a number of computational studies of the coupling between the membrane and embedded multi-TM proteins identified specific residues of the GPCRs where the hydrophobic matching remains incomplete in a membrane of specific composition, thus highlighting specific regions of the GPCR where unfavorable hydrophobic-polar interactions carry a significant energy cost (Mondal et al. 2011, 2012; Shan et al. 2012; Khelashvili et al. 2012).

Results from CTMD calculations and coarse-grained simulations suggest that some of this energy cost contributes to the modulation of receptor oligomerization by the membrane. GPCRs have been shown to self-associate spontaneously upon reconstitution into liposomes, without the need for any cellular machinery (Mansoor et al. 2006; Harding et al. 2009), with oligomerization occurring preferentially at specific interfaces

(Fung et al. 2009). The oligomerization is a reversible process (Dorsch et al. 2009; Hern et al. 2010), and is driven by hydrophobic mismatch (Botelho et al. 2006). Coarse-grained MD simulations of the interaction of GPCRs in the membrane have provided results consistent with the experimental observations, and offered important insights into the energetics of this oligomerization (Periole et al. 2007, 2012). Such simulations, taken together with the quantitative results from CTMD calculations, suggest that the residual exposure contributes to the organizational role of the hydrophobic mismatch (Mondal et al. 2011, 2012). In particular, these calculations suggest that there is a drive for GPCR oligomerization at specific interfaces that remove the structural context for the residual hydrophobic mismatch, thereby reducing the hydrophobic mismatch and its energy cost (Mondal et al. 2011, 2012).

The detailed quantitative analysis of the incomplete hydrophobic matching associated with the protein-membrane coupling of multi-TM-segment proteins, revealed a persistent local exposure of residues to unmatched hydrophobic environments, which could not be eliminated by the overall membrane remodeling. This was termed "residual exposure" (Mondal et al. 2011). Calculations in several different GPCR-membrane systems (Mondal et al. 2011, 2012) showed that this residual exposure is due to the radial asymmetry of the GPCR's hydrophobic surface. Specifically, if polar and hydrophobic residues, e.g., in two different TMs, are adjacent in space, membrane deformation would need to produce a match to both polar and hydrophobic residues in a small neighborhood, and this may be energetically unfavorable. Therefore, adjacencies of hydrophobically disparate residues mark possible sites of residual exposure and thus energetically costly membrane-protein interactions. The CTMD approach that determines the energetic cost of the residual exposure caused by such adjacencies (Mondal et al. 2011), is presented in Sect. 4.3 below.

That adjacencies responsible for the residual exposure of multi-TM proteins are a common feature of GPCRs, is evident from the recently determined crystal structures and validated structural models of a variety of rhodopsin-like GPCRs, including B2AR, KOR, DOR, 5-HT_{2A}R, and D2R in the literature (Mondal et al. 2011, 2012; Shan et al. 2012). Indeed, computational modeling with the CTMD method showed that residual exposure occurs at these adjacencies in different GPCRs and in a number of different membranes (see Tables 4.1 and 4.2) (Mondal et al. 2011, 2012; Shan et al. 2012). This is illustrated in Fig. 4.5a–c showing the residual exposure in TM5 of rhodopsin at a site where Phe and Gln are juxtaposed. Figure 4.5d shows another example of the adjacency, with the polar N1.33 next to the hydrophobic V2.66 and V2.67 in a structural model of the dopamine D2 receptor. Here too, the calculations show a residual exposure, with an energy cost of ~2 kT at N1.33 in a raft-like membrane.

These results from computational modeling highlight the two key elements of the coupling energy between the membrane and the GPCR, viz. (1)-the radially asymmetric membrane deformation, and (2)-the residual exposure remaining after the membrane deforms to alleviate the hydrophobic mismatch. The nature of these two identifiable elements demonstrates that the coupling between the membrane and GPCR proteins is not a simple result of the difference between the average hydrophobic length of the protein and that of the pure membrane as previously postulated for model peptides (Harroun et al. 1999a, b; Lundbæk and Andersen 1999; Marsh 2008). Rather, the important consequences of the coupling emerge from specific features of the protein surface, both with respect to residue identity and to local structure. Therefore, very different consequences can be expected from even small differences between highly homologous proteins (Mondal et al. 2012), from point mutations, or from structural changes due to ligand binding (see next Section).

Table 4.1 Residual SA_{res} values (in Å^2) and the corresponding ΔG_{res} values (in k_BT) calculated for TMs of rhodopsin in diC$_{14:1}$PC, diC$_{16:1}$PC, diC$_{18:1}$PC, diC$_{20:1}$PC, and 7:7:6 SDPC/POPC/cholesterol membranes

	diC$_{14:1}$PC		diC$_{16:1}$PC		diC$_{18:1}$PC		diC$_{20:1}$PC		SDPC/POPC/cholesterol	
TM	SA_{res}	ΔG_{res}	SA_{res}	ΔG_{res}	SA_{res}	ΔG_{res}	SA_{res}	ΔG_{res}	SA_{res}	ΔG_{res}
1	210	9.9	80	3.8	75	3.5	70	3.3	63	3.0
2	74	3.5	29	1.4	0	0	32	1.5	27	1.3
3	N.D.[a]	N.D.	N.D.	N.D.	N.D.	N.D.	N.D.	N.D.	N.D.	N.D.
4	0	0	77	3.6	90	4.3	79	3.7	134	6.3
5	115	5.4	1	0	9	0.4	7	0.3	17	0.8
6	36	1.7	31	1.5	56	2.7	38	1.8	51	2.4
7	119	5.6	44	2.1	22	1.1	0	0	0	0

[a]*N.D.* Not Determined

Table 4.2 Residual SA_{res} values (in Å^2) and the corresponding ΔG_{res} values (in k_BT) calculated for TMs of 5-HT$_{2A}$R in complex with 5-HT, LSD, and KET, respectively, in an SDPC/POPC/cholesterol lipid bilayer

	5-HT$_{2A}$R with 5-HT		5-HT$_{2A}$R with LSD		5-HT$_{2A}$R with KET	
TM	SA_{res}	ΔG_{res}	SA_{res}	ΔG_{res}	SA_{res}	ΔG_{res}
1	0	0	0	0	107	5.1
2	30	1.4	0	0	0	0
3	N.D.[a]	N.D.	N.D.	N.D.	N.D.	N.D.
4	25	1.2	81	3.8	103	4.9
5	94	4.4	81	3.8	69	3.3
6	40	1.9	0	0	0	0
7	1	0	0	0	0	0

[a]*N.D.* Not Determined

4.2.2 Properties of GPCR-Membrane Interactions Determined by Ligand Binding

Of particular significance to GPCR signaling is the observation that both the membrane deformation profile and the residual exposure pattern, can be specific to the ligand that acts on the receptor. For example, analysis of ligand-dependent conformational states of a 5-HT$_{2A}$R receptor model showed that membrane remodeling around the receptor bound to the agonist 5-HT is different from that observed around 5-HT$_{2A}$R in complex with either the partial agonist LSD, or the inverse agonist Ketanserin (see Fig. 4.6) (Shan et al. 2012). The membrane is thinner near TM4 and TM6 when either the agonist 5-HT, or the partial agonist LSD, is bound, compared to the structure stabilized by the inverse agonist Ketanserin. The difference in membrane deformation profile reflects the different conformational rearrangements of the receptor, e.g., the different tilt angles of TM6 (Shan et al. 2012), induced by the three ligands. The patterns of residual exposure calculated for the receptor complexes with the three ligands and the host membrane are shown in Table 4.2 (Mondal et al. 2011). Incomplete hydrophobic matching with an energy penalty substantially larger than 1 kT is predicted for TMs 1, 4, and 5 for the Ketanserin complex, but only for TMs 4 and 5 for the LSD-bound receptor. These calculations illustrate how ligand binding to GPCRs can affect membrane-protein interactions, and consequently the patterns of oligomerization (both extent and interface). Indeed, in addition to being membrane-driven, GPCR oligomerization has been reported to be ligand-sensitive for the dopamine D2, serotonin 5HT2C, and B2AR receptors (Guo et al. 2005; Mancia et al. 2008; Fung et al. 2009). The findings reviewed here suggest that the modulation of oligomerization

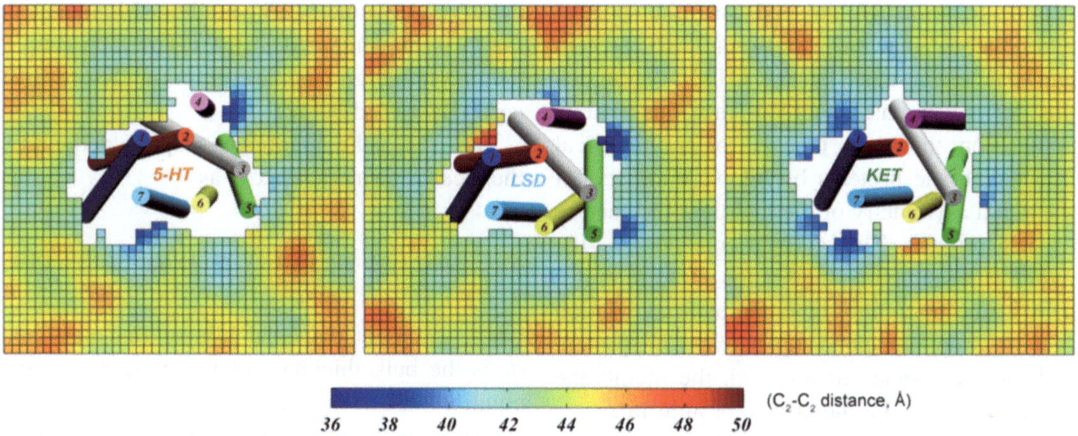

Fig. 4.6 Hydrophobic thickness profiles of the membranes around 5-HT_{2A}R in complex with 5-HT, LSD, or KET. The structures of the various ligand-bound receptor models averaged over the last 100 ns of MD simulations are shown in cartoon, with only the helices depicted (in different colors) with corresponding TM numbers. The colored fields represent distances (in Å) between lipid backbone C_2 atoms from the opposing leaflets. For this analysis, the membrane plane was divided into square 2 Å × 2 Å bins, and the average C_2–C_2 distances in each bin were collected by scanning the last 100 ns of trajectory (Reprinted with permission from Shan et al. 2012)

in the membrane by a specific ligand involves a contribution from its effect on the hydrophobic mismatch. The CTMD approach presented next, in Sect. 4.3, can be used to obtain a quantitative prediction regarding the effects of binding of different ligands to a given GPCR, specifically whether the different effects might result in a substantial difference in the energetics of membrane-protein interactions.

4.3 Continuum-Molecular Dynamics (CTMD) Approach

The hybrid Continuum-Molecular Dynamics (CTMD) approach we developed evaluates the energy cost of hydrophobic mismatch for multi-TM proteins, such as GPCRs (Mondal et al. 2011). Taking into consideration the energy cost of membrane deformations (Sect. 4.3.1) and the energy penalty due to residual exposure at specific residues (Sect. 4.3.2), the CTMD is, to our knowledge, the only method currently available that takes into account the radial asymmetry of the protein hydrophobic surface in evaluating the energy cost of hydrophobic mismatch. In view of the central role that the radial asymmetry of the hydrophobic surface plays for GPCRs (Sect. 4.2), CTMD appears particularly suited for, and was in fact first developed for, the analysis of GPCR-membrane interactions. We note, however, that the CTMD formulation is general enough to be applied to any transmembrane protein.

4.3.1 Membrane Deformations

CTMD employs the well-established continuum, elastic theory of membranes (Nielsen et al. 1998; Huang 1986; Lundbaek et al. 2010; Andersen and Koeppe 2007) to evaluate the energy cost of membrane deformation (ΔG_{def}). Formulations with elastic terms representing the compression-extension and the splay-distortion of the membrane have explained the functional effects of hydrophobic mismatch in model lipid-protein systems at a quantitative level (Huang 1986; Nielsen et al. 1998; Goforth et al. 2003). To date, typical quantitative applications of the continuum, elastic theory of membranes have been limited to single transmembrane helical proteins because the calculations for such model peptides could be simplified by assuming the peptide to be a cylinder and assuming perfect hydrophobic matching between the protein and the membrane. In the CTMD approach the formalism has enabled the quantitative application of this theory to multi-helical transmembrane proteins by taking into account the radial asymmetry of the hydrophobic surface of such proteins. Thus CTMD considers (1)-the radially asymmetric membrane deformations, and (2)-the possibility of incomplete alleviation of the hydrophobic mismatch between the protein and the membrane. To this end, the membrane deformations at the membrane-protein interface are determined from cognate MD trajectories, and the membrane deformations away from the protein are solved at the continuum level without the typical simplifying assumption of cylindrical symmetry. The method (Mondal et al. 2011) is implemented in the CTMDapp (http://memprotein.org/resources/servers-and-software) and is described below.

In CTMD, the membrane shape is defined by the local deformation $u(x,y)$ as

$$u(x,y) = \frac{1}{2}\left(d(x,y) - d_0\right), \tag{4.1}$$

where $d(x,y)$ is the local bilayer thickness and d_0 is the bulk thickness of the bilayer (i.e., the equilibrium thickness away from the protein).

ΔG_{def} is defined as the sum of contributions from compression-extension, splay-distortion, and surface tension (Nielsen et al. 1998; Huang 1986):

$$\Delta G_{def} = \frac{1}{2}\int_{\Omega}\left\{K_a\frac{(2u)^2}{{d_0}^2} + K_c\left(\frac{\partial^2 u}{\partial x^2} + \frac{\partial^2 u}{\partial y^2} - C_o\right)^2 + \alpha\left(\left(\frac{\partial u}{\partial x}\right)^2 + \left(\frac{\partial u}{\partial y}\right)^2\right)\right\} d\Omega, \tag{4.2}$$

where K_a and K_c are the elastic constants for compression-extension and splay-distortion respectively, C_0 is the monolayer spontaneous curvature, and α the coefficient of surface tension. The phenomenological constants K_a, K_c, C_0, and d_0 characterize the membrane property at the continuum level. Experimental measurements for these constants are available for typical lipid types, see (Rawicz et al. 2000) for example.

The local membrane deformations $u(x,y)$ at the continuum level are calculated by solving the Euler-Lagrange equations corresponding to Eq. 4.2. Following the Euler-Lagrange formulation, the ΔG_{def} definition in Eq. 4.2 leads to the boundary value problem,

$$K_c\,\nabla^4 u - \alpha\nabla^2 u + \frac{4K_a}{{d_0}^2}u = 0,\ u\big|_{\Gamma_{in}} = u_o(x,y),$$
$$u\big|_{\Gamma_{out}} = 0,\ \nabla^2 u\,\big|_{\Gamma_{in}} = v_o(x,y),\ \nabla^2 u\big|_{\Gamma_{out}} = 0 \tag{4.3}$$

where Γ_{in} and Γ_{out} represent the boundary at the GPCR-membrane interface and the outer boundary of the simulation box respectively. To evaluate the membrane deformations and the corresponding energy costs without invoking the typical assumption of cylindrical protein, we next perform the following two steps:

1. The membrane-protein interface is determined from cognate MD simulations, with Γ_{in} and $u_0(x,y)$ being obtained from the time-averaged position of the P-atoms around the protein. Specifically, the trajectory is first centered and aligned on the protein, keeping the Z-axis along the membrane normal. Then, the time-averaged z-position of the P-atoms is calculated on a rectangular grid with a 2 Å by 2 Å mesh to obtain the membrane-protein interface as well as the membrane thickness on this interface. This allows for a *non-cylindrical* Γ_{in} corresponding to the membrane-protein interface. Additionally, the membrane deformations $u_0(x,y)$ is *not* simply the difference between the hydrophobic length of the protein and that of the unperturbed membrane, thus allowing for residual exposure. Note that the pattern of local membrane deformations at the membrane-protein interface from MD accounts for

various interfacial interactions at the atomistic level, including interactions between single lipid molecules and specific sites of the GPCRs and any tilting of the TMs within the steric constraint of the whole TM-bundle.

2. The resulting boundary value problem in Eq. 4.3 is numerically solved without simplifying the fourth order partial differential equation in $u(x,y)$ to an ordinary differential equation by assuming radial symmetry. This is achieved by first expressing the partial differential equation in u as coupled partial differential equations in u and v,

$$K_c \nabla^2 v - \alpha v + \frac{4K_a}{{d_0}^2} u = 0,$$

$$\nabla^2 u = v, u\Big|_{\Gamma in} = u_o(x, y),$$

$$u\Big|_{\Gamma out} = u_1(x, y), v\Big|_{\Gamma in} = v_0(x, y),$$

$$v\Big|_{\Gamma out} = v_1(x, y) \qquad (4.4)$$

This set of simultaneous equations can be solved on a rectangular grid using standard finite difference schemes for Poisson equations. As implemented in the CTMDapp, Eq. 4.4 is discretized using the central 5-point approximation of the Laplacian operator, and the system is then solved using the iterative Gauss-Siedel algorithm. As a guideline, we mention that in our experience the procedure converged within 130 iterations for GPCR monomers in a 100 Å by 100 Å box.

While Γ_{in}, Γ_{out} and $u_0(x,y)$ are obtained from MD trajectories, the boundary condition on the curvature is obtained self-consistently using an optimization procedure that searches for $v_0(x,y)$ to minimize ΔG_{def} globally. The complexity of this optimization procedure is made numerically tractable by reducing the size of the $v_0(x,y)$ vector by expressing it as a truncated Fourier series

$$v_o(x, y) : v_0(\theta) \sim \sum_{n=0}^{7} \{a_n \cos(n\theta) + b_n \sin(n\theta)\}, \qquad (4.5)$$

where θ is the polar angle corresponding to the coordinates (x,y), and a_n and b_n are the Fourier coefficients. This step is required to make the method feasible for multi-TM proteins with long membrane-protein interface, e.g., the 7 TM-bundle of GPCRs has a diameter on the order of ~50 Å, thus sharing a long circumferential boundary with the membrane. A given pair of $\{a_n, b_n\}$ in Eq. 4.5 defines a particular $v_0(x,y)$ vector, for which Eq. 4.4 yields the membrane deformations $u(x,y)$ and Eq. 4.2 the ΔG_{def}. In the CTMDapp, the non-linear optimization problem is solved with the objective function $\Delta G_{def} = f\{a_n, b_n\}$ using the BFGS optimization algorithm, which is a standard global, quasi-Newtonian optimization procedure.

We note that the procedure actually involves two nested minimization procedures. The inner minimization occurs via the Euler-Lagrange formalism to give the membrane deformation shape that minimizes the energy cost, given u_0 and v_0. The outer minimization obtains the v_0 that minimizes the energy cost from the Euler-Lagrange formalism, thus obtaining the curvature at the membrane-protein interface for which the membrane deformations are least costly. To verify that membrane shapes calculated from CTMD and MD are similar, the macroscopic $u(x,y)$ obtained by the nested optimization procedure can be compared to $u(x,y)$ calculated directly from the cognate microscopic MD simulation. The calculations for a number of GPCR-membrane systems have shown that the macroscopic $u(x,y)$ indeed converges to a profile similar (within ~1 Å) to the $u(x,y)$ obtained from the cognate MD trajectories. This is illustrated for rhodopsin in di(C14:1)PC in Fig. 4.7. Importantly, the CTMD calculation evaluates the ΔG_{def} using the well-tested elastic continuum theory of membrane deformations.

4.3.2 Residual Exposure

The residual exposure of each membrane-facing residue ($SA_{res,i}$) is quantified in terms of the surface area involved in unfavorable hydrophobic-polar interactions. $SA_{res,i}$ is obtained in terms of

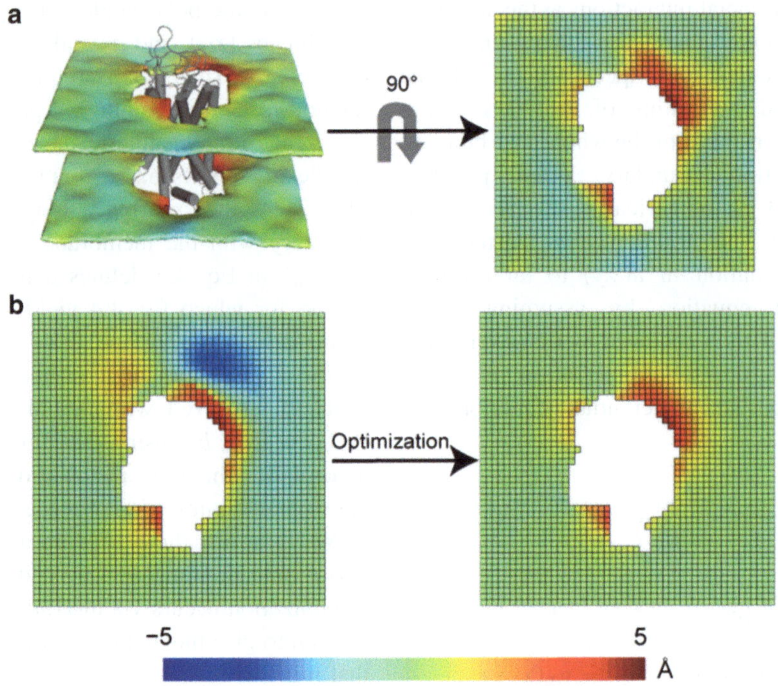

Fig. 4.7 **Protocol for the CTMD approach, illustrated schematically for rhodopsin in a $diC_{14:1}PC$ lipid bilayer.** (**a**) The membrane-deformation profile ***u*(*x,y*)** is calculated directly from the MD simulations. The *left panel* is the deformation profile shown as a color map projected onto the surface defined by fitting a grid (spacing 2 Å) to the positions of the phosphate atoms in the two leaflets during the trajectory, followed by time averaging and spatial smoothing. The *right panel* shows the same deformation color map on the ***x-y*** plane. (**b**) The membrane-deformation profile ***u*(*x,y*)** on a 100 × 100 Å patch, calculated with CTMD. The *left panel* represents the membrane shape calculated with the deformation boundary condition at the membrane-protein interface from the MD profile in panel *A* and a random curvature boundary condition to produce the starting point for the free-energy-based optimization (Eqs. 4.4 and 4.5). The right panel is the membrane-deformation profile calculated with the natural boundary condition, which minimizes the membrane-deformation energy penalty. Note the agreement between the profiles in *A*, calculated using a microscopic theory, and *B*, calculated using the continuum theory (they are within 0.5 Å RMSD of each other). The CTMD approach also allows evaluation of the protein-induced membrane-deformation energy penalty, which is 4.7 $k_B T$ in this case (Reprinted with permission from Mondal et al. 2011. Copyright by Elsevier)

the Solvent Accessible Surface Areas (SASAs) calculated from cognate MD trajectories, as follows:

For hydrophobic residues, $SA_{res,i}$ is calculated as the SASA with the solute taken to be the protein and hydrophobic core of the bilayer. This gives the surface area of the hydrophobic residues exposed to the membrane headgroup or water.

For hydrophilic residues, $SA_{res,i}$ quantifies the surface area of the residue embedded in the hydrophobic core of the membrane. It is calculated as the difference of the SASA with the solute taken to be the protein only and the SASA with the solute taken to be the protein and the hydrophobic part of the membrane.

The corresponding energy penalty ($\Delta G_{res,i}$) can be approximated as linearly proportional to $SA_{res,i}$,

$$\Delta G_{res,i} = \sigma_{res}\, SA_{res,i} \tag{4.6}$$

with the constant of proportionality σ_{res} taken to be 0.028 kcal/(mol. Å^2) (Choe et al. 2008; Ben-Tal et al. 1996).

In these calculations, Lys, Arg, Trp, Tyr, Ser, and Thr need to be treated with special attention to the structural context. If the polar Lys or Arg residues reside within the hydrophobic part of the membrane but near the water-membrane interface, their long side chains can extend towards the phosphate headgroups so that the polar part of their side-chains interacts with the polar phosphate headgroups (Strandberg and Killian 2003; Sankararamakrishnan and Weinstein 2002). This conformational preference is known as "snorkeling", and it has been estimated that as a result, only a small energy penalty is associated with hydrophobic matching by means of snorkeling (Strandberg and Killian 2003). Therefore, Lys and Arg residues near the water-membrane interface are not considered in the calculations of residual exposure if they are found in the MD simulations to snorkel to the lipid headgroup region. *Interfacial* Trp and Tyr are also not considered as candidates for residual exposure, as their commonly observed interfacial location is considered favorable (Yau et al. 1998). Finally, Ser and Thr residues are polar, but when they reside within the hydrophobic part of the membrane, their polar parts can form H-bonds with the helix backbone (Gray and Matthews 1984). In such situations, the Ser and Thr are interacting directly with the protein and not the membrane so they would not be candidates for residual exposure.

4.4 Perspective on "Specific" vs. "Non-specific" GPCR-Membrane Interactions

Functionally relevant interaction of lipids with GPCRs has historically been viewed as being of two different types: (1)-specific lipid-protein interactions, e.g., the binding of individual cholesterol molecules to specific residues of the GPCR; and (2)-non-specific interactions, such as hydrophobic mismatch (Botelho et al. 2002, 2006; Brown 1994; Gibson and Brown 1991, 1993). For example, in affecting rhodopsin activation, lipids with PE headgroups can compensate for the absence of lipids with PS headgroups and lipids with DHA tails, which has been taken to suggest that the lipid-mediated effect on rhodopsin function is not due to lipids of a specific chemical nature (Botelho et al. 2006). Such "non-specific interactions" have been interpreted to involve energetically costly hydrophobic mismatch.

The computational results described herein identify the nature of both "specific" and "non-specific" interactions and suggest that the "specific interactions" are local and the "non-specific" hydrophobic mismatch interactions are non-local in nature. Notably, however, individual lipid molecules that bind to specific sites of the GPCR can also participate in hydrophobic matching. For example, in the 1.6 μs simulation of rhodopsin (see Sect. 4.2.1), the cholesterol that binds specifically at the extracellular end of TM2-TM3 partly reduces solvent accessibility at the hydrophobic Leu3.27 (Khelashvili et al. 2009). Therefore, what is fundamental to this "specific" interaction is its local nature, involving electrostatic/H-bonding/VdW interactions between particular protein residues and the particular cholesterol molecule. On the other hand, the "non-specific" nature of the hydrophobic mismatch is mitigated by the evidence that the membrane deforms differently near TMs of different hydrophobic lengths, and the residual exposure occurs at specific residues of the receptor depending on the spatial organization of polar and hydrophobic residues. However, the deformation profile involves the membrane around the GPCR as a whole thus substantiating the non-local character of the effect. Consequently, membrane deformation profiles could indeed be obtained in the CTMD approach by treating the entire membrane as an elastic continuum and minimizing the total energetic cost of the membrane deformations (provided the thickness boundary conditions were known from atomistic MD simulations). We showed for a number of membrane-GPCR systems that the membrane deformation profile thus obtained was consistent with results from microscopic MD calculations that treat the membrane at the level of lipid molecules (Mondal et al. 2011).

4.5 Conclusion

The multifaceted role of the membrane in GPCR function has, of late, come to the forefront of GPCR research. This has been driven by the experimental data suggesting that the membrane environment affects oligomerization, stability, and activity of GPCRs and the observation of lipid molecules co-crystallized with some GPCR crystals. A central aspect of GPCR-membrane interactions is how the receptor and the membrane couple to each other at the molecular level. Recent computational modeling has provided significant insight into connecting GPCR-membrane interactions with the receptor structure. In particular, this review brings to light the importance of the radial asymmetry of the membrane-facing surface of GPCRs in their interaction with the surrounding membrane. The radial asymmetry creates a context of adjacent hydrophobic and polar residues at specific sites of the GPCR where hydrophobic matching may remain incomplete. Moreover, it is due to this radial asymmetry that there exist a few specific sites where specific lipid molecules such as cholesterol bind. These interactions have been shown to be involved in various aspects of the function and organization of GPCRs.

We have shown how to take into account the radial asymmetry of the hydrophobic surface of GPCRs in providing quantitative information about their interaction with the membrane. With the new understanding gained from the findings and methods presented here, it becomes possible to investigate how GPCR mechanisms, including ligand-determined states and their functional properties, are affected in particular lipid environments. In particular, it becomes possible to quantify the contribution of the lipid-protein interactions with respect to (1)-the individual lipid molecules interacting locally with specific sites of the GPCR, (2)-the deformation/remodeling of the membrane, and (3)-the residual exposure profile and its energy cost.

Acknowledgement This work was supported by the National Institutes of Health grants DA012923-09 and U54-GM087519 to H.W. We also gratefully acknowledge the allocations of computational resources at (1) the Institute for Computational Biomedicine at Weill Medical College of Cornell University, (2) the New York Blue Gene Computational Science facility housed at Brookhaven National Lab, and (3) NSF Teragrid allocation MCB090022.

References

Andersen OS, Koeppe RE (2007) Bilayer thickness and membrane protein function: an energetic perspective. Annu Rev Biophys Biomol Struct 36:107–130

Ballesteros JA, Weinstein H (1995) Integrated methods for the construction of three-dimensional models and computational probing of structure-function relations in G protein-coupled receptors. In: Sealfon SC (ed) Methods in neurosciences, vol 25. Academic, San Diego, pp 366–428

Ballesteros J, Kitanovic S, Guarnieri F, Davies P, Fromme BJ, Konvicka K, Chi L, Millar RP, Davidson JS, Weinstein H (1998) Functional microdomains in G-protein-coupled receptors. J Biol Chem 273(17):10445–10453

Ballesteros JA, Jensen AD, Liapakis G, Rasmussen SGF, Shi L, Gether U, Javitch JA (2001) Activation of the B2-adrenergic receptor involves disruption of an ionic lock between the cytoplasmic ends of transmembrane segments 3 and 6. J Biol Chem 276(31):29171–29177

Barak LS, Menard L, Ferguson SSG, Colapietro AM, Caron MG (1995) The conserved seven-transmembrane sequence NP (X) 2, 3Y of the G-protein-coupled receptor superfamily regulates multiple properties of the beta 2-adrenergic receptor. Biochemistry 34(47):15407–15414

Ben-Tal N, Ben-Shaul A, Nicholls A, Honig B (1996) Free-energy determinants of alpha-helix insertion into lipid bilayers. Biophys J 70(4):1803–1812

Botelho AV, Gibson NJ, Thurmond RL, Wang Y, Brown MF (2002) Conformational energetics of rhodopsin modulated by nonlamellar-forming lipids. Biochemistry 41(20):6354–6368

Botelho AV, Huber T, Sakmar TP, Brown MF (2006) Curvature and hydrophobic forces drive oligomerization and modulate activity of rhodopsin in membranes. Biophys J 91(12):4464–4477

Brown MF (1994) Modulation of rhodopsin function by properties of the membrane bilayer. Chem Phys Lipids 73(1–2):159–180

Brown MF, Thurmond RL, Dodd SW, Otten D, Beyer K (2002) Elastic deformation of membrane bilayers probed by deuterium NMR relaxation. J Am Chem Soc 124(28):8471–8484

Chelikani P, Hornak V, Eilers M, Reeves PJ, Smith SO, RajBhandary UL, Khorana HG (2007) Role of group-conserved residues in the helical core of beta2-adrenergic receptor. Proc Natl Acad Sci U S A 104(17):7027–7032

Cherezov V, Rosenbaum DM, Hanson MA, Rasmussen SGF, Thian FS, Kobilka TS, Choi HJ, Kuhn P, Weis WI, Kobilka BK (2007) High-resolution crystal structure of an engineered human beta2-adrenergic G protein coupled receptor. Science 318(5854):1258–1265

Chien EYT, Liu W, Zhao Q, Katritch V, Won Han G, Hanson MA, Shi L, Newman AH, Javitch JA, Cherezov V, Stevens RC (2010) Structure of the human dopamine D3 receptor in complex with a D2/D3 selective antagonist. Science 330(6007):1091–1095

Choe S, Hecht KA, Grabe M (2008) A continuum method for determining membrane protein insertion energies and the problem of charged residues. J Gen Physiol 131(6):563–573

Choe HW, Kim YJ, Park JH, Morizumi T, Pai EF, KrauÃŸ N, Hofmann KP, Scheerer P, Ernst OP (2011) Crystal structure of metarhodopsin II. Nature 471(7340):651–655

Dorsch S, Klotz KN, Engelhardt S, Lohse MJ, Bunemann M (2009) Analysis of receptor oligomerization by FRAP microscopy. Nat Methods 6(3):225–230

Eroglu Ç, Brügger B, Wieland F, Sinning I (2003) Glutamate-binding affinity of Drosophila metabotropic glutamate receptor is modulated by association with lipid rafts. Proc Natl Acad Sci U S A 100(18):10219

Fritze O, Filipek S, Kuksa V, Palczewski K, Hofmann KP, Ernst OP (2003) Role of the conserved NPxxY (x) 5, 6F motif in the rhodopsin ground state and during activation. Proc Natl Acad Sci U S A 100(5):2290–2295

Fung JJ, Deupi X, Pardo L, Yao XJ, Velez-Ruiz GA, DeVree BT, Sunahara RK, Kobilka BK (2009) Ligand-regulated oligomerization of 2-adrenoceptors in a model lipid bilayer. EMBO J 28(21):3315–3328

Gibson NJ, Brown MF (1991) Role of phosphatidylserine in the MI-MIII equilibrium of rhodopsin*. Biochem Biophys Res Commun 176(2):915–921

Gibson NJ, Brown MF (1993) Lipid headgroup and acyl chain composition modulate the MI-MII equilibrium of rhodopsin in recombinant membranes. Biochemistry 32(9):2438–2454

Gimpl G, Fahrenholz F (2002) Cholesterol as stabilizer of the oxytocin receptor. Biochimi Biophys Acta Biomembr 1564(2):384–392

Goforth RL, Chi AK, Greathouse DV, Providence LL, Koeppe RE, Andersen OS (2003) Hydrophobic coupling of lipid bilayer energetics to channel function. J Gen Physiol 121(5):477–493

Granier S, Manglik A, Kruse AC, Kobilka TS, Thian FS, Weis WI, Kobilka BK (2012) Structure of the delta-opioid receptor bound to naltrindole. Nature 485(7398):400–404

Gray TM, Matthews BW (1984) Intrahelical hydrogen bonding of serine, threonine and cysteine residues within [alpha]-helices and its relevance to membrane-bound proteins. J Mol Biol 175(1):75–81

Green SA, Liggett SB (1994) A proline-rich region of the third intracellular loop imparts phenotypic beta 1-versus beta 2-adrenergic receptor coupling and sequestration. J Biol Chem 269(42):26215–26219

Grossfield A, Feller SE, Pitman MC (2006a) Contribution of omega-3 fatty acids to the thermodynamics of membrane protein solvation. J Phys Chem B 110(18):8907–8909. doi:10.1021/jp060405r

Grossfield A, Feller SE, Pitman MC (2006b) A role for direct interactions in the modulation of rhodopsin by omega-3 polyunsaturated lipids. Proc Natl Acad Sci U S A 103(13):4888–4893

Guo W, Shi L, Filizola M, Weinstein H, Javitch JA (2005) Crosstalk in G protein-coupled receptors: changes at the transmembrane homodimer interface determine activation. Proc Natl Acad Sci U S A 102(48):17495–17500

Haga K, Kruse AC, Asada H, Yurugi-Kobayashi T, Shiroishi M, Zhang C, Weis WI, Okada T, Kobilka BK, Haga T, Kobayashi T (2012) Structure of the human M2 muscarinic acetylcholine receptor bound to an antagonist. Nature 482:547–551

Han Y, Moreira IS, Urizar E, Weinstein H, Javitch JA (2009) Allosteric communication between protomers of dopamine class A GPCR dimers modulates activation. Nat Chem Biol 5(9):688–695

Hanson MA, Cherezov V, Griffith MT, Roth CB, Jaakola VP, Chien EYT, Velasquez J, Kuhn P, Stevens RC (2008) A specific cholesterol binding site is established by the 2.8 Å structure of the human beta2-adrenergic receptor. Structure 16(6):897–905

Hanson MA, Roth CB, Jo E, Griffith MT, Scott FL, Reinhart G, Desale H, Clemons B, Cahalan SM, Schuerer SC, Sanna MG, Han GW, Kuhn P, Rosen H, Stevens RC (2012) Crystal structure of a lipid G protein-coupled receptor. Science 335(6070): 851–855

Harding PJ, Attrill H, Boehringer J, Ross S, Wadhams GH, Smith E, Armitage JP, Watts A (2009) Constitutive dimerization of the G-protein coupled receptor, neurotensin receptor 1, reconstituted into phospholipid bilayers. Biophys J 96(3):964–973

Harroun TA, Heller WT, Weiss TM, Yang L, Huang HW (1999a) Experimental evidence for hydrophobic matching and membrane-mediated interactions in lipid bilayers containing gramicidin. Biophys J 76(2): 937–945

Harroun TA, Heller WT, Weiss TM, Yang L, Huang HW (1999b) Theoretical analysis of hydrophobic matching and membrane-mediated interactions in lipid bilayers containing gramicidin. Biophys J 76(6): 3176–3185

Hern JA, Baig AH, Mashanov GI, Birdsall B, Corrie JET, Lazareno S, Molloy JE, Birdsall NJM (2010) Formation and dissociation of M1 muscarinic receptor dimers seen by total internal reflection fluorescence imaging of single molecules. Proc Natl Acad Sci U S A 107(6):2693–2698

Huang HW (1986) Deformation free energy of bilayer membrane and its effect on gramicidin channel lifetime. Biophys J 50(6):1061–1070

Innis SM (2008) Dietary omega 3 fatty acids and the developing brain. Brain Res 1237:35–43

Jaakola VP, Griffith MT, Hanson MA, Cherezov V, Chien EYT, Lane JR, Ijzerman AP, Stevens RC (2008) The 2.6 angstrom crystal structure of a human A2A adenosine receptor bound to an antagonist. Science 322(5905):1211–1217

Javitch JA, Fu D, Chen J (1995a) Residues in the fifth membrane-spanning segment of the dopamine D2 receptor exposed in the binding-site crevice. Biochemistry 34(50):16433–16439

Javitch JA, Fu D, Chen J, Karlin A (1995b) Mapping the binding-site crevice of the dopamine D2 receptor by the substituted-cysteine accessibility method. Neuron 14(4):825–831

Javitch JA, Ballesteros JA, Weinstein H, Chen J (1998) A cluster of aromatic residues in the sixth membrane-spanning segment of the dopamine D2 receptor is accessible in the binding-site crevice. Biochemistry 37(4):998–1006

Kahsai AW, Xiao K, Rajagopal S, Ahn S, Shukla AK, Sun J, Oas TG, Lefkowitz RJ (2011) Multiple ligand-specific conformations of the B2-adrenergic receptor. Nat Chem Biol 7(10):692–700

Khelashvili G, Grossfield A, Feller SE, Pitman MC, Weinstein H (2009) Structural and dynamic effects of cholesterol at preferred sites of interaction with rhodopsin identified from microsecond length molecular dynamics simulations. Prot Struct Funct Bioinform 76(2):403–417

Khelashvili G, Albornoz PBC, Johner N, Mondal S, Caffrey M, Weinstein H (2012) Why GPCRs behave differently in cubic and lamellar lipidic mesophases. J Am Chem Soc. doi:10.1021/ja3056485

Kofuku Y, Ueda T, Okude J, Shiraishi Y, Kondo K, Maeda M, Tsujishita H, Shimada I (2012) Efficacy of the B2-adrenergic receptor is determined by conformational equilibrium in the transmembrane region. Nat Commun 3:1045

Lagerstrom MC, Schioth HB (2008) Structural diversity of G protein-coupled receptors and significance for drug discovery. Nat Rev Drug Discov 7(4):339–357

Lavie CJ, Milani RV, Mehra MR, Ventura HO (2009) Omega-3 polyunsaturated fatty acids and cardiovascular diseases. J Am Coll Cardiol 54(7):585–594

Lebon G, Warne T, Edwards PC, Bennett K, Langmead CJ, Leslie AGW, Tate CG (2011) Agonist-bound adenosine A2A receptor structures reveal common features of GPCR activation. Nature 474(7352):521–525

Liapakis G, Ballesteros JA, Papachristou S, Chan WC, Chen X, Javitch JA (2000) The forgotten serine: a critical role for Ser-2035.42 in ligand binding to and activation of the B2-adrenergic receptor. J Biol Chem 275(48):37779–37788

Liu W, Chun E, Thompson AA, Chubukov P, Xu F, Katritch V, Han GW, Roth CB, Heitman LH, Ijzerman AP, Cherezov V, Stevens RC (2012) Structural basis for allosteric regulation of GPCRs by sodium ions. Science 337(6091):232–236

Lundbæk JA, Andersen OS (1999) Spring constants for channel-induced lipid bilayer deformations estimates using gramicidin channels. Biophys J 76(2):889–895

Lundbaek JA, Andersen OS, Werge T, Nielsen C (2003) Cholesterol-induced protein sorting: an analysis of energetic feasibility. Biophys J 84(3):2080–2089

Lundbaek JA, Collingwood SA, Ingólfsson HI, Kapoor R, Andersen OS (2010) Lipid bilayer regulation of membrane protein function: gramicidin channels as molecular force probes. J R Soc Interface 7(44):373–395

Luttrell LM, Ferguson SSG, Daaka Y, Miller WE, Maudsley S, Della Rocca GJ, Lin FT, Kawakatsu H, Owada K, Luttrell DK, Caron MG, Lefkowitz RJ (1999) Beta2-arrestin-dependent formation of beta2 adrenergic receptor-Src protein kinase complexes. Science 283(5402):655–661

Mancia F, Assur Z, Herman AG, Siegel R, Hendrickson WA (2008) Ligand sensitivity in dimeric associations of the serotonin 5HT2c receptor. EMBO Rep 9(4):363–369

Manglik A, Kruse AC, Kobilka TS, Thian FS, Mathiesen JM, Sunahara RK, Pardo L, Weis WI, Kobilka BK, Granier S (2012) Crystal structure of the mu-opioid receptor bound to a morphinan antagonist. Nature 485(7398):321–327

Mansoor SE, Palczewski K, Farrens DL (2006) Rhodopsin self-associates in asolectin liposomes. Proc Natl Acad Sci U S A 103(9):3060

Marsh D (2008) Energetics of hydrophobic matching in lipid-protein interactions. Biophys J 94(10):3996–4013

Mondal S, Khelashvili G, Shan J, Andersen OS, Weinstein H (2011) Quantitative modeling of membrane deformations by multihelical membrane proteins: application to G-protein coupled receptors. Biophys J 101(9):2092–2101

Mondal S, Khelashvili G, Wang H, Provasi D, Andersen OS, Filizola M, Weinstein H (2012) Interaction with the membrane uncovers essential differences between highly homologous GPCRs. Biophys J 102(3):514a

Neuringer M, Connor WE, Van Petten C, Barstad L (1984) Dietary omega-3 fatty acid deficiency and visual loss in infant rhesus monkeys. J Clin Invest 73(1):272

Nielsen C, Goulian M, Andersen OS (1998) Energetics of inclusion-induced bilayer deformations. Biophys J 74(4):1966–1983

Niu S-L, Mitchell DC, Litman BJ (2002) Manipulation of cholesterol levels in rod disk membranes by methyl-beta-cyclodextrin. J Biol Chem 277(23):20139–20145. doi:10.1074/jbc.M200594200

O'Dowd BF, Hnatowich M, Regan JW, Leader WM, Caron MG, Lefkowitz RJ (1988) Site-directed mutagenesis of the cytoplasmic domains of the human beta 2-adrenergic receptor. Localization of regions involved in G protein-receptor coupling. J Biol Chem 263(31):15985–15992

Palczewski K, Kumasaka T, Hori T, Behnke CA, Motoshima H, Fox BA, Trong IL, Teller DC, Okada T, Stenkamp RE (2000) Crystal structure of rhodopsin: a G protein-coupled receptor. Science 289(5480):739–745

Park JH, Scheerer P, Hofmann KP, Choe HW, Ernst OP (2008) Crystal structure of the ligand-free G-protein-coupled receptor opsin. Nature 454(7201):183–187

Periole X, Huber T, Marrink SJ, Sakmar TP (2007) G protein-coupled receptors self-assemble in dynamics simulations of model bilayers. J Am Chem Soc 129(33):10126–10132

Periole X, Knepp AM, Sakmar TP, Marrink SJ, Huber T (2012) Structural determinants of the supramolecular organization of G protein-coupled receptors in bilayers. J Am Chem Soc 134(26):10959–10965

Prioleau C, Visiers I, Ebersole BJ, Weinstein H, Sealfon SC (2002) Conserved helix 7 tyrosine acts as a multistate conformational switch in the 5HT2C receptor. J Biol Chem 277(39):36577–36584

Pucadyil TJ, Chattopadhyay A (2004) Cholesterol modulates ligand binding and G-protein coupling to serotonin 1A receptors from bovine hippocampus. Biochimi Biophys Acta Biomembr 1663(1–2): 188–200

Rasmussen SGF, Choi HJ, Rosenbaum DM, Kobilka TS, Thian FS, Edwards PC, Burghammer M, Ratnala VRP, Sanishvili R, Fischetti RF, Schertler GFX, Weis WI, Kobilka BK (2007) Crystal structure of the human beta2 adrenergic G-protein-coupled receptor. Nature 450(7168):383–387

Rasmussen SGF, Choi HJ, Fung JJ, Pardon E, Casarosa P, Chae PS, DeVree BT, Rosenbaum DM, Thian FS, Kobilka TS (2011) Structure of a nanobody-stabilized active state of the beta2 adrenoceptor. Nature 469(7329):175–180

Rawicz W, Olbrich KC, McIntosh T, Needham D, Evans E (2000) Effect of chain length and unsaturation on elasticity of lipid bilayers. Biophys J 79(1): 328–339

Rosenbaum DM, Rasmussen SG, Kobilka BK (2009) The structure and function of G-protein-coupled receptors. Nature 459(7245):356–363

Rosenbaum DM, Zhang C, Lyons JA, Holl R, Aragao D, Arlow DH, Rasmussen SGF, Choi HJ, DeVree BT, Sunahara RK, Chae PS, Gellman SH, Dror RO, Shaw DE, Weis WI, Caffrey M, Gmeiner P, Kobilka BK (2011) Structure and function of an irreversible agonist-beta 2 adrenoceptor complex. Nature 469(7329):236–240

Ruprecht JJ, Mielke T, Vogel R, Villa C, Schertler GFX (2004) Electron crystallography reveals the structure of metarhodopsin I. EMBO J 23(18):3609–3620

Sankararamakrishnan R, Weinstein H (2002) Positioning and stabilization of dynorphin peptides in membrane bilayers: the mechanistic role of aromatic and basic residues revealed from comparative MD simulations. J Phys Chem B 106(1):209–218

Scheerer P, Park JH, Hildebrand PW, Kim YJ, KrauÃŸ N, Choe HW, Hofmann KP, Ernst OP (2008) Crystal structure of opsin in its G-protein-interacting conformation. Nature 455(7212):497–502

Shan J, Khelashvili G, Mondal S, Mehler EL, Weinstein H (2012) Ligand-dependent conformations and dynamics of the serotonin 5-HT_{2A} receptor determine its activation and membrane-driven oligomerization properties. PLoS Comput Biol 8(4):e1002473

Shi L, Javitch JA (2004) The second extracellular loop of the dopamine D2 receptor lines the binding-site crevice. Proc Natl Acad Sci U S A 101(2): 440–445

Shi L, Liapakis G, Xu R, Guarnieri F, Ballesteros JA, Javitch JA (2002) B2 Adrenergic receptor activation: modulation of the proline kink in transmembrane 6 by a rotamer toggle switch. J Biol Chem 277(43): 40989–40996

Soubias O, Teague WE, Gawrisch K (2006) Evidence for specificity in lipid-rhodopsin interactions. J Biol Chem 281(44):33233–33241

Soubias O, Niu SL, Mitchell DC, Gawrisch K (2008) Lipid- rhodopsin hydrophobic mismatch alters rhodopsin helical content. J Am Chem Soc 130(37):12465–12471

Stahl LA, Begg DP, Weisinger RS, Sinclair AJ (2008) The role of omega-3 fatty acids in mood disorders. Curr Opin Investig Drugs 9(1):57–64

Strader CD, Sigal IS, Register RB, Candelore MR, Rands E, Dixon RA (1987) Identification of residues required for ligand binding to the beta-adrenergic receptor. Proc Natl Acad Sci U S A 84(13):4384–4388

Strandberg E, Killian JA (2003) Snorkeling of lysine side chains in transmembrane helices: how easy can it get? FEBS Lett 544(1–3):69–73

Suryanarayana S, Von Zastrow M, Kobilka BK (1992) Identification of intramolecular interactions in adrenergic receptors. J Biol Chem 267(31): 21991–21994

Visiers I, Ebersole BJ, Dracheva S, Ballesteros J, Sealfon SC, Weinstein H (2002) Structural motifs as functional microdomains in G-protein coupled receptors: energetic considerations in the mechanism of activation of the serotonin 5HT2A receptor by disruption of the ionic lock of the arginine cage. Int J Quantum Chem 88(1):65–75

Warne T, Serrano-Vega MJ, Baker JG, Moukhametzianov R, Edwards PC, Henderson R, Leslie AGW, Tate CG, Schertler GFX (2008) Structure of a beta1-adrenergic G-protein-coupled receptor. Nature 454(7203): 486–491

Weinstein H (2005) Hallucinogen actions on 5-HT receptors reveal distinct mechanisms of activation and signaling by G protein-coupled receptors. AAPS J 7(4):871–884

Wu B, Chien EYT, Mol CD, Fenalti G, Liu W, Katritch V, Abagyan R, Brooun A, Wells P, Bi FC, Hamel DJ, Kuhn P, Handel TM, Cherezov V, Stevens RC (2010) Structures of the CXCR4 chemokine GPCR with small-molecule and cyclic peptide antagonists. Science 330(6007): 1066–1071

Wu H, Wacker D, Mileni M, Katritch V, Han GW, Vardy E, Liu W, Thompson AA, Huang XP, Carroll FI, Mascarella SW, Westkaemper B, Mosier PD, Roth BL, Cherezov V, Stevens RC (2012) Structure of the human kappa-opioid receptor in complex with JDTic. Nature 485(7398):327–332

Xiang Y, Rybin VO, Steinberg SF, Kobilka B (2002) Caveolar localization dictates physiologic signaling of beta2-adrenoceptors in neonatal cardiac myocytes. J Biol Chem 277(37):34280–34286

Xu F, Wu H, Katritch V, Han GW, Jacobson KA, Gao ZG, Cherezov V, Stevens RC (2011) Structure of an agonist-bound human A2A adenosine receptor. Science 332(6027):322–327

Yau WM, Wimley WC, Gawrisch K, White SH (1998) The preference of tryptophan for membrane interfaces. Biochemistry 37(42):14713–14718

5 Coarse-Grained Molecular Dynamics Provides Insight into the Interactions of Lipids and Cholesterol with Rhodopsin

Joshua N. Horn, Ta-Chun Kao, and Alan Grossfield

Abstract

Protein function is a complicated interplay between structure and dynamics, which can be heavily influenced by environmental factors and conditions. This is particularly true in the case of membrane proteins, such as the visual receptor rhodopsin. It has been well documented that lipid headgroups, polyunsaturated tails, and the concentration of cholesterol in membranes all play a role in the function of rhodopsin. Recently, we used all-atom simulations to demonstrate that different lipid species have preferential interactions and possible binding sites on rhodopsin's surface, consistent with experiment. However, the limited timescales of the simulations meant that the statistical uncertainty of these results was substantial. Accordingly, we present here 32 independent 1.6 μs coarse-grained simulations exploring lipids and cholesterols surrounding rhodopsin and opsin, in lipid bilayers mimicking those found naturally. Our results agree with those found experimentally and in previous simulations, but with far better statistical certainty. The results demonstrate the value of combining all-atom and coarse-grained models with experiment to provide a well-rounded view of lipid-protein interactions.

Keywords

Rhodopsin • All-atom simulations • Coarse-graining • Lipid-protein interactions • Protein function

5.1 Lipid-Protein Interactions

Protein structure and dynamics play a major role in function. It is not surprising then that membrane protein function would be strongly influenced by interactions at the protein-lipid interface (Marsh, 2008). There are many reasons that the lipid environment plays a major part

J.N. Horn • T.-C. Kao • A. Grossfield (✉)
Department of Biochemistry and Biophysics, University of Rochester Medical Center, 601 Elmwood Ave, Box 712, Rochester, NY 14642, USA
e-mail: alan_grossfield@urmc.rochester.edu

M. Filizola (ed.), *G Protein-Coupled Receptors - Modeling and Simulation*, Advances in Experimental Medicine and Biology 796, DOI 10.1007/978-94-007-7423-0_5,

in the structure and thus the function of integral membrane proteins. First, lipid composition of membranes is incredibly diverse, meaning that there is no single property that defines a membrane; rather, membranes from different cells (or different parts of the same cell) can differ wildly from each other (Brügger et al., 1997). Moreover, the membrane-water interface is a "region of tumultuous chemical heterogeneity" (Wiener and White, 1992), giving membrane proteins a highly diverse surrounding environment (Engelman, 2005).

A number of interactions are at play in the membrane environment, including hydrophobic mismatch (Mouritsen and Bloom, 1984), lipid fluidity, membrane tension, hydrocarbon chain packing (Fattal and Ben-Shaul, 1993), bilayer free volume (Mitchell et al., 1990), the intrinsic curvature of lipids (Gruner, 1985) and elastic strain resulting from bilayer curvature frustration (Lee, 2004). Furthermore, lipid-protein interactions can be subdivided into bulk interactions, or those that result from bilayer properties, and specific interactions, or those that involve an association with individual lipids (Valiyaveetil et al., 2002). Specific interactions can be further subdivided into those with annular lipids, or the boundary lipids forming the first shell around a protein, and non-annular lipids, which can be described as "co-factor" lipids with unique binding sites (Simmonds et al., 1982; Lee, 2003). In some cases, lipids can even be structural elements of membrane proteins (Lee, 2003). A more complete survey of the suggested lipid-protein interaction models can be found in a number of thorough reviews (Mouritsen and Bloom, 1993; Andersen and Koeppe, 2007).

5.1.1 Model Systems for Lipid-Protein Interactions

A number of model proteins exist that depend heavily on lipid interaction for function. One such class of proteins is that of mechanosensitive channels. The function of these channels depends on the relationship between membrane properties, protein structure and flexibility, and protein function (Haswell et al., 2011). One such channel, MscL (mechanosensitive channel of large conductance), responds to turgor pressure changes and hypotonic shock by opening a 30 Å pore to release osmolytes and solutes (Martinac et al., 1987; Cruickshank et al., 1997). The mechanism by which MscL "senses" these conditions is by responding structurally to tension changes in the lipid bilayer. Turgor pressure, as well as cellular processes like cell division, stretches or compresses the bilayer. Thorough reviews of this interplay between membrane structure and protein function for these channels exist elsewhere (Kung et al., 2010; Martinac, 2011; Haswell et al., 2011).

The bacterial potassium channel, KcsA, is another prototypical example. It has been demonstrated that KcsA requires a lipid bilayer for proper folding, despite being stable and active in experiments using detergent micelles, and requires the binding of a single negatively charged lipid on each monomer for activation (Valiyaveetil et al., 2002). Simulation studies confirmed a Arg64–Arg89 binding motif for acidic lipids that was discovered in the experimental studies (Sansom et al., 2005).

Linear gramicidins are a family of bilayer-spanning antibacterial cation channels that increase the permeability of target membranes (Harold and Baarda, 1967). The natural folding preference is to form intertwined (double-stranded) dimers (Burkhart et al., 1998). In the presence of lipid bilayers they refold into the functional end-to-end (single-stranded) dimers (Andersen et al., 1999), with all four Trp residues in each subunit hydrogen bonding with the bilayer at the membrane surface (O'Connell et al., 1990), effectively anchoring the structure in a bilayer spanning configuration. The preference for this conformation in the presence of a lipid bilayer is driven by these Trp residues at the bilayer-solvent interface, which would create a penalty if they were buried in the hydrophobic core. These effects have been noted for other proteins that show a similar Trp anchoring pattern (Killian and von Heijne, 2000).

5.2 Rhodopsin

Rhodopsin, a G protein-coupled receptor (GPCR) responsible for dim-light vision, is a well-characterized transmembrane protein activated by the isomerization of 11-*cis*-retinal to the all-*trans* configuration via light absorption. The ligand's isomerization initiates a cascade of thermal relaxations in the protein, ending with metarhodopsin I (MI). MI exists in equilibrium with metarhodopsin II (MII), the tranducin-binding (or "active") form of the protein (Liebman et al., 1987). The MI-MII transition features significant structural motions that lead to the activation of the G protein and ultimately results in signal transduction.

Rhodopsin is found in large concentrations in the rod outer segment disks (ROS) of rod cells, making up the vast majority of the protein component of each disk's membrane and occupying about a third of the total area (Molday, 1998; Buzhynskyy et al., 2011). The disk membrane phospholipid distribution is about 44 % phosphatidylcholine (PC), 41 % phosphatidylethanolamine (PE), 13 % phosphatidylserine (PS), and 2 % phosphatidylinositol (Boesze-Battaglia and Albert, 1992), with a high concentration polyunsaturated docosahexaenoic acid (DHA) tails (Boesze-Battaglia and Albert, 1989). The concentration of cholesterol in new disks is high (30 %) and decreases as the disk ages (Boesze-Battaglia et al., 1989). Given that rhodopsin is an integral membrane protein with a cascade of structural changes implicated in its function, it is not surprising that this unique membrane environment has been shown to affect the behavior of rhodopsin, particularly the equilibrium between the MI and MII states; recent reviews of these effects are available (Soubias and Gawrisch, 2012). Given the biomedical importance of GPCRs (Drews, 2000) and studies of polyunsaturated fatty acids in dietary intake (Neuringer, 2000), the implications for bilayer regulation of GPCRs to human diseases are clear. Here, we intend to not only highlight experimental and simulation work that explores these effects, but also present long time-scale coarse-grained simulations that provide near-atomic resolution into the possible general and specific mechanisms by which the lipid bilayer interacts with rhodopsin.

5.2.1 Rhodopsin-Lipid Interactions

A number of studies by the Brown lab, starting in the late 1980s and early 1990s (Wiedmann et al., 1988; Gibson and Brown, 1993; Brown, 1994), focused on MII production as a result of the photoisomerization of rhodopsin reconstituted in membranes with a variety of phospholipid and fatty acid combinations. They showed that the population of MII depended on the lipid headgroup composition as well as the concentration of polyunsaturated acyl chains. Native ROS membranes, as expected, showed the greatest quantities of MII. In membranes of PC with short, saturated acyl chains, for instance di(14:0)PC, rhodopsin is essentially inactive. Using di(22:6)PC, they demonstrated that polyunsaturation increased the degree of activity, but not to native levels. The addition of PE lipids also increased activity, though this increase was minor. The presence of polyunsaturation or PE lipids alone does not recreate native activity. Instead, a mixture of phospholipids containing both polyunsaturated chains (22:6ω3) and PE head groups had the highest activity among non-native systems (Wiedmann et al., 1988).

Exploring the role of chain length, and in turn the hydrophobic mismatch between the bilayer and rhodopsin, has shown that the MII population is maximized with chain lengths around 18 carbons, with the equilibrium shifting back towards MI with chain lengths between 16 and 20 carbons. This is coupled to local bilayer compression and stretching effects (Botelho et al., 2006). This mismatch and the resulting bilayer deformations affect rhodopsin activation by altering the helical content of the protein (Soubias et al., 2008). All of this suggests that mechanical and physical properties of the bilayer, including the bilayer thickness and area per lipid, likely modulate the MI-MII transition. However, these bulk effects do not exclude the possibility of localized lipid binding sites on the surface of the protein.

Some preliminary tests were included exploring the role of headgroups in rhodopsin activation, by comparing the effects of PE and PS headgroups (Gibson and Brown, 1993). Later, the role of the membrane potential at the membrane-water interface due to lipid headgroups was explored more thoroughly. It was demonstrated that lipids with PS headgroups have two contradictory effects: they alter the bilayer's properties in ways that oppose MII formation, but their net charge creates an electrostatic environment rich with H_3O^+ ions that promotes MII formation (Brown, 1997). This is in agreement with other results that showed that MII formation is enhanced in acidic conditions (Delange et al., 1997). Finally, they concluded that both PS headgroups and the combination of PE headgroups and DHA chains are needed to maximize the population of MII. The theory suggests that membranes with only PC and PS headgroups would favor MII based only on electrostatics, but this would be counteracted by the structural unfavorability of a charged bilayer surface. As a result, ROS membranes have high concentrations of lipids with highly negative spontaneous curvature (PE) and polyunsaturated chains to counteract this by providing curvature stress and thus promote MII (Wang et al., 2002). These results led Brown and coworkers to a general model, known as the flexible surface model (FSM), where composition of the lipid matrix actively regulates rhodopsin function (Botelho et al., 2002).

The effects of headgroups on the MI-MII equilibrium are not limited to electrostatics and membrane elasticity. In fact, in work intended to discern the energetic contribution of membrane elasticity to rhodopsin function, Gawrisch and coworkers noted that PE headgroups also induce a shift toward MII that correlates with their hydrogen-bonding ability (Soubias et al., 2010). Furthermore, saturation transfer NMR studies of rhodopsin in mixed PC/PE bilayers showed greater magnetization transfer to PE lipids when compared to PC lipids (Soubias et al., 2006), suggesting that in addition to their effects on bulk bilayer properties, there may be a specific role for PE headgroups at the surface of rhodopsin.

5.2.2 Role of Cholesterol

A major component of the rod disk membranes, cholesterol has been shown to have an effect on rhodopsin activation as well. The presence of cholesterol in membranes drives the MI-MII equilibrium towards MI, reducing signaling (Mitchell et al., 1990). However, it is unclear from these experiments whether this is caused by cholesterol's effects on bilayer properties, direct interactions between it and the protein, or some combination of these effects.

Cholesterol's effects on liquid crystalline membrane structure are well documented. The presence of cholesterol causes tighter packing of lipid hydrocarbons (Niu et al., 2002) and increases the thickness of the bilayer, leading to changes in the lateral compressibility (Needham et al., 1988). These bilayer effects may create an environment that inhibits the conformational transition from MI to MII, though cholesterol also promotes negative curvature elastic stress, a property of PE headgroups that tends to promote MII formation (Chen and Rand, 1997).

The alternative means by which cholesterol may affect rhodopsin activation is through direct and possibly specific interactions. Studies of cholesterol, cholestatrienol and ergosterol in ROS membranes have suggested that there is at least one cholesterol binding site on rhodopsin (Albert et al., 1996). Sites of preferential cholesterol interaction have been identified via molecular simulation as well (Grossfield et al., 2006b), although these and other simulations suggested that cholesterol is, on the whole, depleted at the protein surface (Pitman et al., 2005).

5.2.3 Simulation Applied to Rhodopsin

In 2000, the first crystal structure of a GPCR was solved in the form of bovine rhodopsin at 2.8 Å resolution (Palczewski et al., 2000), opening the way for the use of molecular dynamics simulation to probe the atomic-level interactions between rhodopsin and its environment. An early simulation of rhodopsin, performed by Huber et al. (2002), was compared to available NMR data to quantify membrane deformation in the presence of protein and compute cross-sectional protein areas. They found that rhodopsin imposed curvature in the bilayer, which could facilitate selection for polyunsaturated lipids at the surface of rhodopsin. This was seen as further support for the flexible surface model (Huber et al., 2004).

As the available computer power improved, simulations began to explore the interactions between rhodopsin and bilayer constituents via longer molecular dynamics trajectories of the protein embedded in a more realistic bilayer. These systems featured cholesterol, as well as a mixture of lipids with two different headgroups, each of which was linked to one polyunsaturated tail and one saturated tail. Early results showed a preference for the polyunsaturated tails at the surface of rhodopsin and little indication of specific binding sites for cholesterol (Feller et al., 2003; Pitman et al., 2005).

Later, a series of 26 independent 100 ns simulations of rhodopsin in a realistic lipid composition was used to address these questions, with better statistical sampling (Grossfield et al., 2006b). Polyunsaturated tails were again enriched at the surface of the protein. It was also now possible to identify residues that preferentially interacted with cholesterol and each of the lipid components. Although this work required a heroic effort at the time, the sampling was not sufficient to give high confidence in the predictions about specific residues, particularly for cholesterol, since any given interaction was only seen in a small fraction the trajectories (Grossfield et al., 2006b). Nonetheless, the results demonstrated that computational methods could confirm the experimentally suggested trend of preferential interactions with polyunsaturated tails, allowing some to speculate on the roles these flexible lipids could play in the activation of rhodopsin (Feller and Gawrisch, 2005). Preferential sites of interaction between rhodopsin and cholesterol were also identified (Khelashvili et al., 2009).

More recently, the same data (supplemented by an additional 1.6 μs simulation) was reanalyzed to explore the interactions between the palmitoyl moieties attached to a pair of cysteines in the cytoplasmic helix H8 and the helical bundle of rhodopsin (Olausson et al., 2012). There was a high degree of contact between both chains and the protein, but also a significant difference between the two, even though they are attached to consecutive residues. The high level of contact between the palmitate on Cys322 and the protein helices suggests that it may play an important role beyond that of a nonspecific lipid anchor.

Recently, some groups have begun using coarse-grained simulations to explore rhodopsin-lipid interactions as well. For example, Periole et al. (2007) explored the effects of varying the bilayer hydrophobic thickness on the oligomerization of rhodopsin, using the MARTINI force field (Marrink et al., 2007; Monticelli et al., 2008). The results demonstrated that oligomerization is driven by frustration of lipid-protein interactions, something that had already been seen experimentally (Botelho et al., 2006). Rhodopsin significantly altered the local membrane thickness, encouraging oligomerization as a way to reduce unfavorable protein-lipid interactions. While useful for exploring bilayer adaptations attributed to hydrophobic mismatch, each simulation featured homogeneous bilayers, so no new evidence was gleaned about the role for cholesterol and different lipid headgroups in rhodopsin activation.

In this work, we employed simulations featuring membranes with native-like compositions to more thoroughly explore protein-lipid interactions. Our focus in these systems was the role played by the saturation state of the lipids, the headgroups of the lipids, and the presence of cholesterol in modulating lipid-protein interactions.

5.3 Methods

5.3.1 Simulation Systems

The goal of this research was to explore lipid-protein interactions in a system with rhodopsin and a biologically relevant model membrane. We sought to extend previously published work that featured 26 separate 100 ns all-atom simulations of rhodopsin in a membrane (Grossfield et al., 2006b). These systems featured a 2:2:1 ratio of 1-stearoyl-2-docosahexaenoyl-phosphatidylcholine (SDPC), 1-stearoyl-2-docosahexaenoyl-phosphatidylethanolamine (SDPE), and cholesterol, and were intended to mimic those membranes found biologically and used in experimental studies with model membranes (Boesze-Battaglia and Albert, 1989; Wiedmann et al., 1988; Albert et al., 1996). The systems described here mimic these all-atom systems in an attempt to bring together a great deal of experimental work that has been done investigating the role of various membrane properties on rhodopsin activation, taking advantage of the ability of molecular dynamics simulation to explore interactions on the molecular level.

While being a remarkable effort at the time, previous all-atom simulations had some limitations that faster computers and more efficient simulation methods have allowed us to overcome. By employing a coarse-grained model, we simulated larger systems for longer timescales, allowing for better sampling of long time-scale processes critical for bilayer-protein interactions, such as the lateral reorganization of the lipid bilayer. The work described here features systems nearly three times larger than the all-atom work, with timescales an order of magnitude longer. Furthermore, simulations of "dark" and "active" rhodopsin were done to make comparisons about the lipid and cholesterol binding surfaces in these differing structures. Overall, the new results agree with the earlier all-atom work, while providing more robust statistics and allowing us to make some conclusions about specific and general interactions between the lipid environment and rhodopsin.

5.3.2 Construction

Rhodopsin was modeled after the same crystal structure used in the previous work (PDB ID: 1U19 Okada et al. 2004). For comparison, retinal-free opsin was also modeled (PDB ID: 3CAP Park et al. 2008). The MARTINI coarse-grained force field was used (Marrink et al., 2007) with the extension for proteins (Monticelli et al., 2008). Recently, the MARTINI forcefield was applied to study the aggregation properties of rhodopsin (Periole et al., 2012; Knepp et al., 2012). To create the coarse-grained model for each protein, all molecules present other than rhodopsin/opsin were removed from the crystal structure. We then mapped the coarse amino acids onto this structure using the martinize.py script available online at the MARTINI website (MARTINI, 2011). The long C-terminal tail found in the rhodopsin structure, which was not resolved in the opsin structure, was not removed. It makes minimal contact with the bilayer and thus should not affect our results in this context. The acyl chains on the cysteine residues at positions 322 and 323 were manually added with parameters used for palmitoyl chains in lipids. The resulting system was energy minimized. Retinal was not explicitly represented in the rhodopsin model; in the MARTINI protein model, protein fluctuations are dominated by the network of restraints required to stabilize the tertiary structure, and since the retinal itself does not interact with the membrane environment in any significant way, we believe it is sufficient to model the retinal-stabilized dark-state rhodopsin structure of the protein in its absence.

First, we built an SDPC bilayer via self-assembly with a box of randomly placed SDPC lipids and water. For the rhodopsin system, we inserted the protein into this SDPC lipid bilayer using a bilayer expansion and compression technique (Kandt et al., 2007). In this method, the lipids are translated in the plane of the bilayer by a large scaling factor, creating space for the protein insertion without clashes. Then, the lipids are scaled back to the ideal area-per-lipid using a number of cycles of translation and minimization. This reduces lipid-lipid and lipid-protein clashes.

Water and ions were added to solvate and neutralize the system; additional NaCl was added to bring the concentration to 100 mM. Afterwards, we held the protein position fixed and performed a 10 ns simulation to allow the lipid and solvent environment to relax. To generate unique starting states for the bilayer, randomly selected SDPC lipids were converted to SDPE lipids by simply changing the head group beads, while others were swapped with cholesterol. The result was a set of 16 unique rhodopsin systems with 2:2:1 SDPC:SDPE:cholesterol bilayers. We minimized the resulting systems and performed another round of equilibration, again holding the protein fixed.

To make the equivalent opsin model, rhodopsin was replaced with opsin after backbone alignment for each of the rhodopsin systems. These were then equilibrated with a short simulation (100 ps) to ensure that the starting configurations for the opsin systems and the rhodopsin systems were similar.

The final systems included 1 protein molecule (either rhodopsin or opsin), 180 SDPC, 180 SDPE, 90 cholesterol, about 19,000 water beads (each representing 4 water molecules), and about 140 each of Na^+ and Cl^-. This brings the system size to about 26,000 CG beads, which is roughly equivalent to an all-atom system with 105,000 atoms (double the total system size and more than triple the bilayer size of the previous work).

5.3.3 Simulation Protocol

Simulations were performed with version 4.5.4 of the GROMACS molecular dynamics package (van der Spoel et al., 2005; Hess et al., 2008) on a Linux cluster. We used a time step of 10 fs, as suggested for accurate integration (Winger et al., 2009; Marrink et al., 2010), with the neighbor list updated every 5 steps. We held the temperature at 300 K using Nosé-Hoover temperature coupling (Nosé and Klein, 1983; Hoover, 1985) and treated the pressure semi-isotropically with a reference of 1 bar using the Parrinello-Rahman barostat (Parrinello and Rahman, 1981). A shift function was employed for electrostatics with a coulomb cutoff of 12 Å. The Lennard-Jones potential was shifted between 9 and 12 Å.

It is well established that external restraints are required when simulating native proteins using MARTINI. Initially, we performed our simulations with only the parameters necessary to maintain the secondary structure. However, under these conditions the rhodopsin structure moved rapidly away from the crystal structure, reaching a transmembrane alpha carbon RMSD as high as 6 Å (data not shown). This problem in the MARTINI force field with maintaining protein tertiary structure has been noted before and protocols have been developed for overcoming these limitations utilizing distance-based restraints (Periole et al., 2009). To maintain the integrity of our proteins, we included a similar network model, restraining the distances between backbone beads between 2 and 10 Å apart. We tested multiple force constants using short trajectories to try to match the amplitude of fluctuations for the transmembrane helices to previous all-atom simulations; with a force constant of 800 kJ/mol·nm^2, the rhodopsin transmembrane helix RMSD fluctuated between 2.0 and 2.5 Å, consistent with previous all-atom results (Grossfield et al., 2008), and the opsin RMSD fluctuated between 2.5 and 3.0 Å.

We ran 32 independent simulations, 16 each for rhodopsin and opsin. Each simulation was 1.6 μs, for a total of 51.2 μs of simulation time (effective time of about 205 μs if we apply a 4X scaling to the time, as suggested by previous authors to account for the enhanced kinetics of the coarse-grained model Marrink et al. 2007). All times we report here are the actual simulation times, without the 4X scaling.

5.3.4 Analysis

All simulation analyses were performed using tools developed using the LOOS library. LOOS is an object-oriented library implemented in C++ and Boost for rapidly creating new tools for analyzing molecular dynamics simulations (Romo and Grossfield, 2009, 2012). All analysis was performed on trajectories with 1 ns time resolution.

5.3.4.1 Bound Lipid Lifetimes

In order to measure the time required for lipids to exchange off the surface of the protein, we calculated a conditional survival probability.

Specifically, we identified those lipids within 6 Å of any protein bead as "bound", and computed the probability that if a lipid is bound at time t it will also be bound at time $t + \Delta t$. Error bars are estimated by treating each simulation as an independent measure of the decay curve and computing the standard error in the means. This calculation was implemented using LOOS.

5.3.4.2 Lateral Radial Distribution Function

We computed the lateral radial distribution function (RDF) of various bilayer species relative to the center of mass of the protein structure in the membrane plane using the xy_rdf tool distributed with LOOS tools. Each molecule was treated as a single unit, located at its centroid.

5.3.4.3 Density Maps

We created 2D density maps to show average density of each lipid component in the plane of the bilayer. We aligned the protein structure of each frame of the simulation to the initial structure. Then we binned the centroid of each component in a grid on the plane parallel to the membrane, with 1 Å^2 bins. The resulting density histogram is displayed as a heat map.

To probe the 3-dimensional distribution of lipid components about the membrane, we first aligned our trajectories using transmembrane C_α's. Then we used a 1 Å^3 grid superimposed over the protein's bounding box, padded by 30 Å. Each atom is then placed in the nearest bin and the resulting histogram is convolved with a gaussian for a smoother visualization. The gaussian smoothing is only applied to the 3D visualizations, not the 2D density maps.

5.3.4.4 Residue-Based Binding Scans

To highlight interactions between lipid components and individual residues, we computed a residue binding score for each residue of the protein to each lipid component. The residue score R for residue n and lipid component m can be expressed as:

$$R_{nm} = \frac{1}{N} \sum_{i}^{N} \sum_{j}^{M} \frac{1}{r_{ij}^{6}} \tag{5.1}$$

Where N is the number of atoms in residue n, M is the sum of all atoms for all molecules of lipid component m in the system, r_{ij} is the distance between atoms i and j. The normalized residue score is then simply the residue score divided by the average residue score for all transmembrane alpha helix residues:

$$R^{\circ}_{nm} = \frac{R_{nm}}{\frac{1}{N_{\mathrm{tm}}} \sum_{k}^{N_{\mathrm{tm}}} R_{km}} \tag{5.2}$$

5.3.4.5 Statistical Significance

Because we have multiple independently constructed trajectories of each system, we have attempted to assess the statistical significance of our results. This was done using a Welch's T-test, treating the average result (e.g. the radial distribution density at a particular distance) from each trajectory as a single data point. We typically plot the p-values on the same axes as the results themselves. To compute these p-values, we used statistical tools available in SciPy (Jones et al., 2001–).

5.4 Results and Discussion

Before beginning any kind of analysis of lipid binding, it is critical that we show thorough sampling of the exchange between surface bound lipids and bulk lipids; otherwise, the results would be dominated by the starting configuration of the lipids. To accomplish this we calculated the probability if a lipid is found at the protein surface at time t it will also be there at time $t + \Delta t$. The resulting decay curves for SDPE and SDPC are shown in Fig. 5.1. These curves can be fit to a double exponential, resulting in two characteristic decay times. The fast time, accounting for roughly half the amplitude, is below 1 ns. Given that the analysis was performed using 1 ns samples, this is most likely the result of flickering at the edge of the "bound" state. The slow decay times for SDPE and SDPC in the rhodopsin system are 78.3 ns (±2.88) and 63.2 ns (±1.31), respectively.

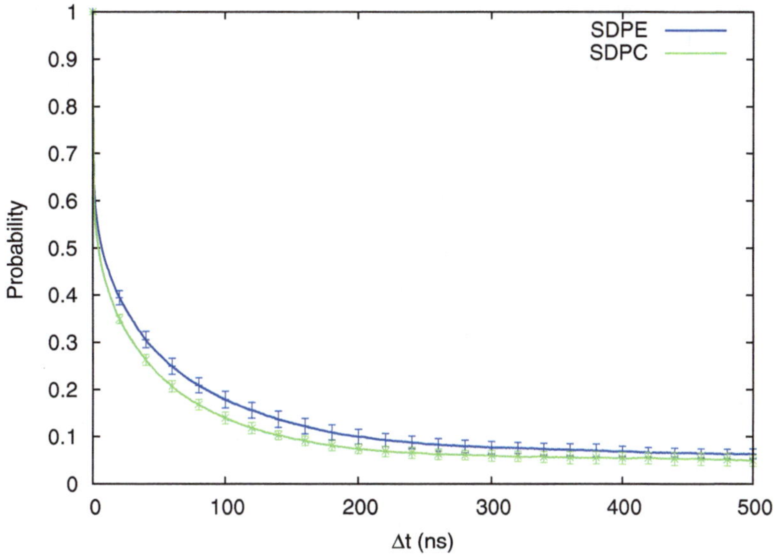

Fig. 5.1 Survival probability for lipids at the surface of rhodopsin. Error bars are the standard error of the means of the 16 trajectories

5.4.1 RDFs Demonstrate Surface Preferences for DHA

To begin our analysis, we used a two-dimensional radial distribution function (RDF) in the plane of the membrane, as was done in previous work (Feller et al., 2003), to assess the packing of the different members of the bilayer against rhodopsin. In Fig. 5.2, we show RDFs for the two lipid tail types and cholesterol. Here we can see a drastic enrichment of docosahexaenoic acid (DHA) between 15 and 20 Å from the center of the protein, with the peaks for cholesterol and stearoyl beyond 20 Å. Interestingly, while cholesterol is not enriched at the surface, it has significant density deeper into the protein than stearoyl, and nearly as deep as DHA. Given its smaller size, this may be indicative of regions accessible only to cholesterol.

To accurately assess the significance of the difference between RDF curves for opsin and rhodopsin, we calculated p-values for each point. Given that each point is the mean of a set of 16 independent samples, we have a fairly large set of data from which to do this assessment (unusual in the simulation community). The resulting p-values are plotted below the RDF curves with the same x-axis, with confidence levels of 0.01 and 0.05 shown for reference. Panels C, D and E show these p-value plots for cholesterol, DHA and stearoyl, comparing the means between rhodopsin and opsin. Statistically significant ($P < 0.01$) differences appear between all analyzed bilayer constituents, but some for only brief stretches of the RDF curves.

The short region of cholesterol significance coincides with a peak in the opsin curve that is not present in rhodopsin, indicating a region of bulk density at the very surface of opsin where the DHA density is the highest. The significant regions are much more substantial for the lipids tails. For stearoyl, large stretches of the curve show very significant differences ($P < 0.001$). Visual inspection of the RDFs shows greater penetration of the lipid tails between 10 and 20 Å. This can be explained to some degree by the greater flexibility and more "open" structure for the opsin system; there is a greater area accessible to the lipid tails between the helices and in the protein interior.

5.4.2 Density Maps Show DHA Preference

The above results are consistent with previous simulation and experimental results. However, simple lateral radial distribution functions

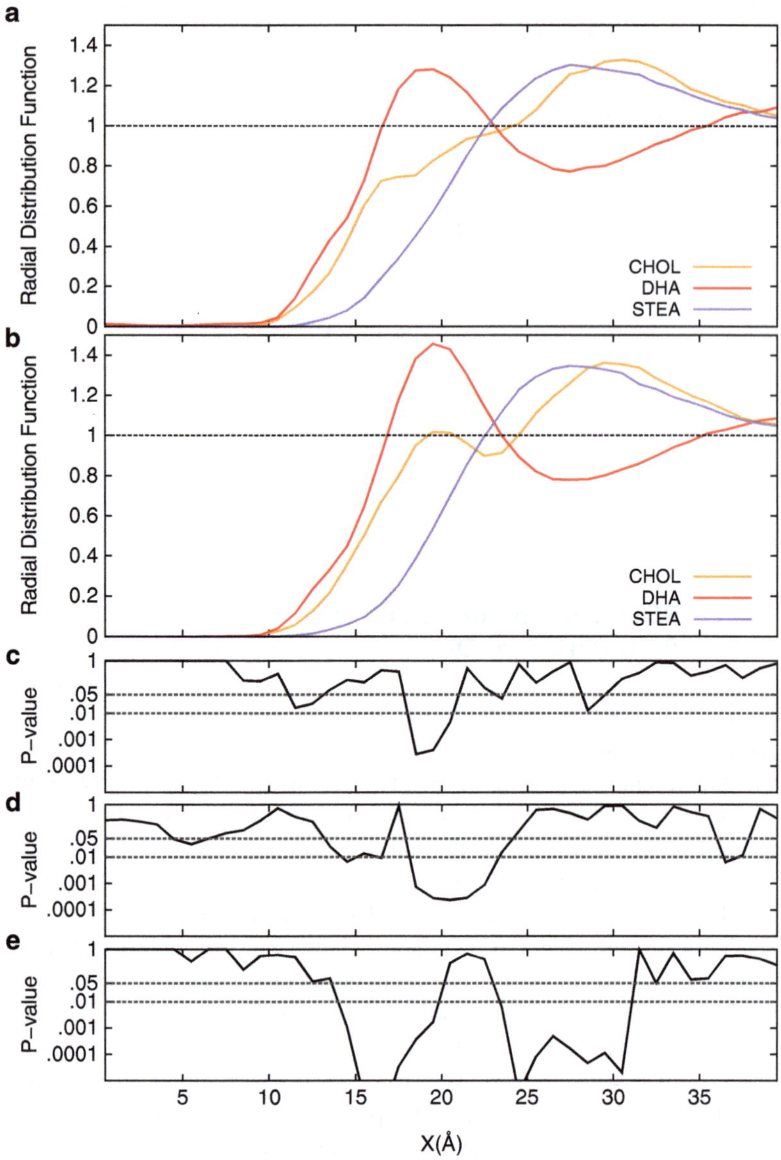

Fig. 5.2 Lateral radial distribution functions of each of the lipid tails and cholesterol for the (**a**) rhodopsin and (**b**) opsin systems. Comparison of means tests between rhodopsin and opsin for (**c**) cholesterol, (**d**) DHA, and (**e**) stearoyl

contain limited information, because they treat rhodopsin as a featureless cylinder, integrating out the distinctions between different portions of the protein surface. Moreover, in these plots, both leaflets were treated together, again averaging away potentially valuable information. Accordingly, we instead project our results along 2 dimensions, using lateral density heat maps.

Figure 5.3 shows density maps for the different lipid components for both rhodopsin and opsin, in both the upper and lower leaflets. The dark region in the center of each frame represents the excluded volume of the helix bundle, as we look down from extracellular side. Opsin and rhodopsin were aligned by their backbones so that they are oriented the same way in all of the heat maps for comparison. These images represent the average of all 16 simulations for each system.

In the plots of DHA density, we see a bright, thin ring tracing the protein space. In contrast with the low densities for stearoyl

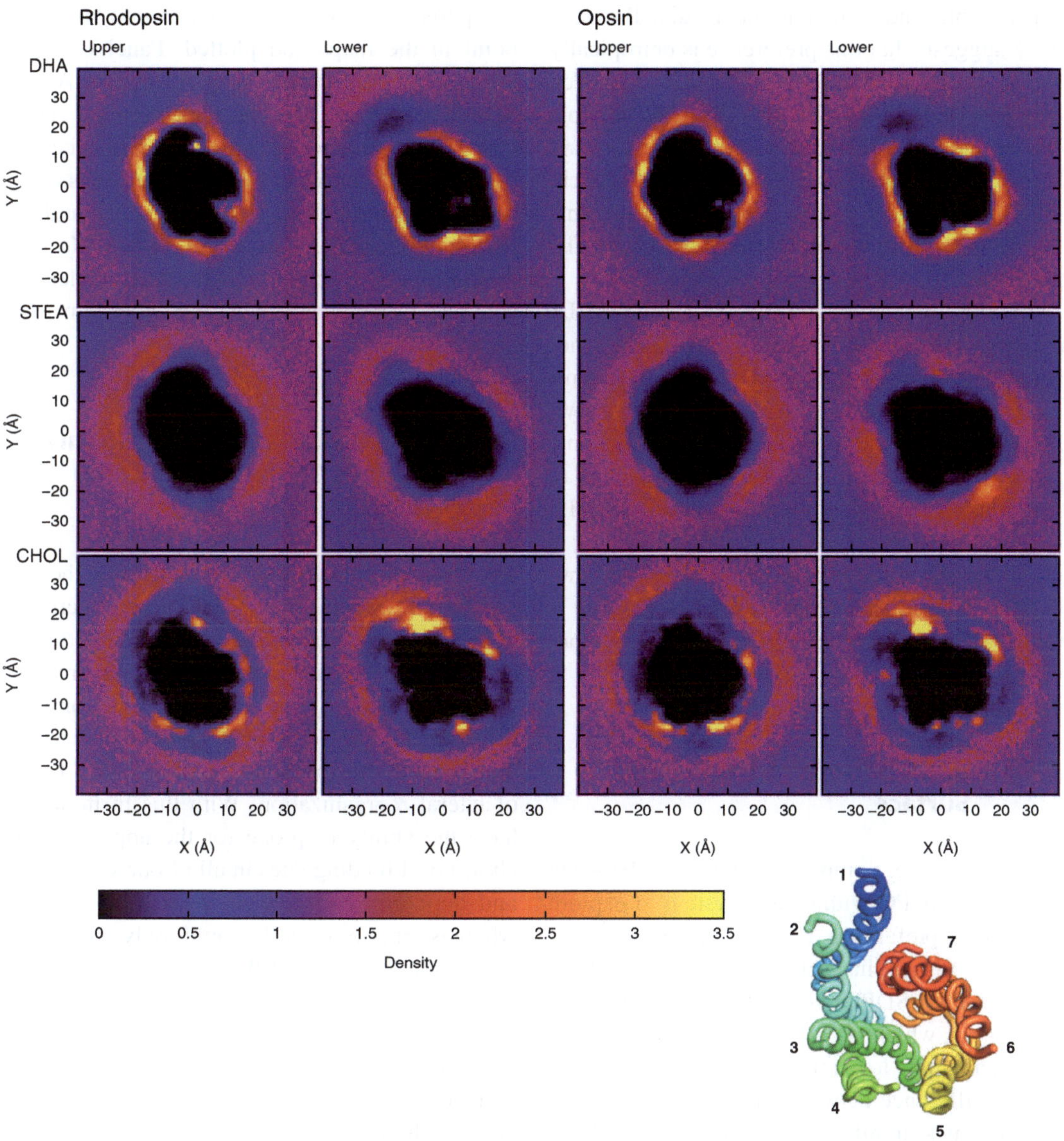

Fig. 5.3 Density maps of each bilayer component, for each leaflet, in each system (rhodopsin or opsin). Density is reported as lipid components per Å^2. All images are viewed from the extracellular side of the protein. The *upper leaflet* refers to the leaflet on the extracellular side. For every map, rhodopsin or opsin was centered at the origin and aligned against a reference structure so that the maps can be directly compared. The orientation of the helices is shown in the *small panel* in the *bottom right corner*, also viewed from the extracellular side of the protein

and cholesterol, this indicates that DHA is preferentially packed against the surface of the protein, with the exception of a bright cholesterol spot next to helices H1 and H7. The corresponding stearoyl densities show rings as well, immediately outside the DHA ring. The stearoyl rings are dimmer and, in general, more diffuse.

The lateral radial distribution functions, coupled with the density heat maps, suggest a strong preference for DHA at the surface of both opsin and rhodopsin, in agreement with previous experimental and computational results. Previous work suggests that this preference is entropically driven (Grossfield et al., 2006a). It has been demonstrated that DHA is extremely flexible (Feller et al., 2002) and rapidly interconverts between conformations (Soubias and Gawrisch, 2007), making it ideal for packing against the relatively rigid but uneven surface of the protein.

The region just beyond the first shell of DHA chains is enriched in stearate. This result is not surprising, because the lipids used in these simulations each have one DHA and one stearoyl. For every lipid with a DHA tail packed against rhodopsin, there is also a stearoyl facing away from the protein, accounting for the inner DHA ring and the outer stearoyl ring. This outer ring is not as bright in the heat maps as the DHA ring because the accessible surface area in this ring is far greater, so the motion of these tails is more diffuse.

5.4.3 SDPE is Preferred at Protein Surface

In Fig. 5.4, we compare preferences between SDPE and SDPC within each system to explore a possible preference for one headgroup over another. Visual inspection suggests a slight preference for SDPE over SDPC at the surface of the protein, which is confirmed to be statistically significant; the lower panels show the p-value for the difference between SDPC and SDPE and demonstrates significance at the 0.01 level for the entire first "solvation" shell of rhodopsin, and most of that region for opsin. Panel E shows that the differences in the SDPC RDF between rhodopsin and opsin are marginally significant at best, but the SDPE RDF (Panel F) does show a significant difference in the location of this initial rise at the protein surface.

This is in agreement with experimental results that suggest that PE is a preferred partner for rhodopsin (Soubias et al., 2006, 2010). To explore the possibility of specific PE interaction sites, we generated density maps of SDPE and SDPC for rhodopsin, found in Fig. 5.5. For quantitative comparison, p-values were computed for every point in the maps and plotted. Panel C shows the p-value plot comparing SDPE (panel A) and SDPC (panel B) for the upper leaflet. There is a large region of statistical significance that indicates differences between the densities of the two headgroups along helices H3, H4 and H5 on the extracellular side of the protein. No such regions of interest are seen in the other leaflet, and it is less pronounced for the opsin system (data not shown).

5.4.4 Mapping Density to Structure Probes Cholesterol Binding Sites

The present simulations, by accessing the microsecond timescale, are much more effective at allowing for sampling of the lateral motion of cholesterol and lipids in the bilayer than previous work. Considering the enhanced rate of lipid diffusion in the MARTINI forcefield, each of our simulations arguably samples nearly 6.5 μs of lateral reorganization. With this in mind, we have the ability to probe for the appearance of cholesterol binding sites in all of our simulations and hopefully converge on representative sites, whereas previous work could only note sites where some fraction of the simulations had seen strong contacts.

In the lateral radial distribution plots, there is evidence that cholesterol can pack inside the helical bundle (deeper than either lipid tail), albeit not with high abundance. Indeed, integrating the RDF suggests the presence of roughly 1 cholesterol immediately at the protein surface (data not shown). The heat maps clarify this result; visual examination shows one site of clear cholesterol preference, next to helices H1 and H7, approximately where the palmitoyls attach to helix H8. DHA lipids are excluded from this location, as seen by a patch of low density. Whether this site represents a true competitive binding site, where binding to cholesterol is preferred over other bilayer constituents, or simply a deep protein pocket or crevice that is only accessible to cholesterol, is unclear.

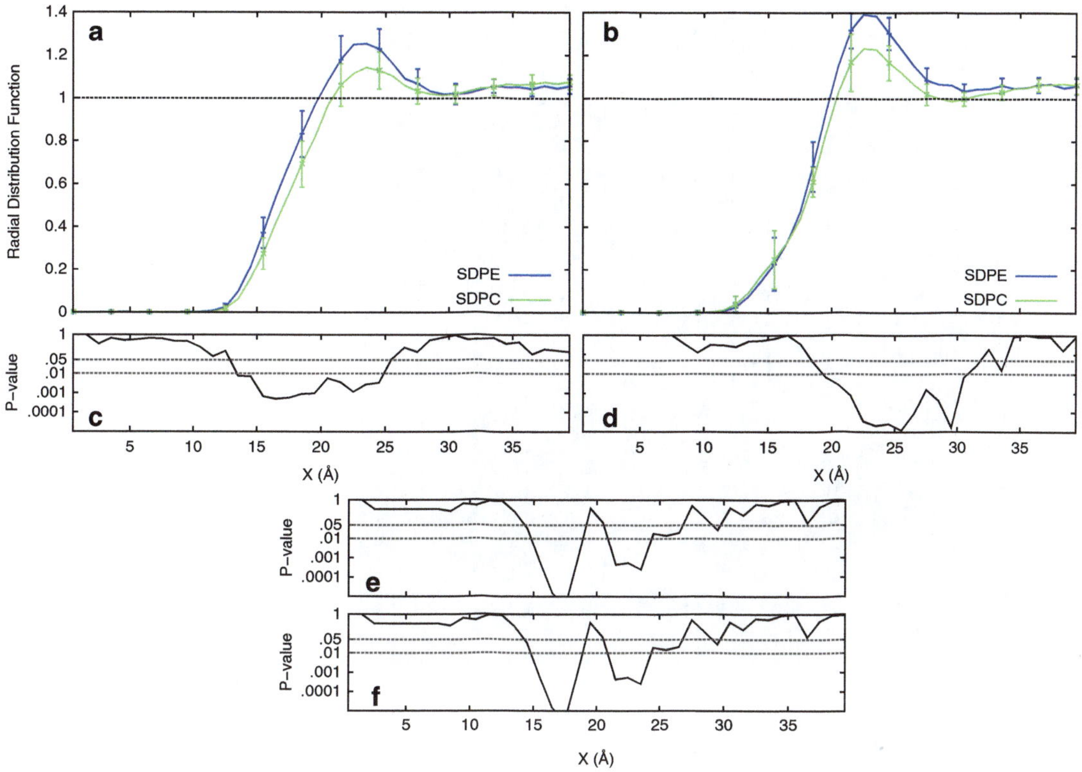

Fig. 5.4 Lateral radial distribution functions of SDPE and SDPC for the (**a**) rhodopsin and (**b**) opsin systems. Comparison of means tests were performed comparing SDPE to SDPC within the (**c**) rhodopsin and (**d**) opsin systems. P-values comparing differences in mean between rhodopsin and opsin were plotted for (**e**) SDPC and (**f**) SDPE

In order to better understand the details of the cholesterol binding, we further expanded the data to look at the full 3-dimensional distribution. Figure 5.6 shows contours of regions with high average cholesterol density superimposed onto the structures of rhodopsin and opsin. The brightest cholesterol spot in the 2D maps, which as mentioned excludes other lipids, corresponds with the region of high density beside H8, the intracellular helix lying parallel to the bilayer, and packed against H1 and H7 of the protein. The density surface packs neatly behind the pair of post-translational palmitates on Cys322 and Cys323. In these simulations, H8 is embedded in the bilayer interface, creating a pocket in the hydrophobic core that is too small for lipids to comfortably diffuse, leading us to conclude that in this case, cholesterol is able to pack where other lipids cannot penetrate efficiently. This cholesterol "hot spot" is found in both the opsin and rhodopsin simulations, as is a second region found on the opposite side of the protein, between the cytoplasmic end of helix H3 and helices H4 and H5. This corresponds with a cholesterol interaction site predicted from a pair of 800 ns simulations of the adenosine A_{2A} receptor (Lee and Lyman, 2012).

Lastly, at the density contour level chosen, a third high density cholesterol region is present in the opsin system packed against helices H5 and H6. At lower contour thresholds, this region appears in the rhodopsin systems as well, but its presence here indicates a possible difference in cholesterol packing interfaces between the two structures. Given that the greatest structural changes between the chosen opsin and rhodopsin structures are the orientations of helices H5 and H6, as well as the elongation of helix H5

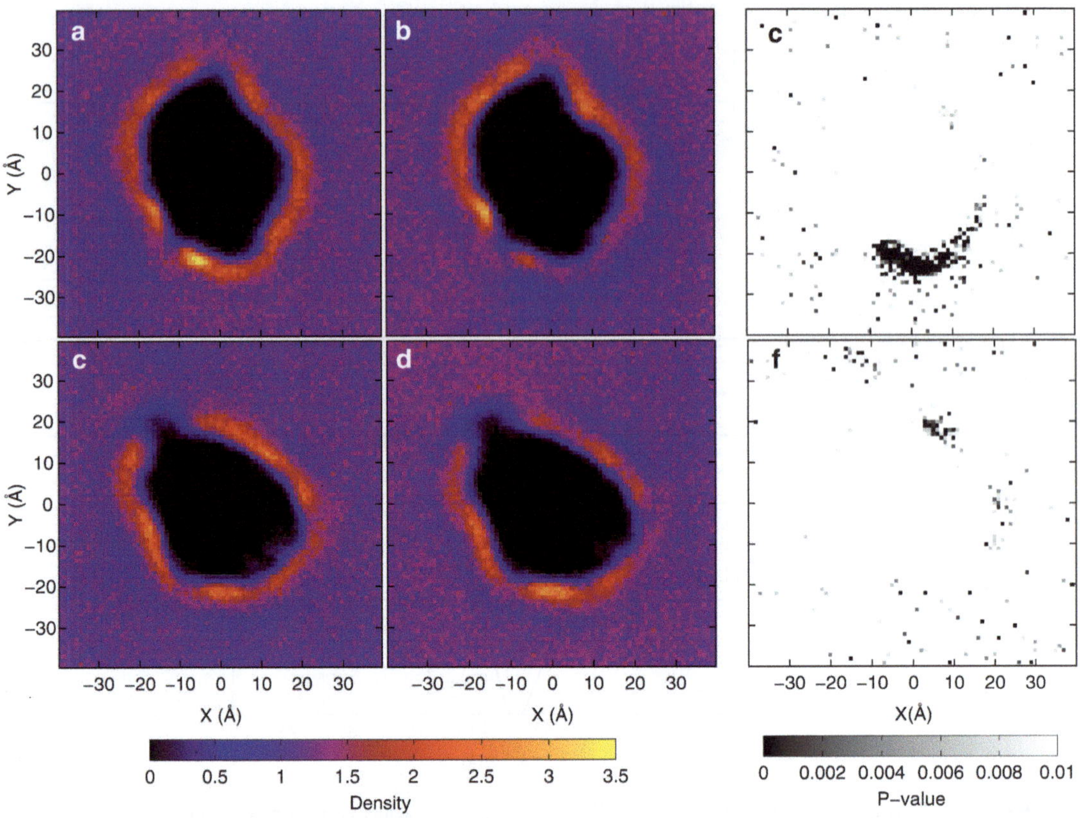

Fig. 5.5 Density maps of (**a**) SDPE in the *upper leaflet* of rhodopsin, (**b**) SDPC in the *upper leaflet* rhodopsin, (**d**) SDPE in the *lower leaflet* of rhodopsin and (**e**) SDPC in the *lower leaflet* of rhodopsin. Density is reported as lipid components per Å^2. Maps of p-values comparing the means for SDPE and SDPC for (**c**) the *upper leaflet* and (**f**) the *lower leaflet* are also shown. The protein orientation is identical to that in Fig. 5.3

in opsin, these changes in cholesterol packing suggest a role for helix-helix interactions and arrangement in cholesterol preferences. The presence of the extra cholesterol binding spot is consistent with Fig. 5.2, which shows greater cholesterol binding at the surface of the opsin system. This is indicative of a general trend that the opsin structure is more amenable to cholesterol binding, again either because the surface has greater preference for cholesterol or the more "open" opsin structure provides a greater number of cholesterol accessible pockets.

The above visualizations are convenient, but do not in themselves tell us precisely which protein residues are involved in the "binding sites." Accordingly, we decided to track specific lipid-residue contacts, using a variant of the packing score applied in previous work (Grossfield et al., 2006b), as discussed in the methods section. Unlike that work, here we account for the size of the residues, and report the ratio of the packing score to that of the average score of all residues in the transmembrane region (Eq. 5.1). Figure 5.7 shows the protein structures colored by residue score. The residues clustered into three well-defined groups based on physical location; Table 5.1 lists the residues with values at least three times greater than the average. These groupings correspond nicely with the three high density regions for cholesterol as seen previously for the opsin structure. Overall, the high-scoring clusters bear a striking resemblance to the groups of residues identified in the previous all-atom work (Grossfield et al., 2006b).

The most dominant cluster is a collection of residues that pack between helices H1 and

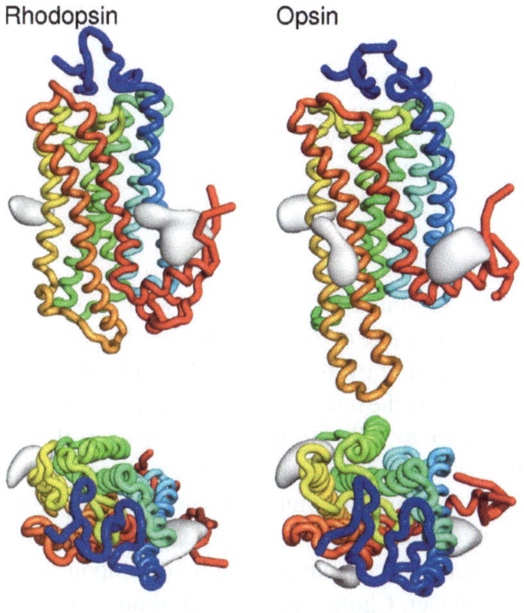

Fig. 5.6 3D images of regions of high cholesterol density (*gray*) for both rhodopsin and opsin. *Bottom panels* are top down views, as seen from the extracellular side. Rhodopsin and opsin are colored with a spectrum from the N-terminus (*blue*) to the C-terminus (*red*)

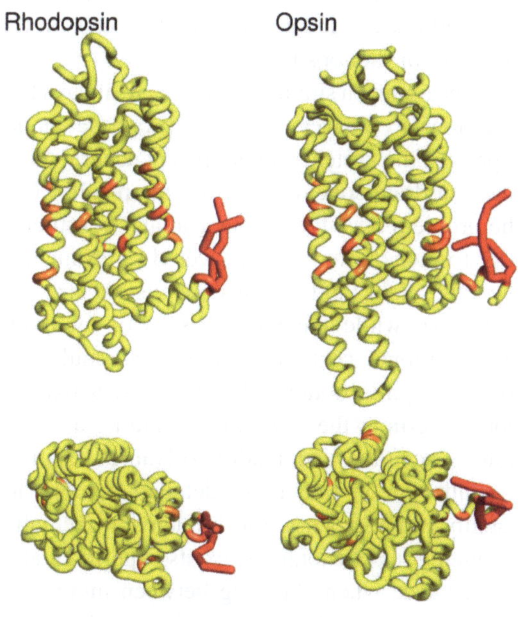

Fig. 5.7 3D images of rhodopsin and cholesterol interactions, color coded from low contact in *yellow* (binding score less than 3) to high contact in *red* (binding score greater than 3)

H8, with the greatest contribution to the cluster score coming from the palmitoyl chains attached to H8. In recent β_2-adrenergic receptor crystal structures, cholesterol and palmitic acids have been resolved in the H1/H8 interface, close to where the palmitoyl post-translational modifications in rhodopsin are located (Cherezov et al., 2007). However, it has been suggested that these cholesterols may exist as an artifact of crystal packing or protein dimerization. Our simulations suggest a real effect exists, showing preferential interactions between cholesterol and rhodopsin at helix H8, despite the presence of only one rhodopsin molecule.

Despite success correlating our results with previous simulation and crystallographic data, we have been unable to detect significant contacts between cholesterol and the groove on the intracellular ends of helices H1, H2, H3 and H4, which had been detected in previous all-atom simulations (Khelashvili et al., 2009) and in structures of the β_2-adrenergic GPCR (Cherezov et al., 2007; Hanson et al., 2008). We were also unable to locate cholesterol binding sites that correspond with three sites identified in a recent high-resolution structure of the adenosine A_{2A} receptor (Liu et al., 2012). One of these sites, between the intracellular ends of helices H1 and H2, was noted in all-atom simulations (Lyman et al., 2009) and crystal structures of the β_2-adrenergic structure (Hanson et al., 2008), and absent in other simulations of A_{2A} (Lee and Lyman, 2012).

5.5 Conclusions

The membrane environment around rhodopsin contains a diverse set of constituents that impact receptor activation, from general bilayer structural properties to specific binding interactions. Utilizing coarse-grained simulation and a combination of radial distribution functions, density representations and quantitative binding scans, we explored and identified a number of bilayer-rhodopsin interactions.

Table 5.1 Clusters of key cholesterol-binding residues identified via binding scan

Structure	Group	Residues	Average score
	1	43, 46, 50, 53, 56, 294, 301, 321, 322, 323	6.0683
Rhodopsin	2	126, 159, 206, 209, 210, 213, 214, 220, 221	3.6042
	3	263	3.3597
	1	50, 53, 54, 56, 318, 321, 322, 323	6.7907
Opsin	2	126, 159, 162, 206, 209, 210, 213, 214, 220	3.9135
	3	256, 259, 263	3.4568

DHA chains were found in higher concentrations at the protein surface, with stearoyl chains excluded to a second solvation shell in the bilayer. There was enrichment of PE headgroups over PC headgroups at the surface of the protein. A region of significant difference was discovered, suggesting a possible specific binding site for lipids with PE headgroups. Possible cholesterol-binding sites were also identified, with the predominant one at the helix H1 and helix H8 interface, behind the palmitoyl chains attached to the protein.

We also found differences between the rhodopsin and opsin systems for these lipid constituents. In the opsin system, the concentrations of cholesterol, stearoyl and DHA reach bulk levels deeper in the protein and the stearoyl and DHA peaks were much higher at the surface of the protein, suggesting a more open structure with greater available surface area.

The use of restraints to maintain protein stability limited the motions available to rhodopsin and opsin, likely preventing any major structural changes that would result from bilayer-protein interactions. We do not feel that this is an issue for our particular simulations, as we are probing preferential interaction sites along the surface of the protein, not the effects of these interactions on protein structure. Our chosen rhodopsin and opsin structures represent distant endpoints along the activation path of rhodopsin, allowing us to explore the effect of these major structural changes on the surface available for interaction.

Coarse-grained models can limit the ability of the simulations to capture specific binding interactions. For example, coarse-grained representations of cholesterol maintain the molecule's hydrophobicity, but cannot capture the chemically distinct faces. Interaction sites that are uniquely suited to interactions with cholesterol are then unable to interact. The use of an all-atom model would overcome these limitation to some degree, but the loss of sufficient sampling remains a barrier to utilizing these models to explore a process as slow as bilayer lateral reorganization.

In the future, a number of other variables could be explored. First, it would be worth exploring variations in concentration of lipid components in the bilayer. Cholesterol concentrations vary with the age of the disc membrane and can result in a drastic change in rhodopsin activity. Also, an improved MARTINI model that accurately maintains tertiary structure would remove the need for restraints and allow us to explore the effect of the lipid bilayer and specific lipid species on the structure of rhodopsin.

In designing a simulation, there are always two questions to answer: Does the proposed model represent the system with sufficient accuracy that one can draw conclusions from the results? Can the problem be solved using available resources? The former questions is answered by evaluating the force field (and perhaps the system size and contents), while the latter has to do with the computational cost of running the simulations long enough to obtain valid statistics. Satisfying both criteria at the same time—finding a model that is accurate and readily solvable—is often a challenge, in that more detailed models are usually more expensive to implement. Actually, an analogous phenomenon exists experimentally, for instance when choosing between more realistic in vivo experiments and simpler (but more interpretable) in vitro ones.

We believe that there is often a certain synergy to combining all-atom and coarse-grained simulations, as we have done here. The all-atom calculations can serve to validate surprising results from the coarse-grained ones; usually, this role is played by comparison to experiment, but sometimes there is no readily available experimental comparison. Alternatively, the all-atom simulations may suggest interesting phenomena, but lack sufficient duration to draw strong conclusions. In this case (as happened here), coarse-grained simulations can be used to flesh out the case.

In a broader sense, if one asks the question "Do I want to use all-atom or coarse-grained simulation or experiment?", the best answer is simply "Yes". All approaches have their own strengths and weaknesses, and the best way to answer the scientific question is to attack from as many directions as possible.

Acknowledgements We would like to thank Nick Leioatts, Dejun Lin and Tod Romo for critical reviews of this manuscript. We would also like to gratefully acknowledge financial support from the U.S. National Institutes of Health (1R01GM095496). We also thank the University of Rochester's Center for Integrated Research Computing for the computing resources necessary to support this work.

References

Albert AD, Young JE, Yeagle P (1996) Rhodopsin-cholesterol interactions in bovine rod out segment disk membranes. Biochim Biophys Acta 1285:47–55

Andersen OS, Koeppe RE (2007) Bilayer thickness and membrane protein function: an energetic perspective. Ann Rev Biophys Biomol Struct 36:107–130. doi:10.1146/annurev.biophys.36.040306.132643

Andersen O,Apell HJ, Bamberg E, Busath D, Koeppe R, Sigworth F, Szabo G, Urry D, Woolley A (1999) Gramicidin channel controversy – the structure in a lipid environment. Nat Struct Mol Biol 6(7):609–609

Boesze-Battaglia K, Albert AD (1989) Fatty acid composition of bovine rod outer segment plasma membrane. Exp Eye Res 49(4):699–701

Boesze-Battaglia K, Albert AD (1992) Phospholipid distribution among bovine rod outer segment plasma membrane and disk membranes. Exp Eye Res 54(5):821–823

Boesze-Battaglia K, Hennessey T, Albert AD (1989) Cholesterol heterogeneity in bovine rod outer segment disk membranes. J Biol Chem 264(14):8151–8155

Botelho AV, Gibson NJ, Thurmond RJ, Wang Y, Brown MF (2002) Conformational energetics of rhodopsin modulated by nonlamellar-forming lipids. Biochemistry 41:6354–6368

Botelho AV, Huber T, Sakmar TP, Brown MF (2006) Curvature and hydrophobic forces drive oligomerization and modulate activity of rhodopsin in membranes. Biophys J 91:4464–4477

Brown MF (1994) Modulation of rhodopsin function by properties of the membrane bilayer. Chem Phys Lipids 73(1–2):159–180

Brown MF (1997) Influence of non-lamellar-forming lipids on rhodopsin. Curr Top Membr 44:285–356

Brügger B, Erben G, Sandhoff R, Wieland FT, Lehmann WD (1997) Quantitative analysis of biological membrane lipids at the low picomole level by nano-electrospray ionization tandem mass spectrometry. Proc Natl Acad Sci 94(6):2339–2344

Burkhart BM, Li N, Langs DA, Pangborn WA, Duax WL (1998) The conducting form of gramicidin a is a right-handed double-stranded double helix. Proc Natl Acad Sci 95(22):12950–12955. doi:10.1073/pnas.95.22.12950

Buzhynskyy N, Salesse C, Scheuring S (2011) Rhodopsin is spatially heterogeneously distributed in rod outer segment disk membranes. J Mol Recognit 24(3):483–489. doi:10.1002/jmr.1086

Chen Z, Rand R (1997) The influence of cholesterol on phospholipid membrane curvature and bending elasticity. Biophys J 73(1):267–276. doi:10.1016/S0006-3495(97)78067-6

Cherezov V, Rosenbaum DM, Hanson MA, Rasmussen SGF, Thian FS, Kobilka TS, Choi HJ, Kuhn P, Weis WI, Kobilka BK, Stevens RC (2007) High-resolution crystal structure of an engineered human beta2-adrenergic G protein-coupled receptor. Science 318(5854):1258–1265. doi:10.1126/science.1150577

Cruickshank C, Minchin R, Dain AL, Martinac B (1997) Estimation of the pore size of the large-conductance mechanosensitive ion channel of Escherichia coli. Biophys J 73(4):1925–1931. doi:10.1016/S0006-3495(97)78223-7

Delange F, Merkx M, Bovee-Geurts PHM, Pistorius AMA, Degrip WJ (1997) Modulation of the metarhodopsin I/metarhodopsin II equilibrium of bovine rhodopsin by ionic strength. Eur J Biochem 243(1–2):174–180. doi:10.1111/j.1432-1033.1997.0174a.x

Drews J (2000) Drug discovery: a historical perspective. Science 287(5460):1960–1964

Engelman DM (2005) Membranes are more mosaic than fluid. Nature 438(7068):578–580. doi:10.1038/nature04394

Fattal DR, Ben-Shaul A (1993) A molecular model for lipid-protein interactions in membranes: the role of hydrophobic mismatch. Biophys J 65:1795–1809

Feller SE, Gawrisch K (2005) Properties of docosahexaenoic acid-containing lipids and their influence on the function of the GPCR rhodopsin. Curr Opin Struct Biol 15:416–422

Feller SE, Gawrisch K, MacKerell AD Jr (2002) Polyunsaturated fatty acids in lipid bilayers: intrinsic and environmental contributions to their unique physical properties. J Am Chem Soc 124(2):318–326

Feller SE, Gawrisch K, Woolf TB (2003) Rhodopsin exhibits a preference for solvation by polyunsaturated docosohexaenoic acid. J Am Chem Soc 125(15):4434–4435. doi:10.1021/ja0345874

Gibson NJ, Brown MF (1993) Lipid headgroup and acyl chain composition modulate the MI-MII equilibrium of rhodopsin in recombinant membranes. Biochemistry 32:2438–2454

Grossfield A, Feller SE, Pitman MC (2006a) Contribution of omega-3 fatty acids to the thermodynamics of membrane protein solvation. J Phys Chem B 110(18):8907–8909. doi:10.1021/jp060405r

Grossfield A, Feller SE, Pitman MC (2006b) A role for direct interactions in the modulation of rhodopsin by omega-3 polyunsaturated lipids. Proc Natl Acad Sci USA 103(13):4888–4893. doi:10.1073/pnas.0508352103

Grossfield A, Pitman MC, Feller SE, Soubias O, Gawrisch K (2008) Internal hydration increases during activation of the G-protein-coupled receptor rhodopsin. J Mol Biol 381(2):478–486. doi:10.1016/j.jmb.2008.05.036

Gruner SM (1985) Intrinsic curvature hypothesis for biomembrane lipid composition: a role for nonbilayer lipids. Proc Natl Acad Sci 82(11):3665–3669

Hanson MA, Cherezov V, Griffith MT, Roth CB, Jaakola VP, Chien EY, Velasquez J, Kuhn P, Stevens RC (2008) A specific cholesterol binding site is established by the 2.8 angstrom structure of the human beta2-adrenergic receptor. Structure 16(6):897–905. doi:10.1016/j.str.2008.05.001

Harold FM, Baarda JR (1967) Gramicidin, valinomycin, and cation permeability of Streptococcus faecalis. J Bacteriol 94(1):53–60

Haswell E, Phillips R, Rees D (2011) Mechanosensitive channels: what can they do and how do they do it? Structure 19(10):1356–1369. doi:10.1016/j.str.2011.09.005

Hess B, Kutzner C, van der Spoel D, Lindahl E (2008) GROMACS 4: algorithms for highly efficient, load-balanced, and scalable molecular simulation. J Chem Theory Comput 4(3):435–447. doi:10.1021/ct700301q

Hoover WG (1985) Canonical dynamics: equilibrium phase-space distributions. Phys Rev A 31(3):1695–1697. doi:10.1103/PhysRevA.31.1695

Huber T, Rajamoorthi K, Kurze VF, Beyer K, Brown MF (2002) Structure of docosahexaenoic acid-containing phospholipid bilayers as studied by 2H NMR and molecular dynamics simulation. J Am Chem Soc 124:298–309

Huber T, Botelho AV, Beyer K, Brown MF (2004) Membrane model for the G-protein-coupled receptor rhodopsin: hydrophobic interface and dynamical structure. Biophys J 86:2078–2100

Jones E,Oliphant T, Peterson P, et al (2001–) SciPy: open source scientific tools for Python. http://www.scipy.org/

Kandt C, Ash WL, Tieleman DP (2007) Setting up and running molecular dynamics simulations of membrane proteins. Methods 41(4):475–488. doi:10.1016/j.ymeth.2006.08.006

Khelashvili G, Grossfield A, Feller SE, Pitman MC, Weinstein H (2009) Structural and dynamic effects of cholesterol at preferred sites of interaction with rhodopsin identified from microsecond length molecular dynamics simulations. Proteins 76(2):403–417. doi:10.1002/prot.22355

Killian J, von Heijne G (2000) How proteins adapt to a membrane-water interface. Trends Biochem Sci 25(9):429–434. doi:10.1016/S0968-0004(00)01626-1

Knepp AM, Periole X, Marrink SJ, Sakmar TP, Huber T (2012) Rhodopsin forms a dimer with cytoplasmic helix 8 contacts in native membranes. Biochemistry 51(9):1819–1821. doi:10.1021/bi3001598

Kung C, Martinac B, Sukharev S (2010) Mechanosensitive channels in microbes. Ann Rev Microbiol 64(1):313–329. doi:10.1146/annurev.micro.112408.134106

Lee AG (2003) Lipid-protein interactions in biological membranes: a structural perspective. Biochimica et Biophysica Acta (BBA) – Biomembranes 1612(1):1–40. doi:10.1016/S0005-2736(03)00056-7

Lee AG (2004) How lipids affect the activities of integral membrane proteins. Biochimica et Biophysica Acta (BBA) – Biomembranes 1666(1–2):62–87. doi:10.1016/j.bbamem.2004.05.012

Lee JY, Lyman E (2012) Predictions for cholesterol interaction sites on the A(2A) adenosine receptor. J Am Chem Soc 134(40):16512–16515. doi:10.1021/ja307532d

Liebman PA, Parker KR, Dratz EA (1987) The molecular mechanism of visual excitation and its relation to the structure and composition of the rod outer segment. Ann Rev Physiol 49(1):765–791. doi:10.1146/annurev.ph.49.030187.004001

Liu W, Chun E, Thompson AA, Chubukov P, Xu F, Katritch V, Han GW, Roth CB, Heitman LH, IJzerman AP, Cherezov V, Stevens RC (2012) Structural basis for allosteric regulation of GPCRs by sodium ions. Science 337(6091):232–236. doi:10.1126/science.1219218

Lyman E, Higgs C, Kim B, Lupyan D, Shelley JC, Farid R, Voth GA (2009) A role for a specific cholesterol interaction in stabilizing the apo configuration of the human A(2A) adenosine receptor. Structure 17(12):1660–1668. doi:10.1016/j.str.2009.10.010

Marrink SJ, Risselada HJ, Yefimov S, Tieleman DP, de Vries AH (2007) The MARTINI force field: coarse grained model for biomolecular simulations. J Phys Chem B 111(27):7812–7824. doi:10.1021/jp071097f

Marrink SJ, Periole X, Tieleman DP, de Vries AH (2010) Comment on "on using a too large integration time step in molecular dynamics simulations of coarse-grained molecular models" by M. Winger, D. Trzesniak, R. Baron and W. F. van Gunsteren, Phys. Chem. Chem. Phys., 2009, 11, 1934. Phys Chem Chem Phys 12(9):2254–2256; author reply 2257–2258. doi:10.1039/b915293h

Marsh D (2008) Protein modulation of lipids, and vice-versa, in membranes. Biochimica et Biophysica Acta

(BBA) – Biomembranes 1778:1545–1575. doi:10.1016/j.bbamem.2008.01.015
Martinac B (2011) Bacterial mechanosensitive channels as a paradigm for mechanosensory transduction. Cell Physiol Biochem 28(6):1051–1060
Martinac B, Buechner M, Delcour AH, Adler J, Kung C (1987) Pressure-sensitive ion channel in Escherichia coli. Proc Natl Acad Sci 84(8):2297–2301
MARTINI (2011) http://mdchemrugnl/cgmartini/
Mitchell DC, Straume M, Miller JL, Litman BJ (1990) Modulation of metarhodopsin formation by cholesterol-induced ordering of bilayer lipids. Biochemistry 29(39):9143–9149. doi:10.1021/bi00491a007
Molday RS (1998) Photoreceptor membrane proteins, phototransduction, and retinal degenerative diseases. The Friedenwald Lecture. Invest Ophthalmol Vis Sci 39(13):2491–2513
Monticelli L, Kandasamy S, Periole X, Larson R, Tieleman D, Marrink S (2008) The MARTINI coarse grained forcefield: extension to proteins. J Chem Theory Comput 4:819–839
Mouritsen OG, Bloom M (1984) Mattress model of lipid-protein interactions in membranes. Biophys J 46:141–153
Mouritsen OG, Bloom M (1993) Models of lipid-protein interactions in membranes. Ann Rev Biophys Biomol Struct 22:145–171
Needham D, McIntosh TJ, Evans E (1988) Thermomechanical and transition properties of dimyristoylphosphatidylcholine/cholesterol bilayers. Biochemistry 27(13):4668–4673. doi:10.1021/bi00413a013
Neuringer M (2000) Infant vision and retinal function in studies of dietary long-chain polyunsaturated fatty acids: methods, results, and implications. Am J Clin Nutr 71(1 Suppl):256S–267S
Niu SL, Mitchell DC, Litman BJ (2002) Manipulation of cholesterol levels in rod disk membranes by methyl-β-cyclodextrin. J Bio Chem 277:20139–20145
Nosé S, Klein ML (1983) Constant pressure molecular dynamics for molecular systems. Mol Phys 50:1055–1076
O'Connell A, Koeppe R, Andersen O (1990) Kinetics of gramicidin channel formation in lipid bilayers: transmembrane monomer association. Science 250(4985):1256–1259. doi:10.1126/science.1700867
Okada T, Sugihara M, Bondar AN, Elstner M, Entel P, Buss V (2004) The retinal conformation and its environment in rhodopsin in light of a new 2.2 angstrom crystal structure. J Mol Biol 342:571–583
Olausson BES, Grossfield A, Pitman MC, Brown MF, Feller SE, Vogel A (2012) Molecular dynamics simulations reveal specific interactions of post-translational palmitoyl modifications with rhodopsin in membranes. J Am Chem Soc 134(9):4324–4331. doi:10.1021/ja2108382
Palczewski K, Kumasaka T, Hori T, Behnke CA, Motoshima H, Fox BJ, Le Trong I, Teller DC, Okada T, Stenkamp RE, Yamamoto M, Miyano M (2000) Crystal structure of rhodopsin: a G protein-coupled receptor. Science 289:739–745
Park JH, Scheerer P, Hofmann KP, Choe HW, Ernst OP (2008) Crystal structure of the ligand-free G-protein-coupled receptor opsin. Nature 454(7201):183–187. doi:10.1038/nature07063
Parrinello M, Rahman A (1981) Polymorphic transitions in single crystals: a new molecular dynamics method. J Appl Phys 52(12):7182–7190. doi:10.1063/1.328693
Periole X, Huber T, Marrink SJ, Sakmar TP (2007) G protein-coupled receptors self-assemble in dynamics simulations of model bilayers. J Am Chem Soc 129(33):10126–10132. doi:10.1021/ja0706246
Periole X, Cavalli M, Marrink SJ, Ceruso MA (2009) Combining an elastic network with a coarse-grained molecular force field: structure, dynamics, and intermolecular recognition. J Chem Theory Comput 5(9):2531–2543. doi:10.1021/ct9002114
Periole X, Knepp AM, Sakmar TP, Marrink SJ, Huber T (2012) Structural determinants of the supramolecular organization of G protein-coupled receptors in bilayers. J Am Chem Soc 134(26):10959–10965. doi:10.1021/ja303286e
Pitman MC, Grossfield A, Suits F, Feller SE (2005) Role of cholesterol and polyunsaturated chains in lipid-protein interactions: molecular dynamics simulation of rhodopsin in a realistic membrane environment. J Am Chem Soc 127(13):4576–4577. doi:10.1021/ja042715y
Romo TD, Grossfield A (2009) LOOS: an extensible platform for the structural analysis of simulations. Conf Proc IEEE Eng Med Biol Soc 2009:2332–2335. doi:10.1109/IEMBS.2009.5335065
Romo TD, Grossfield A (2012) LOOS: a lightweight object-oriented software library. LOOS: Lightweight object oriented structure analysis, Grossfield Lab. http://loos.sourceforge.net
Sansom MS, Bond PJ, Deol SS, Grottesi A, Haider S, Sands ZA (2005) Molecular simulations and lipid-protein interactions: potassium channels and other membrane proteins. Biochem Soc Trans 33(Pt 5):916–920. doi:10.1042/BST20050916
Simmonds A, East J, Jones O, Rooney E, McWhirter J, Lee A (1982) Annular and non-annular binding sites on the (Ca^{2++} Mg^{2+})-ATPase. Biochimica et Biophysica Acta (BBA) – Biomembranes 693(2):398–406. doi:10.1016/0005-2736(82)90447-3
Soubias O, Gawrisch K (2007) Docosahexaenoyl chains isomerize on the sub-nanosecond time scale. J Am Chem Soc 129(21):6678–6679. doi:10.1021/ja068856c
Soubias O, Gawrisch K (2012) The role of the lipid matrix for structure and function of the GPCR rhodopsin. Biochim Biophys Acta 1818(2):234–240. doi:10.1016/j.bbamem.2011.08.034
Soubias O, Teague WE, Gawrisch K (2006) Evidence for specificity in lipid-rhodopsin interactions. J Biol Chem 281(44):33233–33241. doi:10.1074/jbc.M603059200
Soubias O, Niu SL, Mitchell DC, Gawrisch K (2008) Lipid-rhodopsin hydrophobic mismatch

alters rhodopsin helical content. J Am Chem Soc 130(37):12465–12471. doi:10.1021/ja803599x
Soubias O, Teague WE, Hines KG, Mitchell DC, Gawrisch K (2010) Contribution of membrane elastic energy to rhodopsin function. Biophys J 99(3):817–824. doi:10.1016/j.bpj.2010.04.068
van der Spoel D, Lindahl E, Hess B, Groenhof G, Mark AE, Berendsen HJC (2005) GROMACS: fast, flexible, and free. J Comput Chem 26(16):1701–1718. doi:10.1002/jcc.20291
Valiyaveetil FI, Zhou Y, MacKinnon R (2002) Lipids in the structure, folding, and function of the KcsA K+ channel. Biochemistry 41(35):10771–10777. doi:10.1021/bi026215y
Wang Y, Botelho AV, Martinez GV, Brown MF (2002) Electrostatic properties of membrane lipids coupled to metarhodopsin II formation in visual transduction. J Am Chem Soc 124(26):7690–7701
Wiedmann TS, Pates RD, Beach JM, Salmon A, Brown MF (1988) Lipid-protein interactions mediate the photochemical function of rhodopsin. Biochemistry 27:6469–6474
Wiener MC, White SH (1992) Structure of a fluid dioloeoylphosphatidylcholine bilayer determined by joint refinement of x-ray and neutron diffraction data: III. Complete structure. Biophys J 61:434–447
Winger M, Trzesniak D, Baron R, van Gunsteren WF (2009) On using a too large integration time step in molecular dynamics simulations of coarse-grained molecular models. Phys Chem Chem Phys 11(12):1934–1941. doi:10.1039/b818713d

Beyond Standard Molecular Dynamics: Investigating the Molecular Mechanisms of G Protein-Coupled Receptors with Enhanced Molecular Dynamics Methods

6

Jennifer M. Johnston and Marta Filizola

Abstract

The majority of biological processes mediated by G Protein-Coupled Receptors (GPCRs) take place on timescales that are not conveniently accessible to standard molecular dynamics (MD) approaches, notwithstanding the current availability of specialized parallel computer architectures, and efficient simulation algorithms. Enhanced MD-based methods have started to assume an important role in the study of the rugged energy landscape of GPCRs by providing mechanistic details of complex receptor processes such as ligand recognition, activation, and oligomerization. We provide here an overview of these methods in their most recent application to the field.

Keywords

Enhanced methods • Biased simulations • Metadynamics • Umbrella sampling • Free-energy calculations • Coarse-graining • Ligand recognition • GPCR activation • Molecular mechanisms • Oligomerization • Dimers • GPCR structure • GPCR dynamics • GPCR function

6.1 Introduction

GPCRs are components of complex macromolecular machines with multiple ligand-induced 'active' states that can engender different signaling outputs through association with specific accessory proteins. These functionally versatile macromolecular complexes, recently termed GPCR signalosomes (Huber and Sakmar 2011), are suggested to operate through a multitude of dynamic steps and allosteric signaling conduits whose properties may not depend necessarily on their individual elements. Thus, it appears evident that a conceptual framework strongly relying on a combination of high-content experimental platforms and computational approaches is necessary to account for the complexity of these signalosomes and ultimately understand the relationships between their structure, dynamics, and function (Huber and Sakmar 2011).

J.M. Johnston • M. Filizola (✉)
Department of Structural and Chemical Biology, Icahn School of Medicine at Mount Sinai, One Gustave L. Levy Place, New York, NY 10029, USA
e-mail: marta.filizola@mssm.edu

M. Filizola (ed.), *G Protein-Coupled Receptors - Modeling and Simulation*, Advances in Experimental Medicine and Biology 796, DOI 10.1007/978-94-007-7423-0_6,

One way to interpret the functional versatility of GPCRs is in terms of their structural plasticity, through system representation as energy landscapes (Deupi and Kobilka 2010; Choe et al. 2011; Vaidehi and Kenakin 2010). Building upon a concept put forth in 1991 by Frauenfelder and colleagues (1991), the energy landscape of a protein is described as a hyper-surface in a conformational space of very high dimension, with a very large number of valleys (conformational sub-states) and peaks (energy barriers). The valleys and peaks of such a rugged landscape cannot be classified individually, but must be described by distributions. Specific energy levels are no longer relevant, but are more accurately described in statistical terms. At any given time, an individual protein molecule can jump between conformational sub-states as it navigates its own energy landscape, and this exploration is critically dependent on temperature (Frauenfelder et al. 1991). The available X-ray crystal structures of GPCRs (for a review see Katritch 2012) represent single, static sub-states of these proteins. Sections of missing density in many of the structures (e.g. Wu et al. 2012) highlight the inherent flexibility, even at low temperatures, of some regions of the receptor, particularly parts of the intra and extracellular loops and the N- and C-termini, and remind us that the dynamic and adaptable nature of the GPCR structure is a key part of their functionality.

Ligands with different efficacies can modulate the GPCR energy landscape by shifting the conformational equilibrium towards active or inactive conformations, thus eliciting different physiological responses (Deupi and Kobilka 2010; Vaidehi and Kenakin 2010). Examining the molecular basis of this ligand-induced conformational shift is not an easy task using standard MD approaches due to the limited timescales they can access. Although much has been accomplished by standard MD of atomistically-represented GPCRs using some of the largest multiprocessor computing clusters, running the most efficient MD codes currently available (e.g., see the work recently done to simulate ligand binding to the β2-adrenergic receptor (B2AR) (Dror et al. 2011)), these hardware/software configurations are presently the prerogative of a small group of investigators. Undoubtedly, atomistic descriptions provide the most comprehensive and complete representations of a GPCR, but it is arguable that multiple, lengthy stochastic simulations of such systems are not best suited to access so-called "rare events" in the lives of GPCRs. Since the number of calculations required per step of a MD simulation of a GPCR system scales with the square of the number of particles, reducing the system size considerably increases the speed with which simulations can be performed. This size reduction can be achieved by eliminating the explicit representation of a component of the system (e.g. the solvent or the membrane, as in Generalized Born models), or by grouping individual atoms into interaction sites (e.g., coarse-grained bead models). Of course, such decisions must be taken with care and are inevitably dependent on the nature of the question that the simulation is designed to answer. In the cases where the fastest degrees of freedom can be neglected, these approaches have helped to smooth the energy landscape of GPCRs and thereby extend the range of accessible time and length scales.

The reduced representation of a GPCR system may, however, still be insufficient to bridge the gap between the timescales accessible to standard MD simulations and average experimental timescales. We will illustrate here how both simplified physical representations and enhanced MD-based methods have recently proven useful in the study of the rugged energy landscape of GPCRs by providing otherwise inaccessible details of important events in the life cycle of these receptors, such as recognition of ligands, ligand-specific conformational changes, and oligomerization. However, we acknowledge that both the rate at which advances are being made in this field, and the space constraints of this chapter do not permit an entirely comprehensive review here, but rather our goal is to provide an overview by way of a few selected examples.

6.2 Ligand Recognition in GPCRs

A thorough understanding of the mechanisms by which GPCRs recognize their ligands is fundamental to successful drug discovery. While multiple long timescale atomistic standard MD simulations (of the order of hundreds of μs) have recently permitted researchers to visualize, for instance, the binding of different ligands to the B2AR (Dror et al. 2011), biased MD techniques have been successfully employed to enhance the probability of observing such events during shorter simulation timescales. Three studies we discuss in this section focus on the premise that from knowledge of the pathways by which a ligand can exit a receptor, one can infer specific details of the binding mechanism (Gonzalez et al. 2011; Selvam et al. 2012; Wang and Duan 2009). The probability of spontaneous ligand exit from a binding cavity, on the timescale accessible to MD simulations, can be enhanced by addition of an external force imposed upon the ligand, as is the case for steered MD and random acceleration MD. We compare the results of these studies with those obtained by the very long timescale, unbiased simulations from Dror and colleagues (2011). A fourth study we discuss below focuses on a binding event in which the ligand travels along a predetermined pathway from the bulk solvent, towards a bound state within the receptor (Provasi et al. 2009). We also comment on a novel method of altering the orthosteric binding pocket of receptor homology models, to permit successful docking of ligands of differing shape and size to that in the template structure (Kimura et al. 2008).

6.2.1 Exploring the Binding Site in Homology Models

Prior to the application of receptor engineering innovations that have given rise to the recent influx of solved crystal structures of GPCRs, homology modeling was the only tool in the arsenal of the researcher wishing to computationally investigate any receptor other than rhodopsin. However, for drug discovery purposes, a limitation of GPCR homology models has been that the binding pockets are not always big enough to accommodate the variety of ligands that have been demonstrated to bind to a particular receptor through other experimental assays. Indeed, retinal is a small, covalently bound ligand, so ligands of interest that are different in shape or size to retinal, and most importantly, not covalently bound, might not be expected to be a good fit to receptor models generated from a rhodopsin backbone template. Kimura and colleagues (2008) have developed a method for increasing the available space for ligand binding within the orthosteric pocket of a model of a GPCR, which may be readily transferable to other types of binding pockets for other proteins. Once the binding pocket has been expanded using their algorithm, the endogenous ligand can be docked into the pocket, and induced fit docking methods, sidechain rotamer sampling or MD methods can be used to refine the pocket before virtual screening or docking of novel ligands is performed.

Conceptually, the method proposed by Kimura and colleagues is similar to inflating a balloon inside the pocket. By slowly increasing the pressure on the cavity walls during a MD simulation, the pocket can be expanded. To achieve this, the authors placed several (>20) Lennard-Jones (LJ) beads of small radius (0.25 Å) within the cavity, and these particles collided with the cavity walls with velocities consistent with a Maxwell distribution according to the temperature of the simulation. A slow increase in pressure was implemented by increasing the radii of the particles. The particles filled the pocket, and were tethered by weak ($k = 1$ kcal/(mol Å^2)) harmonic forces to their four immediate neighbors, but these bonds were reassigned with every increase of the particle radius. The radius was increased in steps of 0.05–0.1 Å every 300–500 ps. The simulations were performed using NAMD, with explicitly represented solvent and palmitoyl-oleoyl-phosphatidyl-choline (POPC) lipid bilayer. The system was minimized and equilibrated for several hundreds of picoseconds prior to the pocket expansion stages. During the expansion stages, the backbone dihedral angles of the protein were weakly restrained (k = 2.0 kcal/(mol rad^2)) to prevent gross deformations and unwinding of the TM helices.

Kimura and colleagues tested their protocol on three receptors; firstly, in their 2008 paper (Kimura et al. 2008), they generated an apo state of the rhodopsin structure (PDB ID: 1LH9 (Okada et al. 2002)). By removing the retinal ligand and simulating the receptor for a period of ~4 ns, the sidechains were rearranged such that the pocket was too small to accommodate retinal without modification. After the pocket expansion simulations, docking studies were performed with GLIDE (Friesner et al. 2004, 2006; Halgren et al. 2004) and a comparison of the crystal structure, the MD relaxed structure and the structure after expansion with the balloon potential indicate that re-docking the retinal is significantly improved between the relaxed structure and the balloon expansion structure, but the crystal structure pose is not recovered exactly, with an RMSD of 7.2 Å between the re-docked pose and the crystallographic pose.

Secondly, they built a homology model of the chemokine receptor type 2 (CCR2) based on a rhodopsin template, Three potent antagonists, RS-504393 (MacKerell et al. 1998), TAK-779 (Baba et al. 1999), and a Teijin lead compound (Shiota 1999) were docked into the receptor at different stages of the pocket expansion. At r = 0.35 Å, a cluster of poses correlating with the available mutagenesis data was observed. Docking into the pockets formed with r > 0.35 Å yielded many more poses, owing to the increased space available in the pocket, but clustering these poses did not yield any that matched with the known mutagenesis data.

Finally (Krystek et al. 2006), the same authors have built and refined a homology model of B2AR from a rhodopsin template (PDB ID: 1F88 (Okada et al. 2002)) and validated this model according to the paradigm of structure activity relationships (SAR) and mutagenesis data existing at the time (the study was published in 2006, prior to the solution of the B2AR structure (Cherezov et al. 2007)). After pocket expansion, docking of agonists such as propranolol into the r = 0.3 state satisfied available experimental data of agonist ligand binding, although required the re-orientation of some side chains (Strader et al. 1988, 1989). It seems that key advantage provided by the balloon expansion methodology is that it generates an ensemble of models, unbiased by the inherent dependence on the template introduced during the homology modeling process, which may then be evaluated in the context of the available mutagenesis, SAR and small molecule pharmacophoric data. However, these expanded models require further refinement before the results of any docking studies may be considered any more than qualitatively accurate.

6.2.2 Exploring Ligand Egress Pathways with Random Acceleration MD

Random acceleration MD (Wang and Duan 2007, 2009), also called random expulsion MD (Ludemann et al. 2000) uses a randomly directed, externally applied force to encourage unbinding of a ligand from the pocket of a receptor. This ensures thorough exploration of the pocket, and good sampling of possible modes of ligand egress. According to the formulation, the direction of the force applied to the center of mass (COM) of the ligand is chosen at random, and it is maintained for *N* steps. During these steps, the COM of the ligand is expected to move a minimum distance r_{min}, i.e. the average velocity, $<v>$, of the ligand will maintain a threshold value of at least:

$$\langle v \rangle = {r_{\min}}\big/{N\Delta t}\,, \tag{6.1}$$

where Δt is the timestep of the MD simulation.

Upon encountering an unyielding portion of the binding site, the velocity of the ligand falls below the threshold velocity and the trajectory is considered to be complete, thus a new random direction is assigned. For each new trajectory, the direction of the force, and thus the movement of the ligand, is maintained as long as the velocity of the ligand exceeds the threshold value, or until *N* steps of MD have been completed. By using multiple trajectories, the ligand can thoroughly probe the binding pocket until it finds a suitable exit pathway or no egress, if appropriate. The key feature of this technique, which makes it particularly effective for the purpose of exploring ligand binding pathways, as compared to e.g. steered MD, is that no prior knowledge of an exit pathway is required.

In 2007, Wang and Duan (2007) applied the random acceleration MD technique to the rhodopsin crystal structure, (PDB ID: 1U19 (Okada et al. 2004)), embedded in a POPC bilayer and solvated with explicit water. The endogenous ligand, 11-cis-retinal, is covalently bonded to K296$^{7.43}$ within a deep binding pocket formed by the TM domain of rhodopsin. The superscript refers to the Ballesteros-Weinstein generic residue numbering scheme (Ballesteros and Weinstein 1995), where the first number (e.g., 1 in 1.48) indicates the helix, and the second number (e.g., 48 in 1.48) represents the residue position in that helix relative to the most conserved residue in the helix (numbered 50 by definition). EL2 in rhodopsin forms a beta-hairpin fold that completely blocks access to this binding pocket from the extracellular side. Thirty-eight random acceleration MD simulations were performed on the rhodopsin system and the predominant exit pathways were observed to be towards the extracellular side, through interhelical clefts, either between TM4 and TM5, or between TM5 and TM6. These exit pathways involved transient breakage of the interhelical interactions, which reformed immediately upon complete exit of the ligand from the binding site.

In 2009, the same authors conducted a similar investigation on the B2AR (Wang and Duan 2009). A total of 100 random acceleration MD trajectories were performed on the B2AR crystal structure (PDB ID: 2RH1) (Cherezov et al. 2007), which was first crystallized in the inactive conformation, with carazolol (an inverse agonist) located in a binding pocket defined by strong interactions with polar residues in TM3, TM5 and TM7. The second extracellular loop (EL2) forms a short helix in the B2AR crystal structure, and is extended outward, rendering the binding pocket slightly open to the extracellular side. Egress from the binding site is, however, restricted by two bulky aromatic residues (F193 on EL2 and Y308$^{7.35}$ on TM7) and a salt bridge between EL2 and TM7, formed by D192-K305$^{7.32}$.

The random acceleration MD trajectories suggested that the dominant exit pathway of carazolol from the B2AR crystallographic binding pocket was via the extracellular opening of the binding pocket. The salt bridge between EL2 and TM7, (D192-K305$^{7.32}$), was broken during exit along this pathway. Furthermore, if the salt bridge can be thought of as bisecting the extracellular opening of the receptor, exit via this dominant pathway can be subdivided into two, and egress was found to be approximately equally distributed between both the A1 sub-pathway (toward TM 5, 6 and 7, thick red arrow 1 in Fig. 6.1), and A2 sub-pathway (toward TM 2, 3 and 7, thick red arrow 2 in Fig. 6.1), of the salt bridge. Five additional exit pathways were observed through inter-helical clefts. Of these, the most statistically significant offered an exit through transient breakage of the inter-helical interactions between TM4 and TM5. The predominant barrier to ligand egress was presented by the interactions between the ligand and the polar groups within the binding pocket.

During 120 ns of unbiased, standard MD simulation of the receptor in the absence of ligand, the authors found that the D192-K305$^{7.32}$ salt bridge was dynamic in nature, and that a conformational change in F193 caused it to rotate outward toward TM7 and forming a 'hydrophobic cluster' between EL2 and TM7. The authors repeated their random acceleration MD simulations from this state after re-docking the carazolol. Overall, the average ligand egress time was slightly longer from the putative 'ligand-free' conformation than from the crystal structure and this increase was attributed to the barrier formed by the clustering of F193 and the salt bridge connecting TM7 and EL2. Furthermore, the population of the dominant pathway was shifted strongly toward exit via the A2 sub-pathway (arrow 2 in Fig. 6.1), between ECL2 and TM2, 3, and 7. The authors used these results to propose a binding pathway for carazolol. First, the ligand was found to access the receptor via the cleft

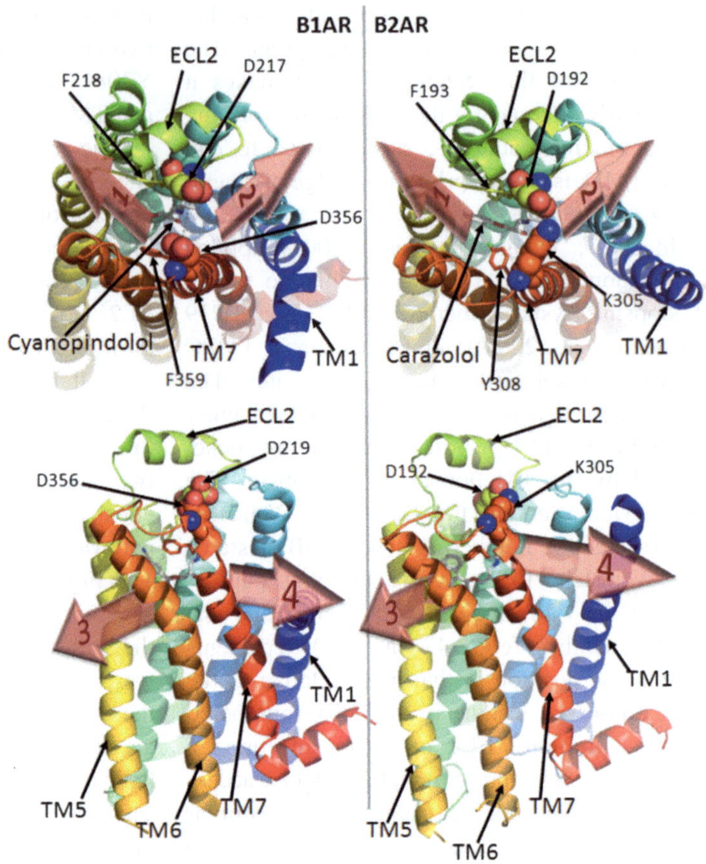

Fig. 6.1 Exit/entry pathways (*red transparent arrows*, labeled 1–4) for the B1AR (*left*) and the B2AR (*right*). *Top* (*top panels*) and *side* (*bottom panels*) views. Receptor is shown in *cartoon representation*, colored *blue* (TM1) to *red* (H8). The crystallized antagonists cyanopindolol (in B1AR) and carazolol (in B2AR) are shown in *stick representation*

between TM2, TM3 and TM7 at the extracellular opening. Subsequently, the ligand interacted with F193 as it passed through the TM7-EL2 junction on the way to the bottom of the binding pocket, where it was oriented and stabilized by the polar interactions with TM3, TM5 and TM7. The difference in results between the B2AR and rhodopsin studies described here, alongside speculative reports of different lipid phase ligand entry pathways for e.g. the cannabinoid receptor (Hurst et al. 2010) and for the PAR1 (Zhang et al. 2012) indicate that ligand binding pathways may not be general across receptor types.

6.2.3 Forced Ligand Unbinding by Steered MD

Under experimental conditions, unbinding of a ligand can be monitored by means of Atomic Force Microscopy (AFM), by applying a time-dependent external force to the system through the atomistically fine tip, and measuring the mechanical resistance properties of the biomolecule. Such a process can be mimicked virtually, using steered MD. Steered MD uses the application of an external force to drive the system towards a desired state, be it via a conformational change, or a ligand-unbinding event. Steered MD is a non-equilibrium technique, during which the MD pathway is irreversible. Analysis of the position and interactions of the dissociating ligand, and the evolution of the applied forces during a forced ligand unbinding by steered MD can provide reliable qualitative insights into the irreversible work required for the unbinding process (Isralewitz et al. 1997; Jensen et al. 2002; Fishelovitch et al. 2009; Yang et al. 2009). Such information may enable researchers to determine structural features of these receptor-ligand complexes that contribute crucially to ligand binding. Unlike random acceleration MD, the direction of the force must be predetermined, though usually this is unknown. The egress pathway can therefore often be arbitrary, and as such the sampling of different possible modes of ligand exit may be less comprehensive than with random acceleration MD.

In steered MD, the elastic force is proportional to the change in the spring extension, relative to its equilibrium length:

$$F(t) = k\left(vt - \vec{r}(t) - \vec{r}_0\right) \bullet \vec{n}\Big) \quad (6.2)$$

where k is the spring constant; v is the constant velocity of pulling, mimicking the retracting cantilever; r_0 and r(t) are the ligand center of mass position at initial and current time t respectively and n is the direction of the pulling vector.

The potential of mean force (PMF) along the reaction coordinate was calculated by the second-order cumulant expansion of the irreversible work measurements (Park et al. 2003) according to:

$$W(t) = v\int_0^t F(t)dt \quad (6.3)$$

$$PMF = \langle W\rangle - \frac{1}{2k_B T}\left(\langle W^2\rangle - \langle W\rangle^2\right) \quad (6.4)$$

where <W> is the mean work averaged from the six trajectories, k_B is Boltzmann's constant and T is the bulk temperature.

Gonzalez and colleagues embedded the crystal structure of the human B2AR (PDB ID: 2RH1) (Cherezov et al. 2007) and a MODELLER (Eswar et al. 2007; Fiser and Sali 2003) generated structure of the human B1AR, based on the turkey B1AR crystal structure (PDB ID: 2VT4) (Warne et al. 2008) into POPC membranes, and then used steered MD to remove the two crystallized ligands, cyanopindolol and carazolol, from the binding sites of B1AR and B2AR respectively, through different channels (Gonzalez et al. 2011). Steered MD simulations were performed at constant velocity of 10 Å/ns and the spring constant was set to 250 pN/Å. Trajectories were repeated six times, and lasted approximately 3 ns each. The force profile, as a function of simulation time, revealed the ease with which the ligand can be extracted from the receptor along the different pathways (Gonzalez et al. 2011). To address the shortcomings of an arbitrary choice of exit pathway from the receptors, CAVER (Petrek et al. 2006) was used to establish feasible egress pathways connecting the orthosteric binding pocket with the surface of the receptor. This showed two possible pathways, depicted in Fig. 6.1, by red arrows labeled 1 and 2. Two additional, lipid phase exit pathways between TM1/7 and TM5/6 inter-helical clefts were also tested (red arrows 3 and 4 in Fig. 6.1). Though not found by the CAVER exploration, they have been implicated as entry/exit pathways for retinal in rhodopsin (Hildebrand et al. 2009), and an exit pathway between TM5/6 was found to be significant in the random acceleration MD study of rhodopsin, from Wang and Duan (2007) (neither of these pathways was found to be significant in their investigation of B2AR (Wang and Duan 2009)). In Gonzalez's investigation, the initial force peaks for extracting ligands through pathways 3 and 4 were found to be twice that for extraction via pathway 1 or pathway 2, so the lipid phase extraction channels were considered to be unfavorable for the adrenergic receptors. For B2AR, extraction of carazolol via pathway 1 (bounded by TM5, 6 and 7) was determined to have two force peaks, indicating breaking of interactions between ligand and receptor along the exit pathway. The first (and highest) peak in the force occurred early in the simulations and represents extraction from the orthosteric binding pocket, breaking key interactions with D3.32, S5.42 and N7.39, while the second peak occurred later, and represents breaking interactions with D192 and N301 in the extracellular loops EL2 and EL3 respectively. Extraction of cyanopindolol from B1AR through the same channel showed similar behavior, showing not only a second barrier, arising from interactions with D217 in EL2, and D7.32, but also some subsequent barriers, as additional interactions in the extracellular loops were broken. Extractions of the ligands through channel 2 (bounded by TM2, 3 and 7) for both receptors demonstrate two retention sites after the initial orthosteric pocket interactions were broken, before being released into the solvent. The potentials of mean force (PMFs) of these extractions indicate channel 1 is favored for both receptors, but only by a small amount (~1 kcal/mol). The PMFs also show that the difference between the bound and the unbound states is positive, indicating that the bound state is favored in all cases.

These pathways suggested by random acceleration MD and steered MD can be directly compared with the recent entry pathways observed for similar ligands (antagonists alprenolol, propranolol and dihydroalprenolol) binding to B1AR and B2AR, by Dror and colleagues, during 82 standard MD simulations, ranging from 1 to 19 μs in length (Dror et al. 2011). During their 82 simulations, Dror and colleagues observe 21 binding events in total. Out of 12 binding events for alprenolol to B2AR, in 6 cases alprenolol replicates the crystallographic pose. The authors note that entry is almost exclusively through the cleft between ECL2 and TM5, 6 and 7 (11 out of 12 binding events) i.e. pathway 1 in Fig. 6.1. In the remaining event, entry was through pathway 2. Similar results were observed for binding dihydroalprenolol to B1AR. These results are largely in agreement with the preference noted by Gonzalez and colleagues (2011), but opposite to that of Wang and Duan (2009). Also, similarly to Gonzalez and colleagues, but again unlike Wang and Duan, Dror and colleagues observed two energetic barriers to ligand binding through pathway 1 for B2AR (Dror et al. 2011). The main barrier is that presented by ligand dewetting and orientation within the binding pocket, and is comparable to that observed during ligand extraction from the orthosteric pocket for both of the biased MD studies. The other barrier observed by Dror and colleagues occurred prior to ligand entry into the binding pocket and was not observed during the random acceleration MD simulations (Wang and Duan 2009), but according to Dror and colleagues, is at least as large, if not larger than the barrier to binding in the pocket (Dror et al. 2011). This barrier is comparable to the second barrier in the force profile observed during Gonzalez and colleagues' steered MD simulations (Gonzalez et al. 2011) and was attributed to dewetting of both the ligand and the binding pocket by Dror and colleagues, noting that alprenolol loses 80 % of its hydration upon binding, and 63 % of this is at the point of entering the pre-binding vestibule (Dror et al. 2011). Gonzalez and colleagues present data showing that the number of water molecules within 3 Å of the ligand in their simulations increases dramatically, from ~10 to ~30 molecules, at the point of encountering this second barrier to ligand egress (Gonzalez et al. 2011).

The biased MD simulations are largely able to reproduce the characteristics of binding to B2AR, i.e. dominant exit through the extracellular opening of B2AR, no or limited exit through interhelical pathways (Dror and colleagues observe that alprenolol partitions into the bilayer but never enters the receptor through an interhelical pathway). The steered MD is also able to replicate the existence of two barriers to ligand binding at stages along the entry/egress pathway comparable to those observed in the unbiased simulations. Furthermore, many of the detailed features of the dynamics of the ligand binding events can be captured by both the random acceleration MD and steered MD, including conformational changes in F218/193 and F/Y7.35 in B1AR/B2AR respectively; changes in ligand hydration and the dynamic nature of the salt bridge between K305 and D192. The key advantage of the biased events is the much shorter simulations that are required to obtain these details.

Following the simulations described above, Selvam and colleagues (2012) have used a similar steered MD methodology to Gonzalez and colleagues, to demonstrate that the PMF of extracting ligands from the B1AR and the B2AR can be used to differentiate selective ligands from their non-selective counterparts. The work performed to extract a B1AR selective antagonist (Esmolol) from the B1AR was greater than that required to extract ICI-118,551 (B2AR selective) and the reverse is also true: more work was required to extract ICI-118,551 from B2AR than Esmolol.

6.2.4 Exploring Ligand Binding Pathways Using Well-Tempered Metadynamics

Well-tempered metadynamics (Barducci et al. 2008) is an enhanced sampling technique that enables more efficient exploration of the multidimensional free energy surfaces of biological systems by adding Gaussian bias to a standard MD simulation (Laio and Parrinello 2002; Leone et al. 2010). The dynamics is biased by a non-Markovian (history-dependent) potential, constructed as a sum of Gaussian "hills" localized along the trajectory in the direction of the reaction coordinate (or collective variable, CV) of interest. The accumulation of this biasing potential enables the simulated system to overcome high energy barriers in order to explore efficiently its free energy landscape as defined by the CVs. The success of a metadynamics study is predicated on a careful *a priori* choice of a set of CVs to provide a satisfactory description of the process of interest. In well-tempered metadynamics, the height of the added Gaussian hills depends both on a temperature scaling factor and the underlying bias, and decreases to zero once a given energy threshold is reached. As a result, convergence of the algorithm to the correct free energy profile can be proven rigorously, and exploration of physically relevant regions of the conformational space is ensured for complex systems.

In 2009, we performed well-tempered metadynamics to elucidate the mechanistic details of flexible-ligand, flexible-protein docking of naloxone (NLX), a non-selective antagonist for opioid receptors, from the bulk water environment into the orthosteric binding pocket of a B2AR-based homology model of δ opioid (DOP) receptor (Provasi et al. 2009). Using the multiple walker method (Raiteri et al. 2006), several well-tempered metadynamics simulations can be simultaneously performed, each contributing to the same history-dependent bias potential; this is a key advantage of the method, enabling the most efficient use of cluster computing resources. Thus, the DOP receptor was simulated for 500 ns, in a hydrated DPPC-cholesterol lipid bilayer, using ten walkers. All simulations were performed using GROMACS 4.0.5 (Van der Spoel et al. 2005) with PLUMED (Bonomi et al. 2009).

To represent the binding event, two CVs were chosen to describe (1) the distance between the center of mass of the heavy atoms of NLX and the center of mass of the heavy atoms of the alkaloid-binding pocket of opioid receptor and (2) the distance between the center of mass of the DOP receptor alkaloid binding pocket and the center of mass of the heavy atoms of the middle residues of the EL2. This second CV was selected to enable enhanced conformational sampling of the EL2 region of the DOP receptor, given its uncertain predicted structure (this work was performed before the crystal structure of the DOP receptor became available (Granier et al. 2012)). Using these two CVs, we reconstructed the free-energy surface of the NLX binding event, determining that the non-selective antagonist NLX exhibits a molecular recognition site on the DOP receptor surface at a cleft formed by EL2 and EL3, and ends in a preferred orientation into the receptor alkaloid binding pocket, after visiting some less stable states in between. The most stable NLX-bound state of DOP corresponded to a conformation in which NLX interacted directly with D128$^{3.32}$ via a salt bridge, and with Y308$^{7.43}$, W274$^{6.48}$, H278$^{6.52}$,in broad agreement with experimental data from mutagenesis and competition binding assays (Befort et al. 1996a, b; Li et al. 1999; Mansour et al. 1997; Bot et al. 1998; Spivak et al. 1997; Surratt et al. 1994). The ligand was stabilized in a specific orientation through interactions with a number of residues on TM3, TM5, TM6 and TM7, in particular: M132$^{3.35}$ as well as F218$^{5.43}$ and F222$^{5.47}$. The recent crystal structure of the DOP bound to naltrindole confirmed some key interactions in the bound state, in particular, the interactions with Y308$^{7.43}$, H274$^{6.52}$ M132$^{3.36}$ and D128$^{3.32}$ were observed. On the pathway to this final binding mode, NLX visited a number of different less stable alternative pockets and in particular, two metastable states characterized by different degrees of opening of EL2. The existence of these metastable states before binding to the orthosteric pocket is reminiscent of the pathways observed for the B2AR using steered MD (Gonzalez et al. 2011) and unbiased MD (Dror et al. 2011), though the intermediate poses visited by NLX along the binding pathway are, as expected, not the same as those observed for ligand binding to B2AR. Some residue positions are, however, implicated to interact along the entry pathway for both receptors, in particular Y/L7.35, H/W6.58 and D300/D290 in the B2AR and DOP receptor respectively.

Sampling of the NLX ligand in the bulk solvent was corrected for, using a methodology put forward by the Roux lab (Roux 1999; Allen et al. 2004). Using the free energy surface from the metadynamics simulation, with the collective variables described above, we obtained a restrained free-energy profile, following the protocol and equations described in (Provasi et al. 2009). Our calculated equilibrium constant, of $K_{eq} = 80 \pm 13$ nM, for the final bound state of NLX at DOP was remarkably close to the majority of reported experimental values (e.g. Toll et al. 1998), and thus, this methodology offers great potential for describing, quantitatively, the binding events of other GPCRs.

6.3 Activation Mechanisms in GPCRs

Activation processes of GPCRs are known to occur on timescales that are inaccessible to current simulations using standard MD. From recent crystal structures, the most pronounced, common conformational rearrangements that mark the activation of a receptor include: breaking of the so-called "ionic lock" between the E/DRY motif in TM3 and acidic residues in TM6, upon a large (i.e. ~11 Å in the case of B2AR (Rasmussen et al. 2011a)) outward movement and slight rotation of the intracellular end of TM6; smaller (i.e. ~6 Å for B2AR (Rasmussen et al. 2011a)) outward movement of TM5; and some smaller slightly inward movements of TM3 and TM7. Here we describe activated GPCR models obtained using reduced system representations and steered MD, as well as combinations of adiabatic biased MD and metadynamics techniques. We also discuss ligand-specific conformations obtained depending on the physiological response of the bound ligand. In particular, we discuss applications to two GPCRs, the B2AR and the $5HT_{2A}$ serotonin receptor, and their perceived success as judged by virtual ligand screening and, in the case of B2AR, when compared with the recently published crystal structures.

6.3.1 Using Guided MD to Build Activated GPCRs

There are very few currently available X-ray crystal structures of GPCRs displaying the characteristics of an active state, and these are: the ligand free opsin crystal structures at either low pH (PDB ID: 3CAP (Park et al. 2008)) or in complex with a synthetic peptide derived from carboxy terminus of the alpha-subunit of the heterotrimeric G protein (PDB ID: 3DQB (Scheerer et al. 2008)); the crystal structure of the proposed active state of the A2A adenosine receptor (PDB ID: 3QAK, 2YDO (Xu et al. 2011; Lebon et al. 2011)); the crystal structures of metarhodopsin II (PDB ID: 3PXO (Choe et al. 2011)) and a constitutively active metarhodopsin II (PDB ID: 4A4M (Deupi et al. 2012)); and finally the crystal structures of B2AR in complex with the Gs protein (PDB ID: 3SN6 (Rasmussen et al. 2011b)), and a camelid nanobody (PDB ID: 3P0G (Rasmussen et al. 2011a)).

In 2011, two groups combined experimentally derived restraints and ligand binding data from extensive literature searches with steered MD to guide the generation of the active states of the B2AR (Simpson et al. 2011) and the $5HT_{2A}$ receptor (Isberg et al. 2011), and have assessed the success of their approaches by virtual screening of test sets of known agonists, antagonists and inverse agonists, diluted among non-binders.

To build an initial model for the active state of the B2AR, Simpson and colleagues have combined the intracellular part of the opsin crystal PDB ID: 3DQB (Scheerer et al. 2008) with the extracellular portion of the inactive B2AR crystal structure at 2.1 Å (PDB ID: 2RH1 (Cherezov et al. 2007)) to generate a template, and used an alignment based on common motifs in the transmembrane regions as input for MODELLER version 9v1 (Fiser and Sali 2003). Two hundred initial models were narrowed down to a single model that scored well according to the MODELLER objective function (measuring spatial restraint violation) and had a low backbone RMSD from the template.

Isberg and colleagues generated a single homology model of the $5HT_{2A}$ based on the B2AR inactive structure, PDB ID: 2RH1 (Cherezov et al. 2007). They modified this structure, to match the active characteristics found in the opsin structure, PDB ID: 3DQB (Scheerer et al. 2008). Specifically, (1) the lower parts of TM5 and TM6 were tilted outwards, manually; (2) the missing density of IL3 in the B2AR structure was replaced by the opsin IL3 structure, and (3) the $G\alpha_i$ peptide was inserted into the structure and mutated to $G\alpha_s$ and then subsequently to the $G\alpha_q$ subtype. Finally, (4), the rotamer of $R131^{3.50}$ was set to their G protein interacting conformation, and $W286^{6.48}$ was set to its presumed active conformation, as predicted by some earlier investigations (see e.g. Holst et al. 2010; Shi et al. 2002, among others), but not yet observed in any active receptor crystals to date (see supplementary figure 7a of Taddese et al. 2012). This modified structure was then used as the input to MODELLER version 9v6 (Eswar et al. 2007).

In agreement with subsequent observations of the activated B2AR during long timescale MD simulations (Rosenbaum et al. 2011), which revealed that interaction with either the G protein or a nano-body G protein surrogate is essential to maintaining an activated receptor conformation, both groups have included interactions with the G protein to complete their activated conformations. The 3DQB opsin structure represents the apo-receptor, stabilized by a peptide derived from the main binding site of the heterotrimeric G protein, the C-terminus of the α subunit (GαCT). Isberg and colleagues included the G protein peptide, mutated to Gαq while Simpson and colleagues modeled the whole heterotrimeric assembly, using structural details from a number of different crystallographic structures (namely PDB IDs: 1AZS (Tesmer et al. 1997), 1GOT (Lambright et al. 1996) and 1GP2 (Wall et al. 1995)).

The initial model of the B2AR from Simpson and colleagues was energy minimized using the CHARMM (Brooks et al. 1983) force field and an implicitly represented membrane/solvent environment described by the Generalized Born algorithm with the simple switching function (GBSW) (Im et al. 2004). In the GBSW model, the influence of the membrane is included as a solvent-inaccessible, infinite planar slab of low dielectric constant. A simple smoothing function is included to approximate the dielectric boundary between the "bulk water" and the "membrane" (Im et al. 2003). For this study, the membrane thickness was set at 35 Å with a smoothing length of 5 Å. The use of the implicit membrane can minimize problems associated with complete sampling of phase space during short standard MD simulations, since the number of explicit particles for which pair-wise interactions are required to be calculated is dramatically reduced. In particular for this application, this meant that the entire G protein assembly could be included without the prohibitive contribution to the number of particles from explicitly represented solvent. The simulation length total was approximately 175 ns. The choice to use an implicit solvent model could be controversial, given previous reports of a structural and functional role for H-bonding interactions of waters entering the rhodopsin structure upon activation (Grossfield et al. 2008), and the internal crystallographic waters noted in many X-ray derived structures of GPCRs. Simpson and colleagues have indirectly incorporated the effects of such solvation, through specific experimentally derived restraints that would open out parts of the structure sufficient to permit entry of water molecules upon activation under explicit solvation conditions.

Isberg and colleagues embedded their model of the $5HT_{2A}$ receptor in an explicitly represented DPPC membrane, with the TIP4P explicit water model and counterions under the OPLS-AA 2001 (Kaminski et al. 2001) forcefield. Their simulations were carried out using the Schrödinger implementation of the DESMOND software package, version 2.4 originally developed by DE Shaw and associates (Chow et al. 2006). Perhaps reflecting the increased number of degrees of freedom, these simulations are shorter in duration, four stages of 5 ns simulations with restraints for a total of 20 ns. The choice of DPPC may not be the best lipid for activation studies of GPCRs, since Vogel and colleagues presented evidence that, at least for rhodopsin, the fully saturated chains of DPPC can inhibit the activation transition to the receptor activated meta II state (Vogel et al. 2004). Furthermore, the melting temperature of DPPC is 41 °C, and so, under physiological conditions, the membrane might be expected to be gel-like, casting some doubt on the choice of this lipid as a valid membrane mimetic. Nevertheless, Isberg and colleagues ensured a fluid membrane environment for their $5HT_{2A}$ model, by maintaining their simulation conditions at $T = 325$ K ($\sim$51 °C).

Both groups derived sets of restraints from previously published experimental information to define their activated states, and in both cases, during the earliest parts of the simulations, harmonic restraints were applied to the backbone of transmembrane regions of the receptor, to prevent distortions of the helical structure. Isberg and colleagues applied pairwise harmonic distance constraints, collated from mutagenesis and X-ray crystallographic data, to inter-helical, ligand-receptor and G protein-receptor distances. These constraints were used to drive the conformation of the $5HT_{2A}$ towards a presumed activated structure, achieved once they are satisfied. Simpson and colleagues applied six sets of harmonic distance restraints to the B2AR receptor model during MD simulations of 150 ns in production length, to drive the conformation towards an activated state. The basis of these restraints was from experimental evidence sourced from the available literature on both B2AR and other class A GPCRs (in particular, the muscarinic receptor and rhodopsin). The sets of restraints were derived from site-directed spin labeling, disulfide cross-linking, engineered zinc binding and site-directed mutagenesis experiments. The restraints were introduced slowly, as the system was heated to simulation temperature, and all but the toggle switch restraints were maintained throughout the 150 ns production simulation.

The success of the models, in both cases, was probed by virtual screening during which it was demonstrated that the predicted active models successfully discriminated known receptor-selective agonists from antagonists. Simpson and colleagues refined their active structure by flexibly docking full agonists epinephrine, isoprotenerol, TA-2005 and salmeterol into the binding pocket, and screened these refined receptors using GLIDE (Friesner et al. 2004, 2006; Halgren et al. 2004). They were able to check all binding poses, because they used a fairly small set of ligands, comprising 29 full and partial B1/B2AR agonists, 38 B1/B2AR antagonists and inverse agonists, and 56 non-peptide ligands selective for opioid, muscarinic, neurokinin, neurotensin, bradykinin and serotonin receptors. Docking scores demonstrated that the active model showed a preference for B1/B2AR agonists over antagonists, inverse agonists and other ligands. Active enrichment curves showed that the enrichment over the first 10 % of ligands was significantly greater than expected at random. Furthermore, enrichment for B2AR-selective agonists was higher than for B1AR-selective agonists, and enrichment for antagonists was only observed once all the agonists had been found. Isberg and colleagues performed a larger scale virtual screen, also using GLIDE. They selected 182 known $5HT_{2A}$ ligands, including 139 agonists (112 phenylethylamine agonists and 27 others), 39 antagonists and 4 inverse agonists. These were diluted with 9257 diverse compounds from the ZINC database (Irwin et al. 2012). GLIDE docking scores demonstrated that the 29 highest ranked hits were phenylethylamine agonists, and these were the best scoring class of molecules by a large margin, with a mean G-score 1.5 lower than other agonists. Phenylethylamine antagonists also scored better than other agonists, other antagonists, and other ligands. The virtual screenings seem to demonstrate the success of modeling the active state of both receptors, using restraints derived from experimental data. However, the subsequent solution of the active state crystal structures of B2AR (Rasmussen et al. 2011a, b; Rosenbaum et al. 2011) has enabled direct comparison between the B2AR crystal and the active model, and Taddese and colleagues (2012) have discussed this comparison in a brief paper in 2012. Overall, they report an average RMSD of 1.3 Å in the invariant TM region (i.e. residues that have the same environment when compared over a number of structures) between the active model from Simpson and colleagues (2011) and the active B2AR structure. The overall RMSD is perhaps less explanatory than the residue-by-residue variation between the B2AR model and the B2AR crystal structure, which ranges from 1.0 Å in the TM region, to 7.2 Å in the loops. The larger value of RMSD for the loop regions is not unexpected owing to their high mobility. The active B2AR model was based on an active, G protein bound state of opsin, and the difference between the opsin template and the active B2AR crystal structure is 1.7 Å over the invariant TM region. The active model shows good agreement with the overall position of the helices in the structure PDB ID: 3SN6 for TM1, TM2 and TM5, but seems more like the opsin template for TM3, TM4, TM6 and TM7. The orthosteric binding pocket appears to have been successfully modeled, with an RMSD of 1.3 Å between the residues in the active model and the crystal structure, and this may be the reason for the strong enrichment results for the virtual screening. The key receptor-G protein interactions were correctly predicted from the opsin-G protein complex template, with 27/38 residues on the receptor and 20/34 residues on the G protein surface correctly predicted. Overall the modeling moderately successfully reproduced the nature of the activated characteristics observed in the crystal structure, but the atomic resolution specificity of the interactions seen in the single conformation of the X-ray structure was not reproduced. This can be thought of as a direct consequence of the simplification or averaging over interactions collated from different structural/biophysical sources introduced by the homology modeling process, but can also be attributed to the static nature of a crystallographic structure versus the dynamic manner in which the model was constructed.

6.3.2 Combined Adiabatic Biased MD with Metadynamics to Explore Activation Pathways

As mentioned earlier in this chapter, and as observed in experimental studies, the value of GPCRs as pharmaceutical targets is enhanced by the concept of "functional selectivity", or "biased agonism", a concept wherein a single ligand might display different efficacies at a single receptor isoform depending on the effector pathway coupled to that receptor (Rives et al. 2012). A simple mechanistic explanation for this phenomenon can be derived from the dynamic equilibrium of GPCRs between different conformational sub-states. Different ligands or changes in protein-protein interactions (such as may be observed in receptor dimerization) can shift the equilibrium population of these sub-states toward conformations of the receptor that are kinetically and structurally distinct. Evidence of different sub-states can be inferred from crystallographic structures: not all of the attributes usually associated with an 'active-state' receptor are present for active-state crystal structures solved in the presence of different agonists (Provasi et al. 2011). We have recently developed a general computational strategy that combines adiabatic biased MD with path CV metadynamics, which we have used to characterize possible metastable active states of rhodopsin (Provasi and Filizola 2010); to discern the effects of ligands with variable efficacies on the conformation of the B2AR receptor (Provasi et al. 2011) and, finally, to offer a mechanistic explanation of allosteric effects of ligand binding in a GPCR dimer of the $5HT_{2A}$ serotonin receptor

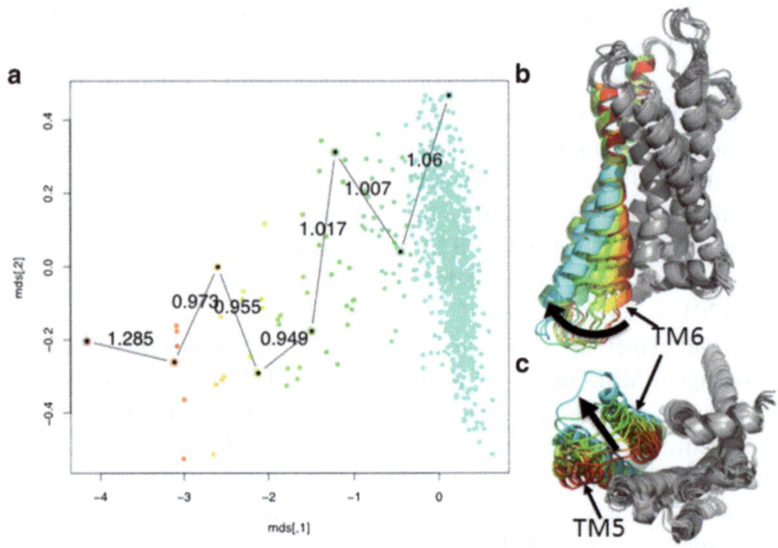

Fig. 6.2 (**a**) Multidimensional scaling representation of multiple adiabatic biased MD trajectories clustered according to RMSD. Representative structures selected from each cluster (shown in (**b**): side view and (**c**): intracellular view) to define a homogenous pathway between the inactive and active states. The receptor is shown in *grey cartoon* representation, in (**b**) and (**c**), except for TM5 and TM6, which show the progress from inactive to active, in colors corresponding to the clusters in (**a**)

and the glutamate receptor (mGluR2) observed by our experimental colleagues (Fribourg et al. 2011).

In the first stage of the method, a monodimensional pathway defining activation of the receptor was derived. Frames depicting the receptor in intermediate states between 'inactive' and 'active' X-ray-derived crystal structures (or models, in the absence of crystal structures) defined the pathway. The frames were derived from multiple (e.g. 5 or 10) statistically independent adiabatic biased MD simulations (Marchi and Ballone 1999). Like targeted MD (Schlitter et al. 1994), adiabatic biased MD drives the system towards the desired final state, by enforcing a reduction in RMSD between the initial and target conformations, to enable observations of a conformational transition during a single trajectory. In contrast to TMD, the adiabatic biased MD algorithm (Marchi and Ballone 1999) ensures exploration of low-energy pathways by keeping the total potential energy of the system constant during the MD run through application of a time-dependent biasing potential. The harmonic biasing potential is only applied when the RMSD is bigger than the minimum value previously achieved during the trajectory. The adiabatic biased MD trajectories were short (e.g. 1–5 ns), with small elastic constants (0.01 kJ/nm^2) and independently drawn Maxwellian initial velocities. All of the independent adiabatic biased MD trajectories were then pooled, and all of the sampled conformations clustered based on the RMSD of the TM domain of the receptor. Representative structures from these clusters were then chosen as reference frames to define the path CVs for further metadynamics simulations. The number of clusters, and hence frames, was between 4 and 10, depending on the nature of application. See Fig. 6.2 for an example of this.

To obtain information about the relative stability of the states populated by the receptor during transition from inactive to active conformations, metadynamics simulations with path collective variables (Branduardi et al. 2007) can be performed. The CVs defining the phase space are the position (s) along and the distance (z) from the pre-determined activation pathway. The reference frames, k, define the position of the progression of the system from inactive ($s \sim 0$) to active

($s \sim 1$) during the pre-determined transition (CV 1), and its distance (z) from it (CV 2), according to the following equation:

$$s\left(R(t)\right) = Z^{-1}\sum_{i=1}^{k}\frac{i-1}{k-1}e^{-\lambda d_{\mathrm{TM}}}\left(R(t), R^{(i)}\right) \tag{6.5}$$

where $\delta = \langle d_{TM}(R^{(i)}, R^{(i\pm1)})\rangle$ is the average distance between two adjacent frames in the transition pathway, the constant exponent, λ, must be chosen in such a way that $\lambda \times \delta \cong 1$, and Z is a normalization factor defined by:

$$Z\sum_{i=1}^{k}e^{-\lambda d_{\mathrm{TM}}}\left(R(t), R^{(i)}\right) \tag{6.6}$$

while

$$z\left(R(t)\right) = -\lambda^{-1}\log Z \tag{6.7}$$

The free-energy of the receptor as a function of s and z can then be reconstructed from converged well-tempered metadynamics simulations of varying length (~80–300 ns) (see Provasi et al. 2011; Provasi and Filizola 2010; Fribourg et al. 2011 for details of the simulation protocols). Convergence of the reconstructed free-energy can be assessed in two ways: firstly, by ensuring minimal variation in the free-energy difference between the minima around the experimental endpoints with time, and secondly, in a converged simulation, exploration of the FES represented by the CVs faces no further energy barriers, therefore one can expect frequent re-crossing of the values of the CVs after convergence is attained.

All simulations in each of the applications of this method to date have been carried out using an explicit atomistic representation of all components, and the receptors were embedded in solvated POPC/10 % cholesterol membranes. The versatility of the method lies in the implementation using Plumed plugin (Bonomi et al. 2009), meaning it has been performed using GROMACS (Van der Spoel et al. 2005) and the OPLS-AA force-field (Kaminski et al. 2001) for rhodopsin and B2AR (Provasi et al. 2011; Provasi and Filizola 2010) and NAMD 2.7 (Phillips et al. 2005) with the Charmm27 force-field (MacKerell et al. 1998) for the $5HT_{2A}$, mGluR2 monomers and dimer (Fribourg et al. 2011).

This combinatorial approach was initially validated on rhodopsin (Provasi and Filizola 2010), where the activation pathway was defined between the crystal structure of a photoactivated, deprotonated intermediate of rhodopsin (PDB ID: 2I37) and the low pH crystal structure of opsin (PDB ID: 3CAP), the only crystal structure displaying characteristics of an active state available at the time of this study. Simulations were performed for wild-type rhodopsin, embedded in a POPC membrane, with either a charged or an uncharged residue $E134^{3.49}$ within the E(D)RY motif, to simulate proton uptake from the bulk occurring in the late stage of rhodopsin activation.

Reconstruction of the rhodopsin free energy landscape indicated three common metastable states between 2I37 and 3CAP along the adiabatic biased MD pre-calculated transition path. Two of the identified minima are comparable to active intermediates of bovine rhodopsin, characterized by different amplitude of the outward movement of TM6. This was revealed by comparing intra-molecular distance analysis of these states with results from double electron-electron resonance spectroscopy (Altenbach et al. 2008). The largest predicted separation between TM3 and TM6 was in line with data obtained for Meta IIb from spectroscopy (Knierim et al. 2007), and for opsin from crystallography (Park et al. 2008).

Key residues, thought to influence the activation pathway, were mutated and additional simulations were carried out on these receptors for comparison to the WT. An interaction between $K231^{5.66}$ and $E247^{6.30}$ appeared to stabilize the predicted active conformation with the largest separation between TM3 and TM6, as also seen in the opsin crystal structure (Park et al. 2008). Furthermore, we carried out metadynamics simulations of the $K231A^{5.66}$ mutant rhodopsin with either a charged or an uncharged residue $E134^{3.49}$ within the E(D)RY motif. We hypothesized that removal of a polar interaction between $K231^{5.66}$ and $E247^{6.30}$ by replacing $K231^{5.66}$ with alanine would switch the equilibrium toward the predicted conformation with a smaller separation between TM3 and TM6. Our simulations showed that the putative active conformation of rhodopsin with the largest separation between TM3 and TM6 was destabilized in the presence of a neutral E134$^{3.49}$ mimicking the activation-dependent proton uptake from the bulk.

In the most comprehensive study using this methodology (Provasi et al. 2011), we were able to present compelling evidence for the concept that ligands of differing efficacy access structurally distinct conformations of the B2AR. Path CV metadynamics simulations were conducted for six ligands, representing a full agonist (epinephrine), weak partial agonists (dopamine and catechol) inverse agonists (carazolol and ICI-118,551) and a neutral antagonist (alprenolol). These different ligands were also compared to the unliganded receptor. Parameters for the ligands were obtained manually, by analogy with existing molecules in the OPLS-AA force-field, while Coulomb point charges were obtained from quantum chemical geometry optimization using Gaussian 03 and restricted Hartree-Fock calculations with the 6-31G* basis set. The ligands were either docked into the receptor using Autodock 4.0 (Morris et al. 1996) (for dopamine, catechol and epinephrine) or positioned according to their crystallographically determined binding modes. The activation pathway was homogenously defined by ten frames, between the inactive state of the B2AR, represented by PDB ID: 2RH1 (Cherezov et al. 2007) and the active state, represented by the nanobody stabilized structure, PDB ID: 3P0G (Rasmussen et al. 2011a). Metadynamics simulations for each receptor/ligand complex were run for approximately 300 ns until the difference between the free energies of the metastable states observed was converged. The activation of the receptor was also measured in terms of other structurally pertinent variables, by reweighting the free energy surface. These variables were: the distance between the sidechains of the R131$^{3.50}$ and E268$^{6.30}$ residues of the ionic lock; the dihedral angle of the sidechain of W286$^{6.48,}$ i.e. the rotamer toggle switch; and the outward displacement of TM6. Comparison of the free energy surfaces of the receptor activation when complexed with different ligands demonstrated selective inactive or active conformations depending on their known elicited functional responses.

In the absence of ligand, the receptor was found to have two low energy conformations, one with s ~0.2, i.e. close to the inactive state, and a second, closer to an active state, with s ~0.6, separated by a low energy barrier. This corresponds well with the known basal activity of the B2AR, and the inherent conformational flexibility of the unliganded B2AR may help to explain the difficulty in obtaining crystallographic structures of the B2AR in the absence of ligand.

The free energy surface for B2AR complexed with alprenolol, ostensibly a neutral antagonist, i.e. a ligand that competitively binds in the orthosteric binding pocket of a receptor and blocks the binding of (and thus cellular responses to) agonists or inverse agonists, also displayed two stable states. One, close to the inactive, s ~0.2, was marginally (~1 kcal/mol) more stable than the second state, at s ~0.6. This type of free energy profile illustrates possible reasons for some studies having found alprenolol to behave in different ways in different assays: most notably, as an inverse agonist (Elster et al. 2007; Hopkinson et al. 2000) or as a partial agonist (Callaerts-Vegh et al. 2004).

The profiles for the inverse agonists (ligands that preferentially stabilize a receptor in an inactive conformational state), ICI-118,551 and carazolol both displayed single conformational states with low values of s ($s < 0.2$), in deep free energy wells. The free energy minimum, in each case, was approximately 4 kcal/mol lower in energy than the next lowest minima. This corresponds with the solution of many crystal structures in inverse agonist bound conformations.

Results for the agonists were less clear-cut. The free energy profiles for both partial (catechol and dopamine) and full (epinephrine) agonists each displayed a deepest free energy minimum at the expected value of s, i.e. s ~0.6 for the partial agonists, and $s > 0.9$ for the full agonist. However, other minima, only marginally less stable than these were found in all cases, at lower values of s. A feasible explanation for this arises from the absence of G protein: as demonstrated by Rosenbaum and colleagues (2011), during very long timescale MD simulations, an active conformation of the B2AR will quickly lose many of the active characteristics in the absence of the G protein or a stabilizing nanobody surrogate.

The results from this study provide an atomic resolution view of structurally distinct receptor conformations stabilized by different ligands, and demonstrate that ligands with varying efficacies might be used to control population shifts toward desirable GPCR conformations. Such knowledge might be useful to aid rational drug design: biased ligands may represent novel, more efficacious therapeutics.

Finally, in conjunction with tissue and animal behavioral studies, we have recently applied this methodology to determine some of the mechanistic details driving the antipsychotic properties of drugs targeting a heteromeric arrangement of the serotonin $5HT_{2A}$ and mGluR2 (Fribourg et al. 2011), in particular, why some $5HT_{2A}$ inhibitors (e.g. clozapine) may exhibit neuropsychological effects while others (e.g. methysergide) do not. Oligomeric receptor complexes have been shown to have distinct signaling properties when compared to their monomeric components, and these receptors have been shown to form a specific hetero-oligomeric complex in mammalian brain tissue that is implicated in schizophrenia. The complex integrates the responses from the Gq-coupled mGluR2 and that of the Gi-coupled $5HT_{2A}$ to modulate signal transduction and influence downstream signaling events. To investigate this computationally, and to provide a structural context for the experimental observations, we built homology models of both receptors: mGluR2 was rhodopsin based, using PDB ID: 1U19 for the inactive state and 3DQB for the active-like state, while $5HT_{2A}$ was based on B2AR, using 2RH1 as the inactive template and 3P0G as the active-like template. To these models, we applied the methodology as in the previous cases, comparing the free energy profiles of activation for $5HT_{2A}$ bound to a strong agonist, DOI, a neutral antagonist, methysergide, and an inverse agonist, clozapine. For the monomeric $5HT_{2A}$, clozapine and methysergide stabilized two different inactive-like conformations, while DOI stabilized an active like conformation. These experiments were repeated for dimeric arrangements of the receptors, with the lowest energy ligand-stabilized conformations of $5HT_{2A}$ complexed with a ligand-free mGluR2. When complexed with the clozapine-stabilized $5HT_{2A}$ conformation, the mGluR2 conformation is shifted toward an activated state. In contrast, when bound to either the DOI-stabilized active-like $5HT_{2A}$, or the methysergide–stabilized inactive-like $5HT_{2A}$ conformation, the conformational equilibrium of mGluR2 is shifted toward an inactive-like state. These results (in combination with the tissue and animal studies) suggest that the formation of the heteromeric complex enables modulation of the mGluR2 response to its endogenous ligand by binding of different ligands to the $5HT_{2A}$. Therefore, the strong $5HT_{2A}$ agonist, DOI, can greatly stimulate Gq signaling, while decreasing Gi signaling, but the inverse agonist, clozapine, abolishes Gq signaling while simultaneously stimulating Gi signaling. These results go some way to explaining the complexities surrounding the disparity between the neuropsychological effects of $5HT_{2A}$ inhibitors, some of which demonstrate antipsychotic effects, while others do not. A drug bound to one receptor of the heteromer was shown to influence the signaling response of the partner to its endogenous ligand and such mechanistic insights are invaluable to therapeutic strategy design for disorders in which oligomeric arrangements of receptors are implicated.

6.4 Oligomerization Mechanisms in GPCRs

In the previous section, we have mentioned the distinct properties of oligomeric arrangements of some GPCRs as compared to those of their monomeric components (Milligan 2009). However, the ability of GPCRs to assemble into stable, heteromeric or homomeric, oligomeric arrangements remains a subject of much debate (Fonseca and Lambert 2009; Lambert 2010). From a computational perspective, understanding the thermodynamics and kinetics of GPCR oligomerization is not a trivial problem owing to (1) the large size of the complex system, and (2) the stochastic nature of the process. In this section we survey the enhanced MD-based methods that have recently been used in combination with reduced representations of the GPCR systems to study the dynamic behavior of dimeric/oligomeric arrangements of these receptors.

6.4.1 Spontaneous Self-Assembly of Coarse-Grained Representations of GPCRs

The number of calculations required by a biological molecule scales as the square of the number of particles included in its representation, and consequently, a lengthy simulation of a large, explicitly represented GPCR, particularly in a dimeric or oligomeric state, becomes particularly difficult at atomistic resolution. This issue has prompted several recent efforts focused on reducing the number of particles included in a simulation of a protein and its environment, and a number of strategies have been proposed. We focus here on the MARTINI coarse-grained (CG) model (Marrink et al. 2008), one of several "bead models" that have been devised to coarsely represent biomolecules (see Tozzini 2005, 2010 for recent reviews).

The MARTINI forcefield has proven to be a successful means of studying oligomerization for GPCRs and has been used in extensive studies of oligomerization for rhodopsin (Periole et al. 2007, 2012) opioid receptors (Johnston et al. 2011; Provasi et al. 2010) and the B1AR and B2AR (Johnston et al. 2012). Developed by Marrink and co-workers, the MARTINI force-field (Marrink et al. 2004, 2007, 2008; Monticelli et al. 2008; Lopez et al. 2009) offers coarse-grained (CG) representations of proteins, several lipid types and cholesterol. Close comparability to both experimental and atomistic MD approaches in residue-level detail is ensured by the extensive calibration of a large library of chemical building blocks against thermodynamic data (in particular, oil/water partition coefficients). These versatile building blocks permit the force field to flexibly represent a large range of biomolecules without the need to re-parameterize the model in each case. The MARTINI force field is compatible with GROMACS (Van der Spoel et al. 2005) and its CG mapping of heavy atoms (including water molecules) to beads is a 4:1 ratio, except in the case of molecules containing rings, where a mapping of 2:1 is used. There are four main types of beads, P = polar, N = non-polar, C = apolar and charged = Q, with subtypes depending on hydrogen bonding capacity and polarity. Each CG bead has a mass of 72 atomic mass units (amu), equivalent to the mass of four water molecules, except for ring structure beads, which have a mass of 45 amu. This improves efficiency and enables a simulation timestep of 20–40 fs, approximately ten times that possible with an all-atom GROMACS simulation. Based on comparison of diffusion rates for atomistic and MARTINI CG models, the CG model affords a scaling factor of 4 for the effective time sampled, though there has been some recent debate surrounding the best time step (Winger et al. 2009; Marrink et al. 2009).

The earliest application of the MARTINI CG model to GPCRs was the self-aggregation of rhodopsin monomers in four different explicit, CG, lipid membrane environments (Periole et al. 2007). Systems with 16 CG rhodopsin monomers (based on the 1L9H structure (Okada et al. 2002)) at a protein:lipid ratio of 1:100, were simulated for up to 8 μs. Such time scales would be very difficult to reach in an atomistic representation. The results of these simulations indicate a localized membrane adaptation to the presence of the receptor, supposedly driven by the motivation to overcome the hydrophobic mismatch between the length of the hydrophobic part of the monomeric receptor and the equilibrium hydrophobic bilayer thickness. Hydrophobic mismatch has been shown to promote self-assembly of rhodopsin reconstituted in membrane bilayers (Kusumi and Hyde 1982; Ryba and Marsh 1992). The details of such phenomena could not be modeled in a simulation using an implicit representation of a membrane, like the generalized Born method (Im et al. 2004) described earlier in this chapter.

Periole and colleagues simulated rhodopsin in different phosphocholine (PC) lipid environments under non-biased conditions, and they observed a clear dependence on chain length. Bilayer adaptation was manifest as local thickening near helices TM2, TM4, and TM7 in $(C12{:}0)_2PC$, $(C16{:}1)_2PC$, and $(C20{:}1)_2PC$ bilayers and as local thinning near helices TM1/H8 and TM5/TM6 in $(C20{:}1)_2PC$ and $(C20{:}0)_2PC$. The results indicated a higher propensity for protein:protein contact interactions in $(C16{:}1)_2PC$. Clear reorganization and increase in scope of the rhodopsin dimer interfaces was observed during the simulations, indicating a search for a shape complementarity that maximized the buried accessible surface area. The number of contact interfaces was higher in $(C12{:}0)_2PC$ and $(C16{:}1)_2PC$, where strong forces were introduced by the hydrophobic mismatch, than in $(C20{:}1)_2PC$ and $(C20{:}0)_2PC$, where the forces were more balanced. Three contact interfaces were clearly visible on the 6–8-μs time scale in $(C20{:}1)_2PC$; these included previously suggested homo- and heterodimerization interfaces in rhodopsin and other GPCRs, involving the exposed surfaces of the helices TM1/TM2/H8, TM4/TM5, and TM6/TM7. The authors concluded that hydrophobic mismatch drives self-assembly of rhodopsin into liquid-like structures with short-range order, and that the interactions may not be dictated by specific residue-residue contacts at the contact surfaces.

6.4.2 Combining CG Representation and Biased MD to Investigate GPCR Dimerization

By combining the MARTINI reduced representation of the system with biased MD techniques, it has been possible to make predictions about the relative strength of dimers formed at different interfaces in an explicit membrane environment.

We pioneered the use of a free energy approach to characterize oligomerization of GPCRs. Two studies of dimerization of the DOP receptor were performed using established methodologies, firstly umbrella sampling, from which we derived the PMF of a dimerization event (Provasi et al. 2010), and secondly, a metadynamics study in which we established the most favorable orientation of the individual protomers involved in different dimeric arrangements (Johnston et al. 2011), i.e. comprised of different contact interfaces, of the DOP receptor. The computational results of this study compared favorably with inferences from cysteine crosslinking experiments, supporting the role of specific residues at the interfaces.

To characterize dimerization for the DOP receptor, we performed biased MD simulations of a CG representation of a homo-dimeric arrangement of this receptor in an explicitly CG represented, POPC:10 % cholesterol-water environment. Since these studies were conducted prior to the crystallographic solution of the opioid receptor structures in 2012 (Wu et al. 2012; Granier et al. 2012; Thompson et al. 2012; Manglik et al. 2012), an all-atom structural model of the DOP receptor protomer from *Mus musculus*, was generated by homology modeling, using the crystal structure of the B2AR at 2.4 Å resolution (PDB ID: 2RH1) (Cherezov et al. 2007) in MODELLER 9v3 (Eswar et al. 2007), using the same strategy we recently reported in the literature for the human DOP receptor (Provasi et al. 2009). The loop regions were built *ab initio* using ROSETTA 2.2 (Wang et al. 2007). A pair of the resultant DOP receptor models was placed facing one another at a putative symmetrical interface, involving residue 4.58, inferred from cysteine cross-linking data on this and other GPCRs (see e.g. Guo et al. 2005, 2008).

In an effort to improve the stability of the secondary structure of the CG representation of the receptor, we combined an elastic network model (ENM) with the MARTINI CG representation, using a method developed and termed ELNEDIN by Periole and colleagues (2009). ENMs are ideally suited to preserve the secondary or tertiary structure of biomolecules, since they are structure derived, and therefore introduce an intrinsic bias toward the structure

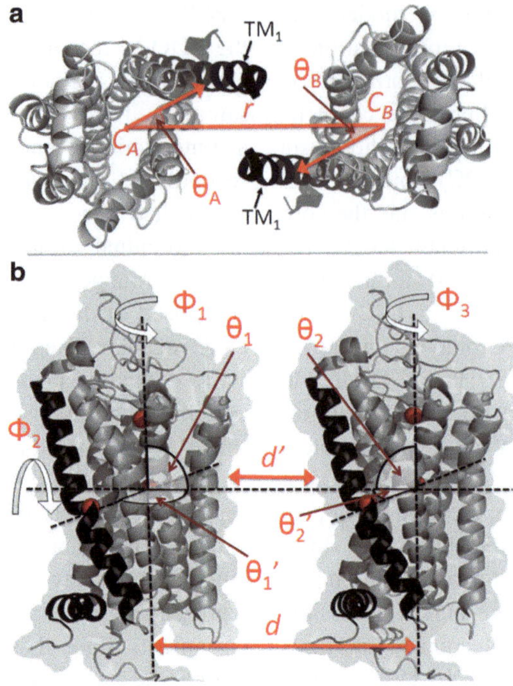

Fig. 6.3 (**a**) Definition of collective variables (CVs) used in studies of dimers from the Filizola lab (Johnston et al. 2011, 2012; Provasi et al. 2010). CV1 is the distance r between the COMs of the protomers (C_A) and (C_B); CV2 is the rotational angle θ_A, defined by the projection on to the plane of the membrane of the COM of the TM_i of A, the C_A, and the C_B, and CV3 corresponded to the equivalent rotational angle θ_B. Shown for the B2AR at a TM1/H8 interface, TM1 and H8 are colored *black*. The CVs are defined according to the interface of interest. (**b**) Definitions of the virtual bond algorithm used by Periole and colleagues in their recent publication (Periole et al. 2012). The axes (*black dotted lines* in Fig. 6.3b) are anchored by the backbone CG beads of Cys 187, Gly 121 and Gly51 (shown as *red spheres*) on each protomer, thus the variables are consistent across different interface arrangements. They describe the distance between the receptors, d; the tilt of long axis of each receptor relative to the receptor-receptor direction θ1 and θ2; the rotation of the receptors around their long axis (parallel to the membrane normal), ϕ1, ϕ3 or θ1′ and θ2′, the relative orientation of the receptor's long axis, ϕ2. d′ is the interfacial receptor distance and is defined as the distance, d, between the receptors from which the distance at the minimum of the PMFs was subtracted, see supplementary material from (Periole et al. 2012). Once again, TM1 and H8 are colored *black*

upon which they are established. In a novel extension to the ELNEDIN method, we applied a secondary-structure-dependent construct to models of the DOP receptor dimer (Provasi et al. 2010). The strength of the force constant for the harmonic restraint, K_{SPRING}, was determined by the secondary structure of each of the residues. If the residue was determined to have a defined secondary structure (by DSSP (Kabsch and Sander 1983)), e.g. α-helix as in the case of the TM regions of the DOP receptor, then a force constant of $K_{SPRING} = 1{,}000$ kJ/mol/nm^2 was applied. For a sequence of >2 residues with undefined secondary structure (e.g. coil, bend, or turn), a force constant of $K_{SPRING} = 250$ kJ/mol/nm^2 was applied. A comparison of the RMSF for the Cα beads for the CG representation with an equivalent atomistic simulation indicated a qualitative agreement, permitting the secondary structure of the helices to be maintained, without compromising the flexibility of the loop regions.

We used the CVs illustrated in Fig. 6.3a to describe the relative position of interacting DOP receptor protomers A and B during the simulations: CV1 represented the distance r between the COM of protomers A (C_A) and B (C_B); CV2 described the rotational angle θ_A, defined by the projection on to the plane of the membrane of the COM of the TM4 of A, the C_A, and the C_B, and CV3 corresponded to the equivalent rotational angle θ_B. To allow exploration of an experimentally-supported TM4 interface of DOP receptor homodimers involving position 4.58 in a reasonable timescale, we limited the sampling of the two rotational angles θ_A and θ_B to a ~25° interval, using steep repulsive potentials.

The system setup was the same for both of the studies (Johnston et al. 2011; Provasi et al. 2010). All simulations were performed using GROMACS (Van der Spoel et al. 2005) incorporating the Plumed plugin (Bonomi et al. 2009). In the first of these two, we used umbrella sampling to define the PMF of dimerization as a function of the separation between the centers of mass of the two protomers.

Umbrella sampling was introduced in 1977 by Torrie and Valleau (1974), and has been considered the prototypic biased MD technique for improving sampling of a PMF as defined by the Kirkwood equation (Kirkwood 1935) (6.8), from the average distribution, $<\rho(\zeta)>$, along a predefined reaction coordinate, ζ:

$$W(\zeta) = W(\zeta *) - k_B T \ln \left[\frac{\langle \rho(\zeta) \rangle}{\langle \rho(\zeta *) \rangle} \right] \quad (6.8)$$

where k_BT is the Boltzmann factor, and $\zeta *$ and $W(\zeta *)$ are arbitrary constants. In this method, the reaction coordinate is divided into windows, and a biasing potential ($w(\zeta)$) is introduced to tether the system to the centre of each window. The biasing potential acts to restrict the variations of the variable in order to ensure enhanced configurational sampling within each of the independent windows, and as a result, along the entirety of the reaction coordinate when all of the windows are combined. The biased simulations have potential energy $[U(\mathbf{R}) + w_i(\zeta)]$, (where $\mathbf{R}$ is the system coordinates) and $w_i(\zeta)$ is a harmonic potential of the form:

$$w_i(\zeta) = {}^1\!/_2\, K(\zeta - \zeta_i)^2 \quad (6.9)$$

centered on the successive values of ζ_i for each window. K represents the force constant of the potential. The advantage of umbrella sampling, considered to be key from a computational perspective, is that the relative independence of the windows from one another (though this should be considered and confirmed when recombining the windows) allows the biased simulations along the length of ζ to be conducted in parallel on large computer clusters.

To obtain the PMF, ($W(\zeta)$) for the range of ζ that is of interest, the simulations from each of the window must be unbiased and combined. Thus, it is crucial that the histogrammed variations in the measured variable obtained from each of the windows overlap sufficiently such that the whole of the coordinate space is covered.

Roux offers a thorough description of the derivation in (Roux 1995), but briefly, the biased distribution function, $<\rho(\zeta)>_{(i)}$ as obtained from the ith biased ensemble, is:

$$\langle \rho(\zeta) \rangle_{(i)} = e^{-w_i(\zeta)/k_B T} \langle \rho(\zeta) \rangle \langle e^{-w_i(\zeta)/k_B T} \rangle^{-1} \quad (6.10)$$

where k_BT is the Boltzmann factor and the unbiased PMF from the ith biased ensemble is:

$$W_i(\zeta) = W(\zeta *) - k_B T \ln \left[\frac{\langle \rho(\zeta) \rangle_{(i)}}{\langle \rho(\zeta *) \rangle} \right] - w_i(\zeta) + F_i \quad (6.11)$$

$\zeta *$ and $W(\zeta *)$ are arbitrary constants. F_i is an undetermined constant representing the free energy associated with introducing the biasing potential:

$$e^{-F_i/K_B T} = \langle e^{-w_i(\zeta)/k_B T} \rangle \quad (6.12)$$

Most commonly, the Weighted Histogram Analysis Method, or WHAM (Kumar et al. 1992, 1995) is used to derive the PMF from the output of the simulations. This probably reflects the fact that the Grossfield lab (University of Rochester, NY) has created a very user-friendly and freely distributed implementation of WHAM, but other unbiasing techniques e.g. MBAR (Shirts and Chodera 2008) have proven to be equally useful. WHAM focuses on optimizing the estimate of the coordinate-dependent unbiased distribution function as a weighted sum over all of the data from the biased simulations, and determining the functional form of the weight factors that minimize the statistical error. The derivation of the WHAM equations can be found in (Kumar et al. 1992) but here, in short, the key expressions can be distilled down to:

$$\langle \rho(\zeta) \rangle = \frac{\sum_{i=1}^{N_w} n_i \langle \rho(\zeta) \rangle_i}{\sum_{j=1}^{N_w} n_j e^{-[w_j(\zeta)-F_j]/k_B T}} \tag{6.13}$$

$$F_j = -k_B T \ln \left(\int \langle \rho(\zeta) \rangle e^{-w_j(\zeta)/k_B T} d\zeta \right) \tag{6.14}$$

where N_w is the number of windows; n_i is the number of independent data points, used to construct the biased distribution function, in the *i*th window at specific value of ζ; $<\rho(\zeta)>_i^b$ is the biased distribution function in the *i*th window; $w_j(\zeta)$ is the window potential of the *j*th window, at specific ζ; F_j is the (unknown and to be determined) free energy constant. The unbiased distribution function, $<\rho(\zeta)>$ is dependent on $\{F_j\}$, and thus the simultaneous Eqs. (6.13) and (6.14) must be solved iteratively, until a consistent solution is obtained for both of them. This unbiased distribution function may then be substituted into the Kirkwood equation (Kirkwood 1935), (6.8), to obtain the PMF along the coordinate ζ.

For our first umbrella sampling simulations of GPCRs (Provasi et al. 2010), 43 independent windows with values of the reaction coordinate in question, i.e. protomer separation, *r*, between $r_1 = 3.0$ nm and $r_{43} = 4.90$ nm were simulated for 300 ns, using harmonic restraints on the distance. The distribution *p(r)* of the separation was harvested for the last 250 ns, and the resulting probability distributions were combined using WHAM to derive the free energy as a function of the separation between the protomers.

The free energy surface identified two different, yet energetically very close, homodimeric states of the DOP receptor (D_1 and D_2) involving the TM4 interface, energetically stabilized relative to the monomeric state, which were separated from each other by a transition state at $r_{TS1} = 3.28$ nm, and from the monomeric state at large values of the separation ($r \geq 4.90$ nm), by a transition state at $r_{TS2} = 3.75$ nm. In D_1, the structure indicated that TM4 from each protomer inserts into a groove on the surface of the opposite protomer, formed by helices TM2, the C-terminal half of TM3, and TM4. As expected from the energetic similarity, the D_2 structure was similar to D_1 in the overall orientation of the protomers, but corresponded to a slightly less compact ($r = 3.40$ nm), and slightly asymmetric arrangement of the protomers.

Using the derived free-energy surface of DOP receptor homodimers, and the formalism described by Roux and co-workers in Allen et al. (2004) and Roux (1999), we were able to calculate several observable quantities from our simulations. The dimerization constant for the identified lowest-energy DOP homodimer was $K_D = 1.02\ \mu m^2$ (see details of the derivation in Johnston et al. 2012). From this calculated K_D, and in combination with a diffusion coefficient D_T value of 0.08 mm^2/s determined experimentally for the mu-opioid (MOP) receptor (Sauliere-Nzeh et al. 2010), and also in line with a value of 0.1 mm^2/s obtained for several other GPCRs (Barak et al. 1997; Hegener et al. 2004; Henis et al. 1982; Poo and Cone 1973), we obtained an estimate of a few seconds for the half-time of DOP receptor dimers at a contact interface comprised of TM4 in a lipid bilayer mimetic. We repeated these calculations in a second simulation (Johnston et al. 2011), for a contact interface simultaneously involving contact at both TM4 and TM5 of each protomer. The comparison revealed that the TM4 interface was marginally more stable than the interface involving both TM4 and TM5, indicating that the stability of the dimer pair is dependent on the region of contact between the protomers. These calculated lifetimes of DOP receptor homodimers are consistent with the transient, sub-second to millisecond association inferred by recent single-molecule studies of GPCRs (Hern et al. 2010; Kasai et al. 2011). Whether such a lifetime for DOP receptor homodimers and other GPCRs within the membrane has implications on the functional role of receptor complexes and/or the specificity of their interactions requires more in-depth investigation.

In a second study (Johnston et al. 2011), using this same system setup, we performed a well-tempered metadynamics study, to better explore the TM4-TM4 contact interface of the protomers in this same dimeric arrangement, and compare it with the interface comprising contact at both TM4 and TM5 of each protomer. The well-tempered metadynamics implementation is identical to that described earlier in this chapter. We restricted exploration of CV1, r, to $r = 3.80$ nm, i.e. the value of the transition state observed in the umbrella sampling simulations (Provasi et al. 2010), using a steep repulsive potential, and applied a Gaussian biasing potential to ensure thorough exploration of CV2 and CV3, angles θ_A and θ_B, within a ~25° interval for the contact interface involving TM4 only, and a ~17° interval for the more symmetric interface involving TM4 and TM5 simultaneously. The FES as a function of angles θ_A and θ_B, for each of the two contact interfaces (TM4, and TM4/5) was reconstructed from the history of the applied bias. We extracted a structure from the minima of each of the resulting 2-dimensional free energy surfaces, and proceeded to characterize the residues representing symmetric contacts between the two protomers in the minimum energy dimeric structures. The results of the calculated relative stability of the two interfaces were in line with inferences from cysteine cross-linking studies. The key residue defining the TM4 interface was V181$^{4.58}$, and was found to have a Cβ-Cβ separation of <7 Å after conversion to an atomistic resolution representation in the OPLS-AA forcefield (Kaminski et al. 2001; Jorgensen et al. 1996), minimization and short equilibration (1 ns) simulation in GROMACS (Van der Spoel et al. 2005). For the TM4/TM5 interface, the cysteine cross-linked residue, T213$^{5.38}$, was found within a Cβ-Cβ distance of 11 Å between the opposing protomers.

6.4.3 Relative Stability of Dimer Interfaces in GPCRs

In 2012, two papers more comprehensively addressing dimerization of rhodopsin (Periole et al. 2012) and the B1/B2ARs (Johnston et al. 2012) were published almost simultaneously. We have recently combined the two techniques that we have used in the past to investigate the DOP receptor, i.e. umbrella sampling and metadynamics, into a single method to comprehensively characterize dimerization of the B1AR and B2AR, at contact interfaces previously documented to have physiological relevance for a number of GPCRs by experimental techniques. We have focused on contact interfaces involving transmembrane helices TM1/H8 and TM4, and have used umbrella sampling to explore the FES as a function of protomer:protomer separation as in (Provasi et al. 2010), while simultaneously employing metadynamics as in (Johnston et al. 2011) to ensure complete sampling of the angle space available to the protomers at each interface. The missing segments of the crystal structure of the B2AR (PDB ID: 2RH1 (Cherezov et al. 2007)) were generated using Rosetta (Wang et al. 2007), and a homology model of the human B1AR was built using the turkey B1AR crystal structure (PDB ID: 2VT4 (Warne et al. 2008)). The system for simulation was constructed in an identical fashion to that for the DOP receptor simulations, with both the protein and the environment represented in the CG MARTINI forcefield.

Periole and colleagues (2012) extended their long-timescale simulations of rhodopsin in $(C20{:}1)_2PC$ bilayer (Periole et al. 2007), to multiple repetitions on the 100's of μs timescale, and with a larger bilayer, encompassing 64 rhodopsin receptors, rather than the original 16 used in the previous study. The dimerization events were clustered and suggested that the most commonly occurring events involved contact interfaces also comprised of TM4, TM4/5 and TM1/H8.

Despite the long timescale and the multiple receptors, Periole and colleagues found that the number of individual dimerization events was insufficient to provide reliable statistics from which to calculate free energy estimates for the dimerization of rhodopsin. Accordingly, they included umbrella sampling simulations to establish five PMFs for this dimerization event at different interfaces observed from the self-assembly simulations, using the virtual bond algorithm (VBA) defined in Fig. 6.3b. Their VBA defined 8 restraints, but in general, the ones found to be most useful were dihedral angles $\phi 1$and $\phi 3$ and distance, *d*. Periole and colleagues use these CVs to investigate PMFs for dimerization at symmetric interfaces involving TM4, TM5, TM4 and TM5, and TM1 and H8. They also considered an asymmetric interface composed of TM4 interacting with TM6 on the adjacent protomer. We only investigated symmetric interfaces, comprised of TM4 and TM1/H8 that have been routinely suggested to have physiological relevance through biophysical experiments. It must also be noted that the arrangement of the helices in TM1/H8 is similar but not identical between the two studies.

The combination of simulation length per window and the number of windows was different between the two studies. Periole and colleagues used a small number of windows, with a weak force constant (500–5,000 kJ mol^{-1} nm^{-2} on *r*, and 300 kJ mol^{-1} nm^{-2} on $\phi 1$ and $\phi 3$), and a long simulation length, reaching up to 20 μs (effective timescale) in some cases. In our work, we used many shorter windows, centered at intervals of *r* of only 0.05 nm. We also used a large force constant, k = 10,000 kJ mol^{-1} nm^{-2} for the harmonic restraint to tether the measured value of the distance to the center of each window. The angles θ_A and θ_B as defined in Fig. 6.3a were restricted by steep harmonic potentials, as described in previous studies (Johnston et al. 2011; Provasi et al. 2010), but were explored using the metadynamics techniques as described in (Johnston et al. 2011). The simulation length of each window was 1 μs, which was longer than in the previous umbrella sampling study (in which it was only 300 ns) (Provasi et al. 2010), in order to permit a thorough exploration of the angle range using the Gaussian bias. This novel combination methodology was validated by favorable comparison with the individual methods, on a simple two-dimensional toy system (Johnston et al. 2012).

The similarities between the results of these two independent studies are striking. Most notably, both studies find that the most stable of the interfaces investigated, by a significant margin, was that involving TM1 and H8. It is difficult to compare the exact nature of the interfaces, since firstly, the protomeric arrangement at this interface is not identical between the two investigations, and secondly, the measurement of the separation between the protomers is defined by the interfacial distance in the study of rhodopsin, while it is between the centers of mass of the helical bundles in the study of the B1/B2ARs. Nevertheless, the similarity of the result between the two independent investigations is compelling, particularly when the small contact surface area between the protomers is considered. The TM4 interface was found to be the "weakest" or least stable for both rhodopsin and the B1/B2ARs (no PMF was presented for TM4/5 for the B1/B2ARs). The PMF in the rhodopsin simulation (Periole et al. 2012) shows remarkable qualitative similarity to that found for the same interface in our first umbrella sampling study of the DOP receptor (Provasi et al. 2010). Both PMFs show two minima, separated by a small transition state, with the minimum at larger separation being slightly shallower than that in which the protomers are tightly bound together. Periole and colleagues suggest that the shallower minimum corresponds to a protomeric arrangement in which a small number of lipids "lubricate" the contact surface between the two protomers. Both PMFs show a small barrier for dimerization, between the monomeric state and the shallower minimum for the TM4 interface for both rhodopsin and DOP receptor. No such barrier is noted for TM1/H8 for rhodopsin or the B1/B2ARs, (no TM1/H8 interface has been investigated yet for the DOP receptor). Periole and colleagues suggest this barrier may be related to difficulties in delipidation of the surfaces between the protomers. A significant difference between the studies from these two labs is the choice of lipid membrane mimetic for the environment of the proteins. This is particularly pertinent because

Periole and colleagues found that de-lipidation of some of the contact interfaces was very challenging, even on the (effective) time scale of 20 μs. Following on from their earlier studies (Periole et al. 2007), Periole and colleagues chose the longer-tailed lipid mimetic of those tested, i.e. $C(20{:}1)_2$ PC, since this was found to influence dimerization the least. Atomistically, this would represent 1,2, di(D-cis-eicosanoyl)-sn-glycero-3-phosphocholine. In contrast, we used a shorter-tailed lipid membrane mimetic, representing a POPC lipid, with a 10 % cholesterol component. While it is acknowledged that the presence of cholesterol might reduce the mobility of lipids and proteins in the membrane particularly in terms of exchange at the contact interface, plots depicting the residency time at the interface (in terms of percentage of the trajectory spent within 15 Å of the interface helices) of all lipids and cholesterol in every simulation we ran indicated that 75 % of the cholesterol spent less than 60 % of the trajectory within 15 Å of the interface and 75 % of lipids spent less than 20 % of their time at the interface, while the median time at the interface for both lipids and cholesterol was much lower than 10 % of the trajectory. These statistics suggested adequate lipid exchange.

In both studies, the contact interface between the two protomers at TM4 is more extensive than for the TM1/H8 interface, and Periole and colleagues suggest that this means that the strength of the interaction is not dependent on the buried surface area between the protomers for these receptors, as is the case for soluble proteins.

Several post-translational modifications are known to occur for rhodopsin, and in particular, palmitoylation at cysteine residues 322 and 323, has been predicted to play a role in anchoring the receptor to its native membrane environment and specific interactions between the palmitoylate chain and the protein may have a functional relevance for the dark state of rhodopsin (Olausson et al. 2012). The location of these palmitoylated residues, towards the end of the amphipathic helix 8, means that they could feasibly form part of a contact interface between two protomers involving TM1 and H8. Periole and colleagues have accounted for this by performing their umbrella sampling experiments for this interface both with and without palmitoylation. They found that there was no significant difference between the PMF profiles calculated for both cases and thus speculate that the palmitoylation does not contribute to the strength of the contact interface between TM1 and H8 on adjacent protomers.

Both works offer putative physiological explanations for the striking difference in relative strength of interfaces involving different transmembrane regions. We have proposed that the strength of the TM1/H8 interface, and its subsequently calculated "long" lifetime (of the order of minutes), might be reconciled with the sub-second lifetimes reported for the muscarinic receptor (Hern et al. 2010) and the N-formyl peptide receptor (Kasai et al. 2011), through a model wherein a stable dimer is formed at a contact interface involving TM1/H8, and this diffuses through the membrane interacting with other stable TM1/H8 dimers at other, weaker, interfaces. The interactions at the other weaker interfaces are shorter lived, and these may account for the observations of the single molecule experiments. A dimer/tetramer model of oligomerization would also be comparable to the inferences the Scarlata lab in collaboration with us derived (Golebiewska et al. 2011) from the diffusion rates and number and brightness measurements from fluorescence imaging of tagged MOP and DOP receptors in live cells after chronic morphine treatment.

Periole and colleagues compare their results to 2-dimensional rows of dimers of rhodopsin, as suggested by AFM studies of rhodopsin in native disk membranes (Fotiadis et al. 2003). To further test this hypothesis, they built and simulated a system of rows of TM1/H8 dimers interacting in a lipid-mediated or "lubricated" manner with adjacent rows of receptors, matching the structural information derived from the AFM study (Fotiadis et al. 2003). Their system remained stable after 16 μs of simulation (Periole et al. 2012).

Other information from structural studies supports the TM1/H8 interaction as a contact interface with putative relevance. Two-dimensional electron crystallography of rhodopsin (Schertler and Hargrave 1995) and of metarhodopsin I (Ruprecht et al. 2004) and X-ray crystallography of rhodopsin (Salom et al. 2006), opsin (Park and Schulten 2004) and metarhodopsin II (Choe et al. 2011) have hinted at this interaction, as have the most recent KOP (Wu et al. 2012) and MOP (Manglik et al. 2012) receptor X-ray crystal structures, although the physiological relevance of these arrangements remain to be demonstrated, in spite of recent experimental studies (Knepp et al. 2012) leaning in the same direction.

6.5 Conclusions

Throughout this chapter, we have discussed the ways in which biased MD methods are increasingly becoming invaluable tools to investigate, at the detailed molecular level, key events in the lifetime of GPCRs that take place on timescales in nature that are not routinely accessible to standard approaches. As these techniques become more theoretically comprehensive and computational power becomes greater, the combination of these two factors will progressively reveal important mechanistic details of events in the lives of GPCR signalosomes.

Acknowledgements The authors' work on GPCRs is currently supported by NIH grants DA026434 and DA034049. Their computations are run, in part, on resources available through the Scientific Computing Facility at Icahn School of Medicine at Mount Sinai, and in part on advanced computing resources provided by Texas Advanced Computing Center through MCB080077.

References

Allen TW, Andersen OS, Roux B (2004) Energetics of ion conduction through the gramicidin channel. Proc Natl Acad Sci USA 101:117–122

Altenbach C, Kusnetzow AK, Ernst OP, Hofmann KP, Hubbell WL (2008) High-resolution distance mapping in rhodopsin reveals the pattern of helix movement due to activation. Proc Natl Acad Sci USA 105:7439–7444

Baba M, Nishimura O, Kanzaki N, Okamoto M, Sawada H, Iizawa Y, Shiraishi M, Aramaki Y, Okonogi K, Ogawa Y et al (1999) A small-molecule, nonpeptide CCR5 antagonist with highly potent and selective anti-HIV-1 activity. Proc Natl Acad Sci USA 96:5698–5703

Ballesteros JA, Weinstein H (1995) Integrated methods for the construction of three-dimensional models and computational probing of structure-function relations in G protein-coupled receptors. In: Sealfon SC, Conn PM (eds) Methods in neurosciences, vol 25. Academic, San Diego, pp 366–428

Barak LS, Ferguson SSG, Zhang J, Martenson C, Meyer T, Caron MG (1997) Internal trafficking and surface mobility of a functionally intact beta(2)-adrenergic receptor-green fluorescent protein conjugate. Mol Pharmacol 51:177–184

Barducci A, Bussi G, Parrinello M (2008) Well-tempered metadynamics: a smoothly converging and tunable free-energy method. Phys Rev Lett 100:020603

Befort K, Tabbara L, Bausch S, Chavkin C, Evans C, Kieffer B (1996a) The conserved aspartate residue in the third putative transmembrane domain of the delta-opioid receptor is not the anionic counterpart for cationic opiate binding but is a constituent of the receptor binding site. Mol Pharmacol 49:216–223

Befort K, Tabbara L, Kling D, Maigret B, Kieffer BL (1996b) Role of aromatic transmembrane residues of the delta-opioid receptor in ligand recognition. J Biol Chem 271:10161–10168

Bonomi M, Branduardi D, Bussi G, Camilloni C, Provasi D, Raiteri P, Donadio D, Marinelli F, Pietrucci F, Broglia RA et al (2009) PLUMED: a portable plugin for free-energy calculations with molecular dynamics. Comput Phys Commun 180:1961–1972

Bot G, Blake AD, Li SX, Reisine T (1998) Mutagenesis of a single amino acid in the rat mu-opioid receptor discriminates ligand binding. J Neurochem 70: 358–365

Bowers KJ, Chow E, Xu H, Dror RO, Eastwood MP, Gregersen BA, Klepeis KL, Kolossváry I, Moraes MA, Sacerdoti FD, Salmon JK, Shan Y, Shaw DE (2006) Scalable algorithms for molecular dynamics simulations on commodity clusters. In: Proceedings of the ACM/IEEE conference on supercomputing (SC06), Tampa, 11–17 November 2006

Branduardi D, Gervasio FL, Parrinello M (2007) From A to B in free energy space. J Chem Phys 126:054103

Brooks BR, Bruccoleri RE, Olafson BD, States DJ, Swaminathan S, Karplus M (1983) Charmm – a program for macromolecular energy, minimization, and dynamics calculations. J Comput Chem 4:187–217

Callaerts-Vegh Z, Evans KL, Dudekula N, Cuba D, Knoll BJ, Callaerts PF, Giles H, Shardonofsky FR, Bond RA (2004) Effects of acute and chronic administration of beta-adrenoceptor ligands on airway function in a murine model of asthma. Proc Natl Acad Sci USA 101:4948–4953

Cherezov V, Rosenbaum DM, Hanson MA, Rasmussen SGF, Thian FS, Kobilka TS, Choi HJ, Kuhn P, Weis WI, Kobilka BK et al (2007) High-resolution crystal structure of an engineered human beta(2)-adrenergic G protein-coupled receptor. Science 318:1258–1265

Choe HW, Kim YJ, Park JH, Morizumi T, Pai EF, Krauss N, Hofmann KP, Scheerer P, Ernst OP (2011) Crystal structure of metarhodopsin II. Nature 471:651–655

Chow E, Xu H, Dror RO, Eastwood MP, Gregersen BA, Klepeis JL, Kolossvary I, Moraes MA, Sacerdoti FD, Salmon JK et al (2006) In: Scalable algorithms for molecular dynamics simulation on commodity clusters

Deupi X, Kobilka BK (2010) Energy landscapes as a tool to integrate GPCR structure, dynamics, and function. Physiology (Bethesda) 25:293–303

Deupi X, Edwards P, Singhal A, Nickle B, Oprian D, Schertler G, Standfuss J (2012) Stabilized G protein binding site in the structure of constitutively active metarhodopsin-II. Proc Natl Acad Sci USA 109:119–124

Dror RO, Pan AC, Arlow DH, Borhani DW, Maragakis P, Shan Y, Xu H, Shaw DE (2011) Pathway and mechanism of drug binding to G-protein-coupled receptors. Proc Natl Acad Sci USA 108:13118–13123

Elster L, Elling C, Heding A (2007) Bioluminescence resonance energy transfer as a screening assay: focus on partial and inverse agonism. J Biomol Screen 12:41–49

Eswar N, Webb B, Marti-Renom MA, Madhusudhan MS, Eramian D, Shen MY, Pieper U, Sali A (2007) Comparative protein structure modeling using MODELLER. Curr Protoc Protein Sci, Chapter 2:Unit 2 9

Fiser A, Sali A (2003) MODELLER: generation and refinement of homology-based protein structure models. In Macromolecular crystallography, Pt D. Method Enzymol 374: 461-+

Fishelovitch D, Shaik S, Wolfson HJ, Nussinov R (2009) Theoretical characterization of substrate access/exit channels in the human cytochrome P450 3A4 enzyme: involvement of phenylalanine residues in the gating mechanism. J Phys Chem B 113:13018–13025

Fonseca JM, Lambert NA (2009) Instability of a class A G protein-coupled receptor oligomer interface. Mol Pharmacol 75:1296–1299

Fotiadis D, Liang Y, Filipek S, Saperstein DA, Engel A, Palczewski K (2003) Atomic-force microscopy: rhodopsin dimers in native disc membranes. Nature 421:127–128

Frauenfelder H, Sligar SG, Wolynes PG (1991) The energy landscapes and motions of proteins. Science 254:1598–1603

Fribourg M, Moreno JL, Holloway T, Provasi D, Baki L, Mahajan R, Park G, Adney SK, Hatcher C, Eltit JM et al (2011) Decoding the signaling of a GPCR heteromeric complex reveals a unifying mechanism of action of antipsychotic drugs. Cell 147:1011–1023

Friesner RA, Banks JL, Murphy RB, Halgren TA, Klicic JJ, Mainz DT, Repasky MP, Knoll EH, Shelley M, Perry JK et al (2004) Glide: a new approach for rapid, accurate docking and scoring. 1. Method and assessment of docking accuracy. J Med Chem 47:1739–1749

Friesner RA, Murphy RB, Repasky MP, Frye LL, Greenwood JR, Halgren TA, Sanschagrin PC, Mainz DT (2006) Extra precision glide: docking and scoring incorporating a model of hydrophobic enclosure for protein-ligand complexes. J Med Chem 49:6177–6196

Golebiewska U, Johnston JM, Devi L, Filizola M, Scarlata S (2011) Differential response to morphine of the oligomeric state of mu-opioid in the presence of delta-opioid receptors. Biochemistry 50:2829–2837

Gonzalez A, Perez-Acle T, Pardo L, Deupi X (2011) Molecular basis of ligand dissociation in beta-adrenergic receptors. PLoS One 6:e23815

Granier S, Manglik A, Kruse AC, Kobilka TS, Thian FS, Weis WI, Kobilka BK (2012) Structure of the delta-opioid receptor bound to naltrindole. Nature 485:400–404

Grossfield A, Pitman MC, Feller SE, Soubias O, Gawrisch K (2008) Internal hydration increases during activation of the G-protein-coupled receptor rhodopsin. J Mol Biol 381:478–486

Guo W, Shi L, Filizola M, Weinstein H, Javitch JA (2005) Crosstalk in G protein-coupled receptors: changes at the transmembrane homodimer interface determine activation. Proc Natl Acad Sci USA 102:17495–17500

Guo W, Urizar E, Kralikova M, Mobarec JC, Shi L, Filizola M, Javitch JA (2008) Dopamine D2 receptors form higher order oligomers at physiological expression levels. EMBO J 27:2293–2304

Halgren TA, Murphy RB, Friesner RA, Beard HS, Frye LL, Pollard WT, Banks JL (2004) Glide: a new approach for rapid, accurate docking and scoring. 2. Enrichment factors in database screening. J Med Chem 47:1750–1759

Hegener O, Prenner L, Runkel F, Baader SL, Kappler J, Haberlein H (2004) Dynamics of beta(2)-adrenergic receptor – ligand complexes on living cells. Biochemistry 43:6190–6199

Henis YI, Hekman M, Elson EL, Helmreich EJM (1982) Lateral motion of beta-receptors in membranes of cultured liver-cells. Proc Natl Acad Sci USA 79:2907–2911

Hern JA, Baig AH, Mashanov GI, Birdsall B, Corrie JET, Lazareno S, Molloy JE, Birdsall NJM (2010) Formation and dissociation of M-1 muscarinic receptor dimers seen by total internal reflection fluorescence imaging of single molecules. Proc Natl Acad Sci USA 107:2693–2698

Hildebrand PW, Scheerer P, Park JH, Choe HW, Piechnick R, Ernst OP, Hofmann KP, Heck M (2009) A ligand channel through the G protein coupled receptor opsin. PLoS One 4:e4382

Holst B, Nygaard R, Valentin-Hansen L, Bach A, Engelstoft MS, Petersen PS, Frimurer TM, Schwartz TW (2010) A conserved aromatic lock for the tryptophan rotameric switch in TM-VI of seven-transmembrane receptors. J Biol Chem 285:3973–3985

Hopkinson HE, Latif ML, Hill SJ (2000) Non-competitive antagonism of beta(2)-agonist-mediated cyclic AMP accumulation by ICI 118551 in BC3H1 cells endogenously expressing constitutively active beta(2)-adrenoceptors. Br J Pharmacol 131:124–130

Huber T, Sakmar TP (2011) Escaping the flatlands: new approaches for studying the dynamic assembly and activation of GPCR signaling complexes. Trends Pharmacol Sci 32:410–419

Hurst DP, Grossfield A, Lynch DL, Feller S, Romo TD, Gawrisch K, Pitman MC, Reggio PH (2010) A lipid pathway for ligand binding is necessary for a cannabinoid G protein-coupled receptor. J Biol Chem 285:17954–17964

Im WP, Lee MS, Brooks CL (2003) Generalized born model with a simple smoothing function. J Comput Chem 24:1691–1702

Im W, Feig M, Brooks CL (2004) An implicit membrane generalized born theory for the study of structure, stability, and interactions of membrane proteins (vol 85, pg 2900, 2003). Biophys J 86:3330–3330

Irwin JJ, Sterling T, Mysinger MM, Bolstad ES, Coleman RG (2012) ZINC: a free tool to discover chemistry for biology. J Chem Inf Model 52:1757–1768

Isberg V, Balle T, Sander T, Jorgensen FS, Gloriam DE (2011) G protein- and agonist-bound serotonin 5-HT2A receptor model activated by steered molecular dynamics simulations. J Chem Inf Model 51:315–325

Isralewitz B, Izrailev S, Schulten K (1997) Binding pathway of retinal to bacterio-opsin: a prediction by molecular dynamics simulations. Biophys J 73:2972–2979

Jensen MO, Park S, Tajkhorshid E, Schulten K (2002) Energetics of glycerol conduction through aquaglyceroporin GlpF. Proc Natl Acad Sci USA 99:6731–6736

Johnston JM, Aburi M, Provasi D, Bortolato A, Urizar E, Lambert NA, Javitch JA, Filizola M (2011) Making structural sense of dimerization interfaces of delta opioid receptor homodimers. Biochemistry 50:1682–1690

Johnston JM, Wang H, Provasi D, Filizola M (2012) Assessing the relative stability of dimer interfaces in g protein-coupled receptors. PLoS Comput Biol 8:e1002649

Jorgensen WL, Maxwell DS, TiradoRives J (1996) Development and testing of the OPLS all-atom force field on conformational energetics and properties of organic liquids. J Am Chem Soc 118:11225–11236

Kabsch W, Sander C (1983) Dictionary of protein secondary structure – pattern-recognition of hydrogen-bonded and geometrical features. Biopolymers 22:2577–2637

Kaminski GA, Friesner RA, Tirado-Rives J, Jorgensen WL (2001) Evaluation and reparametrization of the OPLS-AA force field for proteins via comparison with accurate quantum chemical calculations on peptides. J Phys Chem B 105:6474–6487

Kasai RS, Suzuki KG, Prossnitz ER, Koyama-Honda I, Nakada C, Fujiwara TK, Kusumi A (2011) Full characterization of GPCR monomer-dimer dynamic equilibrium by single molecule imaging. J Cell Biol 192:463–480

Katritch V, Cherezov V, Stevens RC (2012) Structure-function of the G protein-coupled receptor superfamily. Annu Rev Pharmacol Toxicol 53:531–556

Kimura SR, Tebben AJ, Langley DR (2008) Expanding GPCR homology model binding sites via a balloon potential: a molecular dynamics refinement approach. Proteins 71:1919–1929

Kirkwood JG (1935) Statistical mechanics of fluid mixtures. J Chem Phys 3:300–313

Knepp AM, Periole X, Marrink SJ, Sakmar TP, Huber T (2012) Rhodopsin forms a dimer with cytoplasmic helix 8 contacts in native membranes. Biochemistry 51:1819–1821

Knierim B, Hofmann KP, Ernst OP, Hubbell WL (2007) Sequence of late molecular events in the activation of rhodopsin. Proc Natl Acad Sci USA 104: 20290–20295

Krystek SR Jr, Kimura SR, Tebben AJ (2006) Modeling and active site refinement for G protein-coupled receptors: application to the beta-2 adrenergic receptor. J Comput Aided Mol Des 20:463–470

Kumar S, Bouzida D, Swendsen RH, Kollman PA, Rosenberg JM (1992) The weighted histogram analysis method for free-energy calculations on biomolecules.1. The method. J Comput Chem 13:1011–1021

Kumar S, Rosenberg JM, Bouzida D, Swendsen RH, Kollman PA (1995) Multidimensional free-energy calculations using the weighted histogram analysis method. J Comput Chem 16:1339–1350

Kusumi A, Hyde JS (1982) Spin-label saturation-transfer electron-spin resonance detection of transient association of rhodopsin in reconstituted membranes. Biochemistry 21:5978–5983

Laio A, Parrinello M (2002) Escaping free-energy minima. Proc Natl Acad Sci USA 99:12562–12566

Lambert NA (2010) Gpcr dimers fall apart. Sci Signal 3:pe12

Lambright DG, Sondek J, Bohm A, Skiba NP, Hamm HE, Sigler PB (1996) The 2.0 A crystal structure of a heterotrimeric G protein. Nature 379:311–319

Lebon G, Warne T, Edwards PC, Bennett K, Langmead CJ, Leslie AG, Tate CG (2011) Agonist-bound adenosine A2A receptor structures reveal common features of GPCR activation. Nature 474:521–525

Leone V, Marinelli F, Carloni P, Parrinello M (2010) Targeting biomolecular flexibility with metadynamics. Curr Opin Struct Biol 20:148–154

Li JG, Chen CG, Yin JL, Rice K, Zhang Y, Matecka D, de Riel JK, DesJarlais RL, Liu-Chen LY (1999) Asp147 in the third transmembrane helix of the rat mu opioid receptor forms ion-pairing with morphine and naltrexone. Life Sci 65:175–185

Lopez CA, Rzepiela AJ, de Vries AH, Dijkhuizen L, Hunenberger PH, Marrink SJ (2009) Martini coarse-grained force field: extension to carbohydrates. J Chem Theor Comput 5:3195–3210

Ludemann SK, Lounnas V, Wade RC (2000) How do substrates enter and products exit the buried active site of cytochrome P450cam? 1. Random expulsion molecular dynamics investigation of ligand access channels and mechanisms. J Mol Biol 303:797–811

MacKerell AD, Bashford D, Bellott M, Dunbrack RL, Evanseck JD, Field MJ, Fischer S, Gao J, Guo H, Ha S et al (1998) All-atom empirical potential for molecular modeling and dynamics studies of proteins. J Phys Chem B 102:3586–3616

Manglik A, Kruse AC, Kobilka TS, Thian FS, Mathiesen JM, Sunahara RK, Pardo L, Weis WI, Kobilka BK, Granier S (2012) Crystal structure of the micro-opioid receptor bound to a morphinan antagonist. Nature 485:321–326

Mansour A, Taylor LP, Fine JL, Thompson RC, Hoversten MT, Mosberg HI, Watson SJ, Akil H (1997) Key residues defining the mu-opioid receptor binding pocket: a site-directed mutagenesis study. J Neurochem 68:344–353

Marchi M, Ballone P (1999) Adiabatic bias molecular dynamics: a method to navigate the conformational space of complex molecular systems. J Chem Phys 110:3697–3702

Marrink SJ, de Vries AH, Mark AE (2004) Coarse grained model for semiquantitative lipid simulations. J Phys Chem B 108:750–760

Marrink SJ, Risselada HJ, Yefimov S, Tieleman DP, de Vries AH (2007) The MARTINI force field: coarse grained model for biomolecular simulations. J Phys Chem B 111:7812–7824

Marrink SJ, Fuhrmans M, Risselada HJ, Periole X (2008) The MARTINI forcefield. In: Voth G (ed) Coarse graining of condensed phase and biomolecular systems. CRC Press, Boca Raton

Marrink SJ, Periole X, Tieleman DP, de Vries AH (2009) Comment on "On using a too large integration time step in molecular dynamics simulations of coarse-grained molecular models" by M. Winger, D. Trzesniak, R. Baron and W. F. van Gunsteren, Phys Chem Chem Phys 11:1934. Phys Chem Chem Phys 12:2254–2256; author reply 2257–2258

Milligan G (2009) G protein-coupled receptor hetero-dimerization: contribution to pharmacology and function. Br J Pharmacol 158:5–14

Monticelli L, Kandasamy SK, Periole X, Larson RG, Tieleman DP, Marrink SJ (2008) The MARTINI coarse-grained force field: extension to proteins. J Chem Theor Comput 4:819–834

Morris GM, Goodsell DS, Huey R, Olson AJ (1996) Distributed automated docking of flexible ligands to proteins: parallel applications of AutoDock 2.4. J Comput Aided Mol Des 10:293–304

Okada T, Fujiyoshi Y, Silow M, Navarro J, Landau EM, Shichida Y (2002) Functional role of internal water molecules in rhodopsin revealed by x-ray crystallography. Proc Natl Acad Sci USA 99: 5982–5987

Okada T, Sugihara M, Bondar AN, Elstner M, Entel P, Buss V (2004) The retinal conformation and its environment in rhodopsin in light of a new 2.2 Angstrom crystal structure. J Mol Biol 342:571–583

Olausson BE, Grossfield A, Pitman MC, Brown MF, Feller SE, Vogel A (2012) Molecular dynamics simulations reveal specific interactions of post-translational palmitoyl modifications with rhodopsin in membranes. J Am Chem Soc 134:4324–4331

Park S, Schulten K (2004) Calculating potentials of mean force from steered molecular dynamics simulations. J Chem Phys 120:5946–5961

Park S, Khalili-Araghi F, Tajkhorshid E, Schulten K (2003) Free energy calculation from steered molecular dynamics simulations using Jarzynski's equality. J Chem Phys 119:3559

Park JH, Scheerer P, Hofmann KP, Choe HW, Ernst OP (2008) Crystal structure of the ligand-free G-protein-coupled receptor opsin. Nature 454:183–187

Periole X, Huber T, Marrink SJ, Sakmar TP (2007) G protein-coupled receptors self-assemble in dynamics simulations of model bilayers. J Am Chem Soc 129:10126–10132

Periole X, Cavalli M, Marrink SJ, Ceruso MA (2009) Combining an elastic network with a coarse-grained molecular force field: structure, dynamics, and intermolecular recognition. J Chem Theor Comput 5:2531–2543

Periole X, Knepp AM, Sakmar TP, Marrink SJ, Huber T (2012) Structural determinants of the supramolecular organization of G protein-coupled receptors in bilayers. J Am Chem Soc 134:10959–10965

Petrek M, Otyepka M, Banas P, Kosinova P, Koca J, Damborsky J (2006) CAVER: a new tool to explore routes from protein clefts, pockets and cavities. BMC Bioinformatics 7:316

Phillips JC, Braun R, Wang W, Gumbart J, Tajkhorshid E, Villa E, Chipot C, Skeel RD, Kale L, Schulten K (2005) Scalable molecular dynamics with NAMD. J Comput Chem 26:1781–1802

Poo MM, Cone RA (1973) Lateral diffusion of rhodopsin in the visual receptor membrane. J Supramol Struct 1:354

Provasi D, Filizola M (2010) Putative active states of a prototypic G-protein-coupled receptor from biased molecular dynamics. Biophys J 98:2347–2355

Provasi D, Bortolato A, Filizola M (2009) Exploring molecular mechanisms of ligand recognition by opioid receptors with metadynamics. Biochemistry 48:10020–10029

Provasi D, Johnston JM, Filizola M (2010) Lessons from free energy simulations of delta-opioid receptor homodimers involving the fourth transmembrane helix. Biochemistry 49:6771–6776

Provasi D, Artacho MC, Negri A, Mobarec JC, Filizola M (2011) Ligand-induced modulation of the free-energy landscape of G protein-coupled receptors explored by adaptive biasing techniques. PLoS Comput Biol 7:e1002193

Raiteri P, Laio A, Gervasio FL, Micheletti C, Parrinello M (2006) Efficient reconstruction of complex free energy landscapes by multiple walkers metadynamics. J Phys Chem B 110:3533–3539

Rasmussen SGF, Choi HJ, Fung JJ, Pardon E, Casarosa P, Chae PS, DeVree BT, Rosenbaum DM, Thian FS, Kobilka TS et al (2011a) Structure of a nanobody-stabilized active state of the beta(2) adrenoceptor. Nature 469:175–180

Rasmussen SG, DeVree BT, Zou Y, Kruse AC, Chung KY, Kobilka TS, Thian FS, Chae PS, Pardon E, Calinski D et al (2011b) Crystal structure of the beta2 adrenergic receptor-Gs protein complex. Nature 477:549–555

Rives ML, Rossillo M, Liu-Chen LY, Javitch JA (2012) 6′-Guanidinonaltrindole (6′-GNTI) is a G protein-based kappa-opioid receptor agonist that inhibits arrestin recruitment. J Biol Chem 287:27050–27054

Rosenbaum DM, Zhang C, Lyons JA, Holl R, Aragao D, Arlow DH, Rasmussen SGF, Choi HJ, DeVree BT, Sunahara RK et al (2011) Structure and function of an irreversible agonist-beta(2) adrenoceptor complex. Nature 469:236–240

Roux B (1995) The calculation of the potential of mean force using computer-simulations. Comput Phys Commun 91:275–282

Roux B (1999) Statistical mechanical equilibrium theory of selective ion channels. Biophys J 77:139–153

Ruprecht JJ, Mielke T, Vogel R, Villa C, Schertler GF (2004) Electron crystallography reveals the structure of metarhodopsin I. EMBO J 23:3609–3620

Ryba NJP, Marsh D (1992) Protein rotational diffusion and lipid protein interactions in recombinants of bovine rhodopsin with saturated diacylphosphatidylcholines of different chain lengths studied by conventional and saturation-transfer electron-spin-resonance. Biochemistry 31:7511–7518

Salom D, Lodowski DT, Stenkamp RE, Le Trong I, Golczak M, Jastrzebska B, Harris T, Ballesteros JA, Palczewski K (2006) Crystal structure of a photoactivated deprotonated intermediate of rhodopsin. Proc Natl Acad Sci USA 103:16123–16128

Sauliere-Nzeh AN, Millot C, Corbani M, Mazeres S, Lopez A, Salome L (2010) Agonist-selective dynamic compartmentalization of human Mu opioid receptor as revealed by resolutive FRAP analysis. J Biol Chem 285:14514–14520

Scheerer P, Park JH, Hildebrand PW, Kim YJ, Krauss N, Choe HW, Hofmann KP, Ernst OP (2008) Crystal structure of opsin in its G-protein-interacting conformation. Nature 455:497–502

Schertler GF, Hargrave PA (1995) Projection structure of frog rhodopsin in two crystal forms. Proc Natl Acad Sci USA 92:11578–11582

Schlitter J, Engels M, Kruger P (1994) Targeted molecular-dynamics – a New approach for searching pathways of conformational transitions. J Mol Graph 12:84–89

Selvam B, Wereszczynski J, Tikhonova IG (2012) Comparison of dynamics of extracellular accesses to the beta(1) and beta(2) adrenoceptors binding sites uncovers the potential of kinetic basis of antagonist selectivity. Chem Biol Drug Des 80:215–226

Shi L, Liapakis G, Xu R, Guarnieri F, Ballesteros JA, Javitch JA (2002) Beta2 adrenergic receptor activation. Modulation of the proline kink in transmembrane 6 by a rotamer toggle switch. J Biol Chem 277:40989–40996

Shiota T (1999) Cyclic amine derivatives and their use as drugs. US Patent 1999

Shirts MR, Chodera JD (2008) Statistically optimal analysis of samples from multiple equilibrium states. J Chem Phys 129:124105

Simpson LM, Wall ID, Blaney FE, Reynolds CA (2011) Modeling GPCR active state conformations: the beta(2)-adrenergic receptor. Proteins 79:1441–1457

Spivak CE, Beglan CL, Seidleck BK, Hirshbein LD, Blaschak CJ, Uhl GR, Surratt CK (1997) Naloxone activation of mu-opioid receptors mutated at a histidine residue lining the opioid binding cavity. Mol Pharmacol 52:983–992

Strader CD, Sigal IS, Candelore MR, Rands E, Hill WS, Dixon RA (1988) Conserved aspartic acid residues 79 and 113 of the beta-adrenergic receptor have different roles in receptor function. J Biol Chem 263:10267–10271

Strader CD, Candelore MR, Hill WS, Sigal IS, Dixon RA (1989) Identification of two serine residues involved in agonist activation of the beta-adrenergic receptor. J Biol Chem 264:13572–13578

Surratt CK, Johnson PS, Moriwaki A, Seidleck BK, Blaschak CJ, Wang JB, Uhl GR (1994) Mu opiate receptor – charged transmembrane domain amino-acids are critical for agonist recognition and intrinsic activity. J Biol Chem 269:20548–20553

Taddese B, Simpson LM, Wall ID, Blaney FE, Kidley NJ, Clark HS, Smith RE, Upton GJ, Gouldson PR, Psaroudakis G et al (2012) G-protein-coupled receptor dynamics: dimerization and activation models compared with experiment. Biochem Soc Trans 40:394–399

Tesmer JJ, Sunahara RK, Gilman AG, Sprang SR (1997) Crystal structure of the catalytic domains of adenylyl cyclase in a complex with Gsalpha.GTPgammaS. Science 278:1907–1916

Thompson AA, Liu W, Chun E, Katritch V, Wu H, Vardy E, Huang XP, Trapella C, Guerrini R, Calo G et al (2012) Structure of the nociceptin/orphanin FQ receptor in complex with a peptide mimetic. Nature 485:395–399

Toll L, Berzetei-Gurske IP, Polgar WE, Brandt SR, Adapa ID, Rodriguez L, Schwartz RW, Haggart D, O'Brien A, White A et al (1998) Standard binding and functional assays related to medications development division testing for potential cocaine and opiate narcotic treatment medications. NIDA Res Monogr 178:440–466

Torrie GM, Valleau JP (1974) Monte-Carlo free-energy estimates using Non-Boltzmann sampling – application to subcritical Lennard-Jones fluid. Chem Phys Lett 28:578–581

Tozzini V (2005) Coarse-grained models for proteins. Curr Opin Struct Biol 15:144–150

Tozzini V (2010) Multiscale modeling of proteins. Accounts Chem Res 43:220–230

Vaidehi N, Kenakin T (2010) The role of conformational ensembles of seven transmembrane receptors in functional selectivity. Curr Opin Pharmacol 10:775–781

Van der Spoel D, Lindahl E, Hess B, Groenhof G, Mark AE, Berendsen HJC (2005) GROMACS: fast, flexible, and free. J Comput Chem 26:1701–1718

Vogel R, Ruprecht J, Villa C, Mielke T, Schertler GF, Siebert F (2004) Rhodopsin photoproducts in 2D crystals. J Mol Biol 338:597–609

Wall MA, Coleman DE, Lee E, Iniguez-Lluhi JA, Posner BA, Gilman AG, Sprang SR (1995) The structure of the G protein heterotrimer Gi alpha 1 beta 1 gamma 2. Cell 83:1047–1058

Wang T, Duan Y (2007) Chromophore channeling in the G-protein coupled receptor rhodopsin. J Am Chem Soc 129:6970–6971

Wang T, Duan Y (2009) Ligand entry and exit pathways in the beta(2)-adrenergic receptor. J Mol Biol 392:1102–1115

Wang C, Bradley P, Baker D (2007) Protein-protein docking with backbone flexibility. J Mol Biol 373:503–519

Warne T, Serrano-Vega MJ, Baker JG, Moukhametzianov R, Edwards PC, Henderson R, Leslie AGW, Tate CG, Schertler GFX (2008) Structure of a beta(1)-adrenergic G-protein-coupled receptor. Nature 454:486–491

Winger M, Trzesniak D, Baron R, van Gunsteren WF (2009) On using a too large integration time step in molecular dynamics simulations of coarse-grained molecular models. Phys Chem Chem Phys 11:1934–1941

Wu H, Wacker D, Mileni M, Katritch V, Han GW, Vardy E, Liu W, Thompson AA, Huang XP, Carroll FI et al (2012) Structure of the human kappa-opioid receptor in complex with JDTic. Nature 485:327–332

Xu F, Wu H, Katritch V, Han GW, Jacobson KA, Gao ZG, Cherezov V, Stevens RC (2011) Structure of an agonist-bound human A2A adenosine receptor. Science 332:322–327

Yang LJ, Zou J, Xie HZ, Li LL, Wei YQ, Yang SY (2009) Steered molecular dynamics simulations reveal the likelier dissociation pathway of imatinib from its targeting kinases c-Kit and Abl. PLoS One 4:e8470

Zhang C, Srinivasan Y, Arlow DH, Fung JJ, Palmer D, Zheng Y, Green HF, Pandey A, Dror RO, Shaw DE et al (2012) High-resolution crystal structure of human protease-activated receptor 1. Nature 492:387–392

Part III

GPCR-Focused Rational Design and Mathematical Modeling

From Three-Dimensional GPCR Structure to Rational Ligand Discovery

7

Albert J. Kooistra, Rob Leurs, Iwan J.P. de Esch, and Chris de Graaf

Abstract

This chapter will focus on G protein-coupled receptor structure-based virtual screening and ligand design. A generic virtual screening workflow and its individual elements will be introduced, covering amongst others the use of experimental data to steer the virtual screening process, ligand binding mode prediction, virtual screening for novel ligands, and rational structure-based virtual screening hit optimization. An overview of recent successful structure-based ligand discovery and design studies shows that receptor models, despite structural inaccuracies, can be efficiently used to find novel ligands for GPCRs. Moreover, the recently solved GPCR crystal structures have further increased the opportunities in structure-based ligand discovery for this pharmaceutically important protein family. The current chapter will discuss several challenges in rational ligand discovery based on GPCR structures including: (i) structure-based identification of ligands with specific *effects* on GPCR mediated signaling pathways, and (ii) virtual screening and structure-based optimization of *fragment*-like molecules.

Keywords

Drug design • In Silico methods • Virtual screening • Docking • Crystal structures

7.1 From *In Crystallo* to *In Silico*: GPCR Structures for Rational Ligand Discovery

Knowledge of the three-dimensional structure of G protein-coupled receptors (GPCRs) provides important insights into receptor function and receptor-ligand interactions. This information is key for the *in silico* rational discovery

A.J. Kooistra • R. Leurs • I.J.P. de Esch • C. de Graaf (✉)
Division of Medicinal Chemistry, Faculty of Sciences, Amsterdam Institute for Molecules, Medicines and Systems (AIMMS), VU University Amsterdam, De Boelelaan 1083, 1081 HV Amsterdam, The Netherlands
e-mail: c.de.graaf@vu.nl

M. Filizola (ed.), *G Protein-Coupled Receptors - Modeling and Simulation*, Advances in Experimental Medicine and Biology 796, DOI 10.1007/978-94-007-7423-0_7,

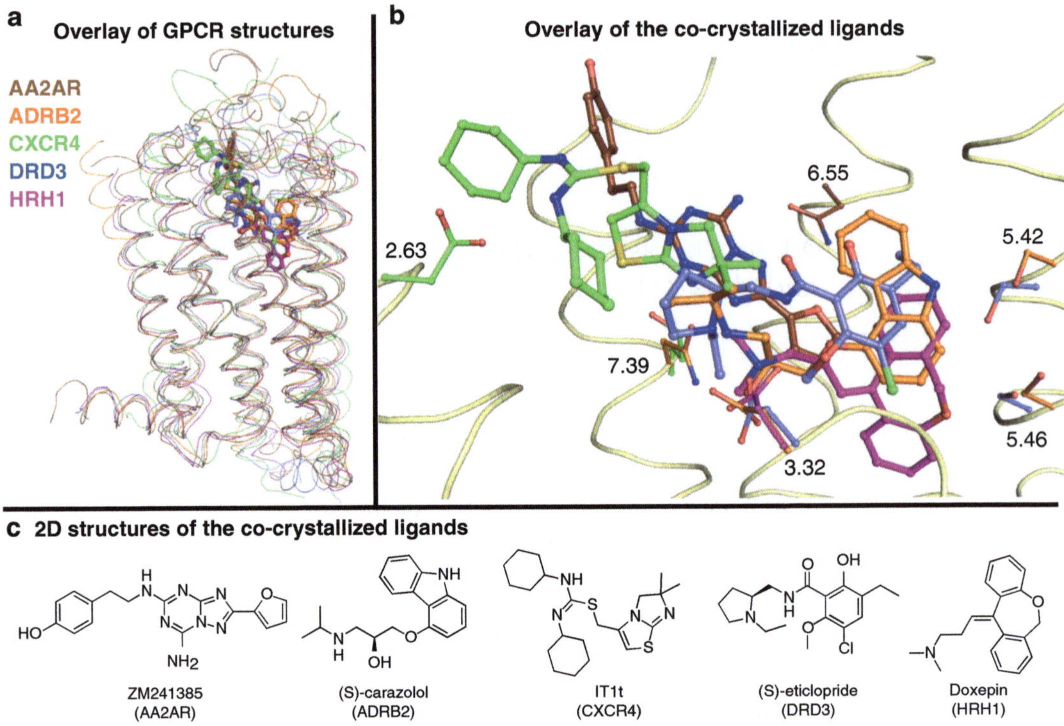

Fig. 7.1 (**a**) Structural alignment of multiple GPCR crystal structures (PDB-codes AA2AR:3EML (Jaakola et al. 2008), ADRB2:2RH1 (Cherezov et al. 2007), CXCR4:3ODU (Wu et al. 2010), DRD3:3PBL (Chien et al. 2010), HRH1:3RZE (Shimamura et al. 2011)) highlight the highly conserved TM fold (**b**) Close-up of the co-crystallized ligands within the conserved binding site. The backbone of carazolol bound ADRB2 (PDB-code: 2RH1 (Cherezov et al. 2010)) is depicted as a light yellow ribbon. Residue positions are indicated using the Ballesteros Weinstein numbering scheme (Ballesteros and Weinstein 1995) and both the ligands and selected residues are colored according to the coloring scheme depicted in **a**. (**c**) 2D structures of the co-crystallized ligands depicted in **b**

of new bioactive molecules that can target this family of pharmaceutically relevant drug targets (Congreve et al. 2011; de Graaf and Rognan 2009). After the first GPCR crystal structure of bovine rhodopsin in 2000 (Palczewski et al. 2000), the first crystal structures of druggable GPCRs have been solved only in the past 6 years (Salon et al. 2011), including beta1 (ADRB1) (Warne et al. 2008) and beta2 (ADRB2) (Cherezov et al. 2007) adrenergic receptors, dopamine D3 receptor (DRD3) (Chien et al. 2010), histamine H1 receptor (HRH1) (Shimamura et al. 2011), and muscarinic M2 (Haga et al. 2012) and M3 (Kruse et al. 2012) receptors, the adenosine A2A receptor (AA2AR) (Jaakola et al. 2008), the chemokine CXCR4 receptor (Wu et al. 2010), the neurotensin 1 receptor (NTR1) (White et al. 2012), the lipid S1PR1 receptor (Hanson et al. 2012), the δ (OPRD) (Granier et al. 2012), μ (OPRM) (Manglik et al. 2012), κ (OPRK) (Wu et al. 2012), and nociceptin (OPRX) (Thompson et al. 2012) opioid receptors, and the protease-activated receptor 1 (PAR1) (White et al. 2012). These GPCR crystal structures (Salon et al. 2011; Katritch et al. 2012) offer unique opportunities to push the limits of structure-based rational ligand discovery and design (Congreve et al. 2011; Salon et al. 2011), and offer higher resolution templates for modeling the structures of GPCRs for which crystal structures have not yet been solved (Katritch et al. 2012; Stevens et al. 2012; Rodriguez and Gutierrez-de-Teran 2013). It should however be noted that modeling of GPCRs with low homology to the currently available GPCR crystal structures (e.g., class B (Miller et al. 2011; Vohra et al. 2013; de Graaf et al. 2011a) and class C GPCRs (Petrel et al. 2004; Malherbe et al. 2006; Gloriam et al. 2011)) still remains a difficult task in which experimental data are of utmost importance to restrict the number of possible models (de Graaf and Rognan 2009; Roumen et al. 2011). The challenges of GPCR modeling has been for example demonstrated in recent community wide competitions to predict GPCR crystal structures (GPCR DOCK 2008 (Katritch et al. 2010a; Michino and Abola 2009), and GPCR DOCK 2010 (Roumen et al. 2011; Kufareva et al. 2011)) and GPCR modeling methods have been described in several reviews (de Graaf and Rognan 2009; de Graaf et al. 2011a; Cavasotto 2011; Kooistra et al. 2013). The current chapter gives an overview of the challenges and opportunities in structure-based discovery of GPCR ligands and focuses in particular on the developments in the period over the past 5 years in which the first druggable GPCR crystal structures have been used for rational GPCR ligand design. Different steps along the virtual screening workflow will be discussed in Sect. 7.2. An overview of several successful structure-based ligand discovery studies in Sect. 7.3 shows that GPCR models, despite structural inaccuracies, can be efficiently used to find novel ligands for GPCRs. Moreover, the recently solved GPCR crystal structures have further increased the opportunities in structure-based discovery of small molecule ligands for this pharmaceutically important protein family. All crystal structures clearly show the conserved heptahelical fold in the TM domain (Fig. 7.1a) as well as for the intracellular helix 8.

The loops on the other hand differ highly between GPCRs in sequence composition, length, and (secondary) structure. Apart from that they have also shown to be harder to resolve in crystal structures as the electron density is distorted in this region, as is the case for ECL2 in, for example, the HRH1 structure as well as some AA2AR crystal structures. Within the conserved 7 TM helices and below the extracellular loops lies the conserved orthosteric binding site of GPCRs (Fig. 7.1b) which can bind a plethora of different small molecules as well as peptides (Fig. 7.1c) (Surgand et al. 2006; Lagerstrom and Schioth 2008).

7.2 Hierarchical Workflow for GPCR Structure-Based Ligand Discovery

This section will describe the steps along a hierarchical virtual screening workflow applied in many GPCR virtual screening studies (Fig. 7.2). Although the different steps can be generally applied to other protein targets, it should be noted that incorporation of target specific information can highly increase the chance on success. In step **1** the initial compound library (Moura Barbosa and Del Rio 2012) can be filtered to remove undesirable compounds that contain chemical moieties that can interfere with experimental validation assays by forming aggregates, chemically reacting with proteins or directly interfere in assay signaling (Baell and Holloway 2010). Furthermore one can decide to exclude com pounds that contain scaffolds associated with toxicity (Muegge 2003) or that have poor oral bioavailability (Lipinski et al. 2001). In most reported GPCR-based virtual screening studies, such pre-filters are applied to construct the chemical library to be screened (Table 7.1). Additional filters, based on the properties (e.g., molecular weight, number of rotatable bonds, number of rings, hydrogen bond donor/acceptor counts, number of positively/negatively charged atoms, etc.) of a set of known actives (step **1** of Fig. 7.2), are applied in most GPCR VS studies as well (Table 7.1) to obtain a more focused chemical database. Substructures (Varady et al. 2003; Evers and Klabunde 2005; Evers and Klebe 2004; Kellenberger et al. 2007), chemical similarity descriptors (Evers and Klebe 2004; Cavasotto et al. 2008; Tikhonova et al. 2008) and 3D-shape similarity or pharmacophore models (de Graaf et al. 2011a; Varady et al. 2003; Evers and Klabunde 2005; Evers and Klebe 2004; Cavasotto et al. 2008) (Table 7.2) derived from known ligands can be used to narrow down the number of compounds to be handled during conformational sam-

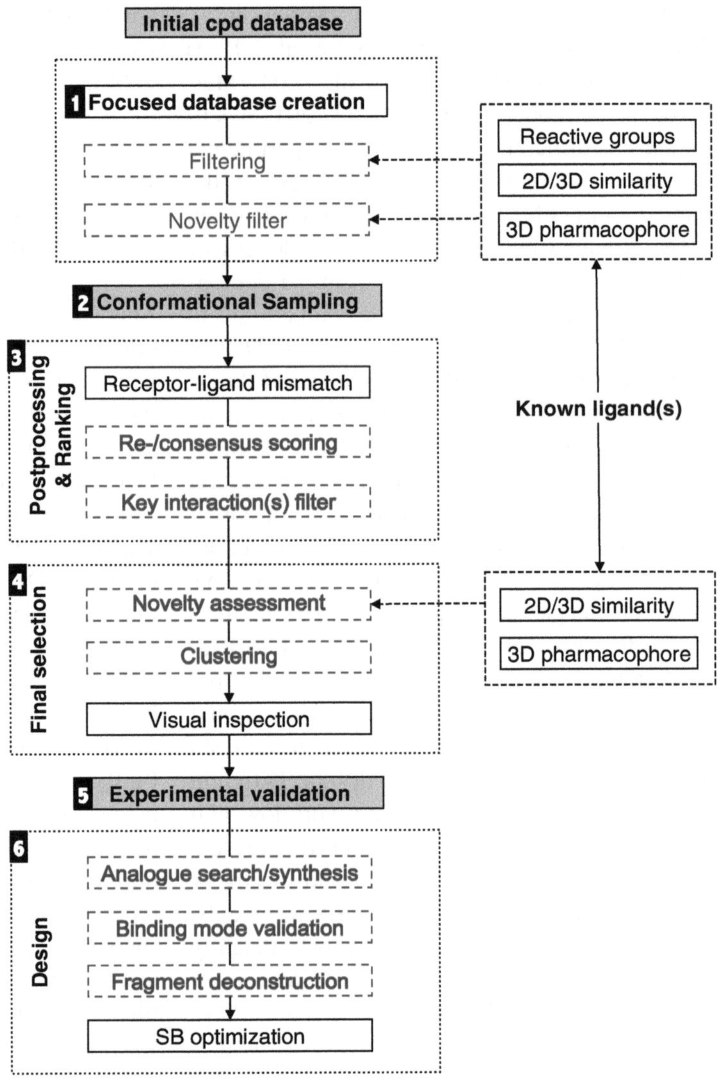

Fig. 7.2 Structure-based virtual screening (*SBVS*) workflow (see Sect. 7.2 for description of the individual steps, Sect. 7.3 and Tables 7.1 and 7.2 for details of SBVS runs against specific GPCRs, and Sect. 7.4 for the discussion of step 6 observed in reported SBVS studies)

pling (step **2**) even further. These methods can also be used inversely in order to limit the focused database to novel ligands, i.e. small molecules that have a low similarity (as assessed by the applied method) compared to known ligands (available in target annotated chemical databases like WomBat (Olah et al. 2007), BindingDB (Liu et al. 2007) and ChEMBL (Gaulton et al. 2012)). However, more often such a filter to identify novel scaffolds is performed (although computationally more expensive) after docking the compounds in the database.

In step **2** the chemical database is automatically docked in the receptor model. Many different automated docking programs and scoring functions based on different physicochemical approximations are available (Moitessier et al. 2008) (Sect. 7.2). In some GPCR SBVS studies docking based screening simulations have been guided by pharmacophore constraints (Evers and Klabunde 2005; Evers and Klebe 2004; Sirci et al. 2012). Pharmacophore models and/or exclusion constraints derived from structural models of receptor-ligand complexes (or the receptor alone) can be used as an alternative structure-based virtual screening approach to molecular docking simulations, as demonstrated in several successful GPCR SBVS campaigns to discover ligands of CA3R (Klabunde et al. 2009), CNR2 (Salo et al. 2005), FFAR1 (Tikhonova et al. 2008), FPR1R (Edwards et al. 2005), MCHR1 (Cavasotto et al. 2008), HRH3 (Sirci et al. 2012).

In step **3**, the docking poses are post-processed and ranked. Many SBVS investigations have only employed docking scoring functions to rank docking poses, but more and more structure-based *in silico* screening protocols include additional filters to post-process docking results (Sect. 7.2), because the scoring accuracy of docking-scoring combinations is very much dependent on physicochemical details of target-ligand interactions and fine details of the protein structure. Strategies to overcome these problems are: (i) the use of consensus scoring strategies (Evers and Klabunde 2005; Bissantz et al. 2003; Chen et al. 2007; Kiss et al. 2008), (ii) topological filters to filter out poses exhibiting steric or electrostatic mismatches between the ligand and the binding site (de Graaf et al. 2011a), (iii) receptor-ligand interaction *post*-processing filters (Tables 7.1 and 7.2) (Carlsson et al. 2010, 2011; Langmead et al. 2012) and/or (iv) receptor-ligand interaction fingerprint (IFP) scoring methods to select and rank poses based on binding mode similarity with reference ligand poses (de Graaf and Rognan 2008, 2009; de Graaf et al. 2008, 2011a, b) (Sect. 7.2, Figs. 7.4 and 7.5).

Before selecting the final compounds (step **4**), the novelty of the discovered hits can be assessed by 2D or 3D similarity searches against previously known ligands (as discussed earlier), as performed in recent virtual screening studies against e.g. AA2AR (Carlsson et al. 2010; Katritch et al. 2010b), ADRB2 (Kolb et al. 2009), DRD3 (Carlsson et al. 2011), and HRH1 (de Graaf et al. 2011b) (Table 7.2). Moreover, 2D and 3D ligand-based similarity searches of the screening database against the reference ligand used to refine the receptor (or present in the receptor co-crystal structure) can be performed to demonstrate the strength of the SBVS approach (de Graaf et al. 2011a, b). When too many ligands are retrieved along the VS funnel, it is generally wise to cluster virtual hits by chemical diversity before visual inspection. Compounds can be classified by their chemical scaffold in order to prioritize scaffolds rather than individual compounds in screening ranking lists. Sampling a few representative analogues for each scaffold usually enables a selection of chemically dissimilar compounds for biological evaluation (Kellenberger et al. 2007). Finally, the selected docking poses and compounds should be visually inspected in step **4** for the ultimate selection: no algorithm yet outperforms the brain of an experienced modeler for such a task.

Although this chapter mainly focuses on SBVS studies (and primarily docking-based virtual screening campaigns) it should be noted, however, that there are alternative (or complementary) screening methods that have been successfully applied for the discovery of novel ligands (e.g. ligand-based similarity (2D and 3D), and receptor/ligand-based pharmacophore searches (Kellenberger et al. 2007; Cavasotto et al. 2008; Tikhonova et al. 2008; Klabunde et al. 2009; Salo et al. 2005; Edwards et al. 2005)). For 3D similarity and receptor-based pharmacophore searches the receptor-bound conformation of reference compounds can be derived from docking simulations in the receptor model. The obtained receptor-ligand complexes are used to derive the information for the 3D-similarity searches or the creation of pharmacophore models. Alternatively, pure receptor-based pharmacophore models, extracted from ligand-receptor interaction hot spots in the binding pocket, can be used even in the absence of a ligand (Sanders et al. 2011, 2012; Barillari et al. 2008), as successfully applied for the discovery of CA3R ligands (incl. **45**) (Klabunde et al. 2009). Frequently GPCR homology-model-based virtual screening studies combine multiple screening methods and filtering steps into a hierarchical workflow. In general, large chemical database are first filtered by application of a 2D and/or 3D pharmacophore model after which the focused library is subjected to molecular docking simulations (Table 7.1, Fig. 7.1). This type of hierarchical SBVS has frequently been applied to bRho-based homology models in order to find novel ligands (de Graaf and Rognan 2009). Apart from the hierarchical approach also integrated approaches are known in which pharmacophore constraints guide the conformational sampling of docking simulations, as applied in virtual screening campaigns against ADA1A (Evers et al. 2005) and NK1R (Evers and Klebe 2004). In general the more recent crystal structure-based SBVS campaigns do not apply ligand-based pharmacophores but instead use physicochemical property filters (e.g., heavy atom count, number of rings, hydrophobicity) and emphasize experimentally supported receptor-ligand interactions (e.g., the essential ionic interaction with $D^{3.32}$ for HRH1 (de Graaf et al. 2011b)) to score and select docking poses (Tables 7.1 and 7.2).

After these final steps of the virtual screening process a highly diminished subset of the original compound database (typically between 0.001 and 0.05 % of the original database, Table 7.1) is obtained and their (commercial) availability is checked. The available compounds are subsequently obtained and experimentally validated (step **5**). Many times the experimentally validated hits of these virtual screening studies are not further investigated, however more and more often the hits are only a starting point and used for further (structure-based) optimization (step **6**, Sect. 7.4).

7.3 *In Silico* Structure-Based GPCR Ligand Discovery

The recent crystal structure determinations of druggable class A GPCRs (Katritch et al. 2013) have now opened up excellent new opportunities to push forward the limit of crystal structure-based GPCR ligand discovery (Carlsson et al. 2010, 2011; Langmead et al. 2012; de Graaf et al. 2011b; Katritch et al. 2010b; Kolb et al. 2009; Topiol and Sabio 2008; Sabio et al. 2008). It should be noted, however, that despite their possible structural inaccuracies GPCR homology models can also be (and have been (de Graaf et al. 2011a, b; Varady et al. 2003; Evers and Klabunde 2005; Evers and Klebe 2004; Kellenberger et al. 2007; Cavasotto et al. 2008; Tikhonova et al. 2008; Sirci et al. 2012; Klabunde et al. 2009; Salo et al. 2005; Edwards et al. 2005; Carlsson et al. 2011; Langmead et al. 2012; Kolb et al. 2012; Lin et al. 2012; Istyastono 2012; Blättermann et al. 2012; Kim et al. 2012; Mysinger et al. 2012; Renault et al. 2012; Costanzi et al. 2012; Becker et al. 2004; Engel et al. 2008; Liu et al. 2008)) efficiently used to find new ligands. Moreover, the new GPCR crystal structures allow for the creation of higher resolution homology models as more templates are available.

As described in Sect. 7.2, many of the reported GPCR-structure-based SBVS campaigns included a customized hierarchical virtual screening workflow. Table 7.1 gives an overview of recent prospective structure-based VS studies against GPCR models (since our review of 2009 (de Graaf and Rognan 2009)). Table 7.2 presents the molecular structures of representative agonists as well as antagonists identified by prospective SBVS studies in GPCR models and crystal structures. Figure 7.3 shows the hit rates, ligand efficiency and size of experimentally validated hits in these *in silico* screening campaigns. It should be noticed that SBVS often yields new chemical scaffolds (Table 7.2) that still contain essential functional groups like positively or negatively charged atoms that are used as substructure or pharmacophore filters/constraints to set up the initial ligand database or score/rank docking poses (Table 7.1). Most of the *in silico* screening studies have focused on bioaminergic receptors (Varady et al. 2003; Evers and Klabunde 2005; Carlsson et al. 2011; de Graaf et al. 2011b; Kolb et al. 2009; Becker et al. 2004), but several successful prospective SBVS campaigns are reported also for other rhodopsin-like GPCRs (adenosine (Carlsson et al. 2010; Langmead et al. 2012; Katritch et al. 2010b), brain-gut peptide (Cavasotto et al. 2008; Engel et al. 2008), chemoattractant (Blättermann et al. 2012), chemokine (Kellenberger et al. 2007; Kim et al. 2012; Becker et al. 2004; Liu et al. 2008), lipid (Salo et al. 2005), peptide (Evers and Klebe 2004; Klabunde et al. 2009; Edwards et al. 2005), purine receptors (Tikhonova et al. 2008)). Only recently the first prospective SBVS studies targeting the allosteric TM cavity of class B GPCRs has been reported (de Graaf et al. 2011a)

Table 7.1 Overview of prospective structure-based virtual screening (SBVS) against GPCR models and crystal structures since 2009[a]

Receptor[b]	Template	Focused database creation[c]	Conf. Sampling[d]	(Re)scoring[e]	Interaction filter[f]	Prospective Initial db[g]	Prediction Hits[h] (tested)	References[i]
Adenosines								
AA1R	AA2AR	ll	ad	Score	$N^{6.55}$	2,200,000	8 ant (39)	Kolb et al. (2012)
AA2AR	X-ray	dl	ad	LE/chem/score	–	4,300,000	23 ant (56)	Katritch et al. (2010b)
AA2AR	X-ray	CNS+dl	ad	Score	$N^{6.55} + E^{45.53}$	14,000,000	7 ant (20)	Carlsson et al. (2010)
AA2AR	ADRB1	dl	ad	Score[m]	$N^{6.55}$	545,000	20 lig (230)	Langmead et al. (2012), Congreve et al. (2012), and Andrews and Benjamin (2013)
Amines								
5HTR2A	ADRB2	fda	ad	Rescoring	D3.32	1,430	1 lig (6)	Lin et al. (2012)
ADRB2	X-ray	dl	ad	Clust+score[m]	$D^{3.32}$	10,000,000	6 ant (25)	Kolb et al. (2009)
DRD3	ADRB1/2	dl	ad	Score[m]	$D^{3.32} + S^{5.43+5.46}$	3,300,000	6 ant (26)	Carlsson et al. (2011)
DRD3	X-ray	dl	ad	Score[m]	$D^{3.32} + S^{5.43+5.46}$	3,300,000	5 ant (25)	Carlsson et al. (2011)
HRH1	HRH1	fl	ad	Score+IFP	$D^{3.32}$	108,790	19 ant (26)	de Graaf et al. (2011b)
HRH3	HRH1	fl	3D	Score (SB+LB)	$D^{3.32}$	771,219	18 lig (29)	Sirci et al. (2012)
HRH4	ADRB2	fl	ad	IFP	$D^{3.32}$	43,326	6 lig (23)	Istyastono (2012)
Chemoattractants								
OXER1	CXCR4	fl	ad	Score	–	1,047	1 lig (10)	Blättermann et al. (2012)
Chemokines								
CXCR4	de novo	none	ad	Score	$D^{4.60}/D^{6.58}/E^{7.39}$	350,000	1 ant (32)	Kim et al. (2012)
CXCR4	bRho ADRB1 ADRB2 AA2AR	ll	ad	Score	$E^{7.39}$	3,300,000	1 lig (24)	Mysinger et al. (2012)
CXCR4	X-ray	ll	ad	Score	$E^{7.39}$	420,000	4 lig (23)	Mysinger et al. (2012)
Lipids								
CNR2	ADRB2	fl	2D+ad	Score	S7.39	250,675	13 lig (97)	Renault et al. (2012)
Peptides								

(continued)

Table 7.1 (continued)

Receptor[b]	Template	Focused database creation[c]	Conf. Sampling[d]	(Re)scoring[e]	Interaction filter[f]	Prospective Initial db[g]	Prediction Hits[h] (tested)	References[i]
C3AR1	AT1R	none	3D	clust.	–	In-house	4 ago (157)	Klabunde et al. (2009)
Purines								
P2RY1	bRho	3D+clust.	3D	3D + clust.		250,675	3 lig (110)	Tikhonova et al. (2008), Sum et al. (2007), and Costanzi et al. (2012)
Secretin								
GLR	CRFR1	ll(+3D)	ad	Score+IFP	$K^{2.53}$	1,900,000	3 lig (26)	de Graaf et al. (2011a)

See de Graaf and Rognan (2009) for overview of successful prospective GPCR structure-based virtual screening campaigns before 2009 (Comprising campaigns against bioaminergic, brain-gut peptide, chemokine, lipid, peptide, and purine receptors)

[a]Only structure-based virtual screening studies targeting the TM domain are included

[b]Receptors are clustered according to Surgand et al. (2006)

[c]Consecutive filters (dl (drug-like physicochemical properties), ll (lead-like physicochemical properties), fl (fragment-like physicochemical properties), 1D (physicochemical properties known ligands), 2D (two-dimensional topological/chemical similarity/pharmacophoric features/sub groups), 3D (three-dimensional pharmacophore)) used to compile database for docking/3D conformer search

[d]Conformer search method: ((H-bond) constr(ained)) automated docking (ad), protein-based or docked ligand-based 3D pharmacophore search (3D)

[e]Method to score, rank and/or filter conformers: clust. (scaffold clustering), (c-)score ((consensus) docking scoring function), 2D (two-dimensional topological/chemical similarity/pharmcoaphoric features/sub-groups), 3D (three-dimensional pharmacophore)

[f]Key interactions with the listed residues were used to filter the docking poses

[g]Prospective validation: initial database (db)

[h]Number of experimentally confirmed hits with detectable affinity/activity (of the total number of tested compounds)

[i]References to homology modeling, virtual screening, and structure-based ligand optimization studies are provided

Table 7.2 Representative ligands obtained in structure-based virtual screening (and design) studies against GPCR homology models and X-ray structures

AA1R (model)(Kolb et al. 2012)

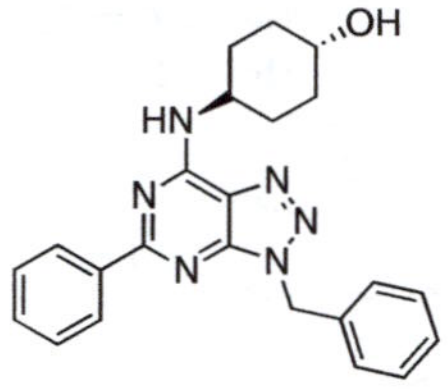

1 (ref)
$pK_i = 8.5$ (ant)

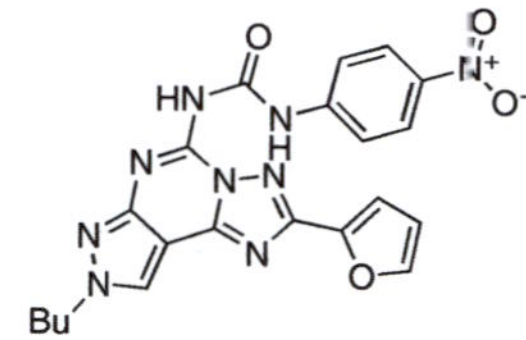

2 (ref)
$pK_i = 6.2$ (ant)

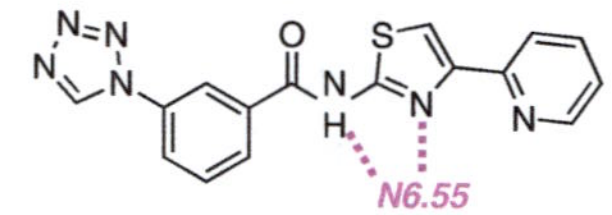

3
$pK_i = 5.5$ (ant)[b] not active on A2A

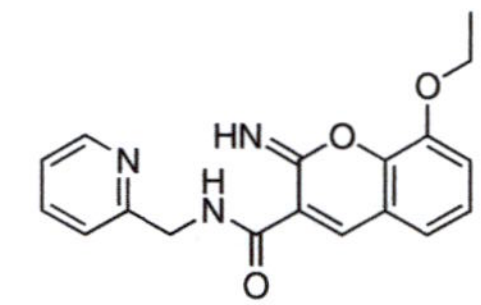

4
$pK_i = 6.4$ (ant)[b] $\Delta pK_{i,\ A1\text{-}A2A} = 0.3$

AA2AR (model/X-ray)

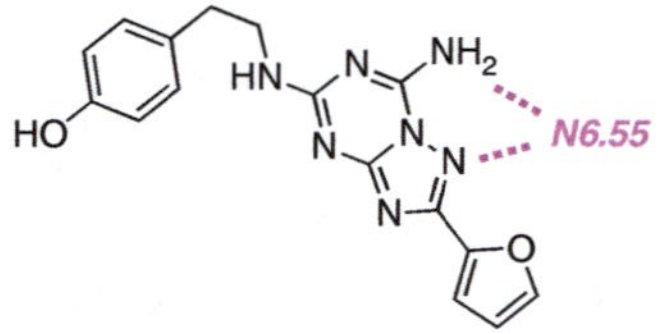

5 (ref) (Carlsson et al. 2010, Katritch et al. 2010b)
$pK_i = 9.0$ (ant)

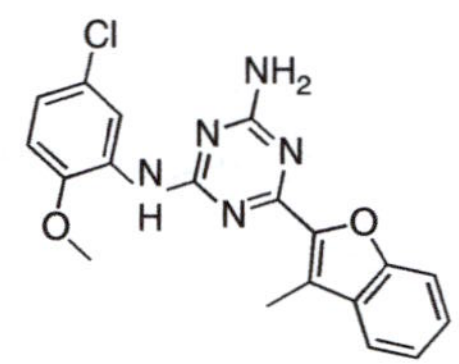

6 (Katritch et al. 2010b)
$pK_i = 7.5$ (ant)[b]

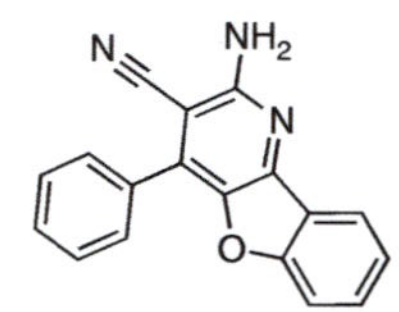

7 (Katritch et al. 2010b)
$pK_i = 7.5$ (ant)[b]

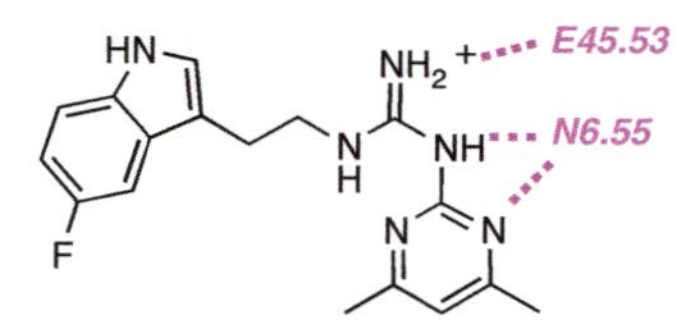

8 (Carlsson et al. 2010)
$pK_i = 6.7$ (ant)[b]

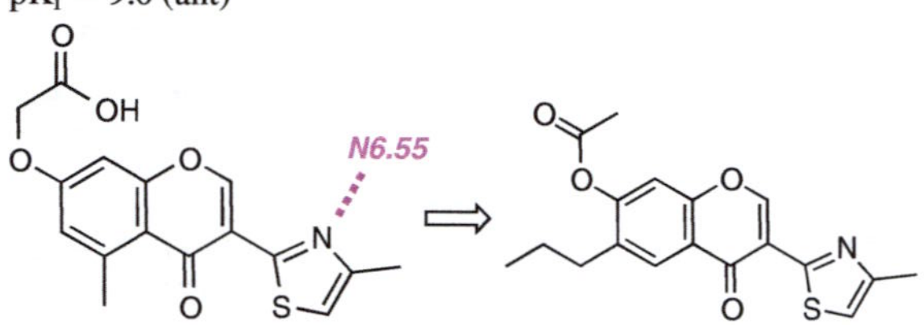

9a (Langmead et al. 2012)
$pK_i = 5.7$ (ant)
–

9b (Langmead et al. 2012)
$pK_i = 8.5$ (ant)
$\Delta pK_{i,\ A2A\text{-}A1} = 1.2$

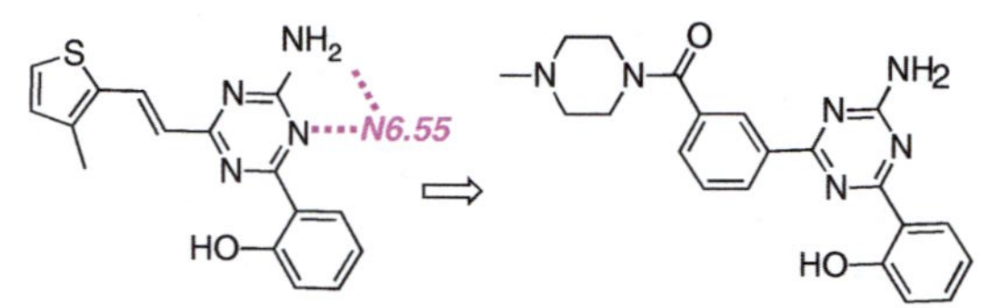

10a (Langmead et al. 2012)
$pK_i = 8.5$ (ant)
$\Delta pK_{i,\ A2A\text{-}A1} = 1$

10b (Langmead et al. 2012)
$pK_i = 9.0$ (ant)
$\Delta pK_{i,\ A2A\text{-}A1} = 1.8$

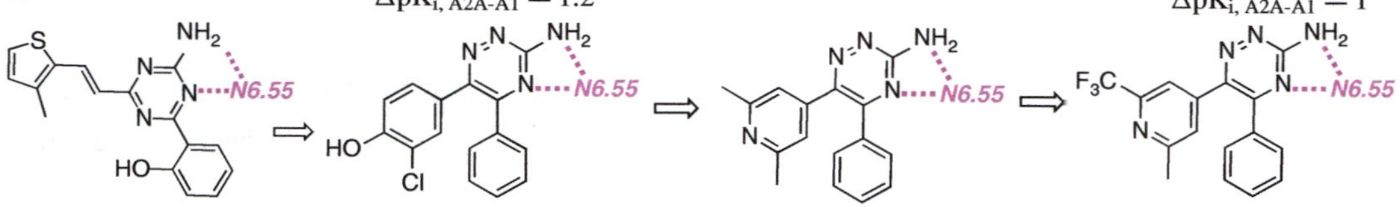

10a (Langmead et al. 2012)
$pKi = 8.5$ (ant)
$\Delta pK_{i,\ A2A\text{-}A1} = 1$

11a (Congreve et al. 2012)
$pK_i = 8.9$ (ant)
$\Delta pK_{i,\ A2A\text{-}A1} = -0.9$

11b (Congreve et al. 2012)
$pK_i = 8.1$ (ant)
$\Delta pK_{i,\ A2A\text{-}A1} = 1$

11c (Congreve et al. 2012)
$pK_i = 8.5$ (ant)
$\Delta pK_{i,\ A2A\text{-}A1} = 1$

(continued)

Table 7.2 (continued)

DRD3 (model/X-ray) (Carlsson et al. 2011)

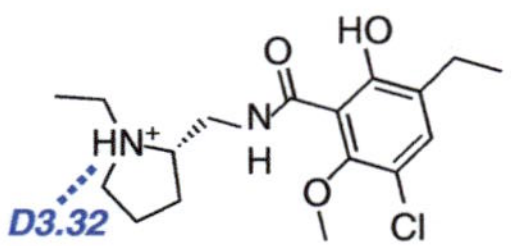

12 (ref)
$pK_i = 9.7$ (ant)[b]

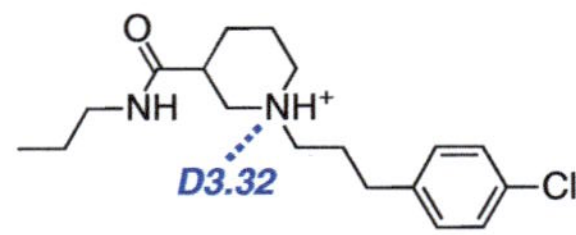

13
$pK_i = 6.5$ (ant)[b]

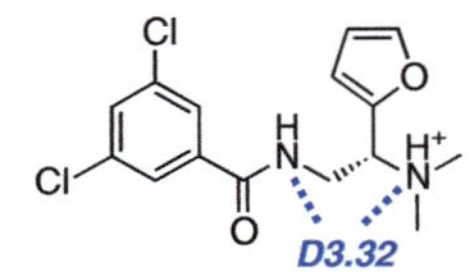

14
$pK_i = 5.7$ (ant)[b]

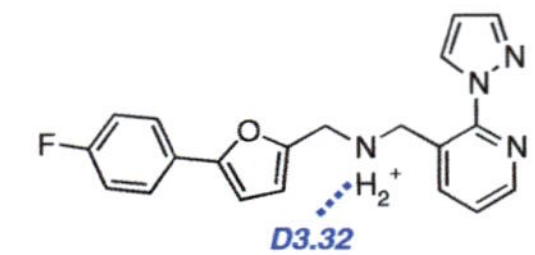

15
$K_i = 5.9$ (ant)[b]

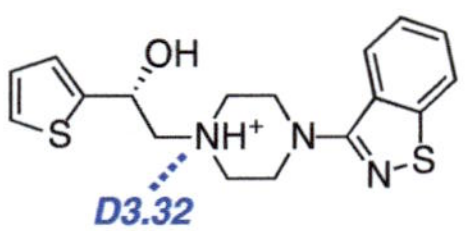

16
$pK_i = 6.7$ (ant)[b]

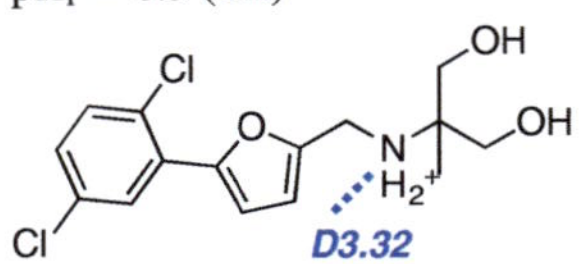

17a
$pK_i = 5.8$ (ant)[b]

⟹

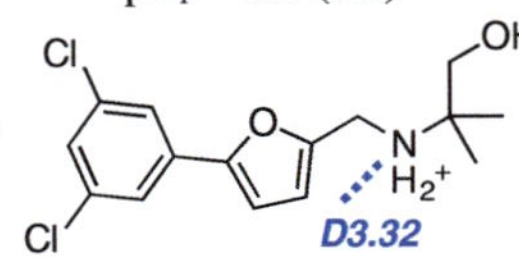

17b
$pK_i = 7.1$ (ant)[b]

⟹

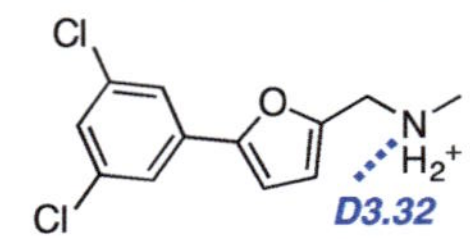

17c
$pK_i = 7.0$ (ant)[b]

ADRB2 (X-ray) (Kolb et al. 2009)

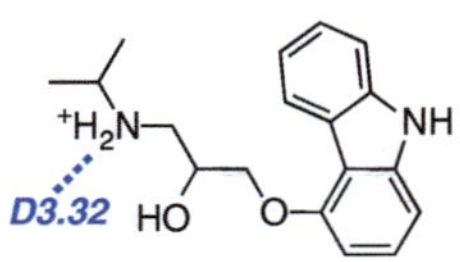

18 (ref)
$pK_i = 10.0$ (ant)

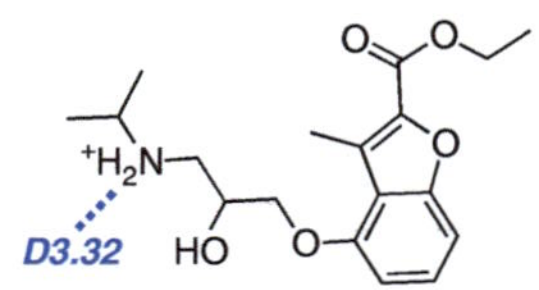

19
$pK_i = 8.0$ (ant)[b]

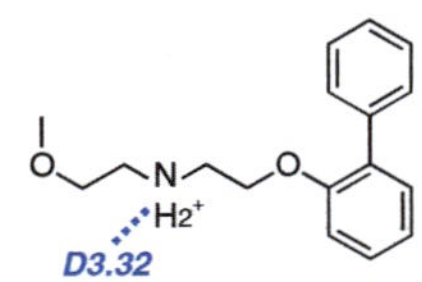

20
$pK_i = 6.3$ (ant)

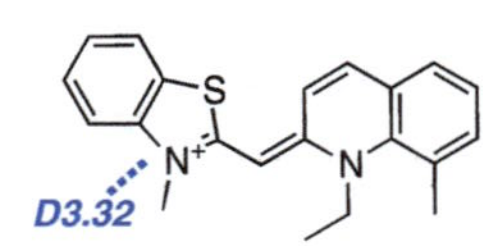

21
$pK_i = 6.0$ (ant)[b]

HRH1 (X-ray) (de Graaf et al. 2011b)

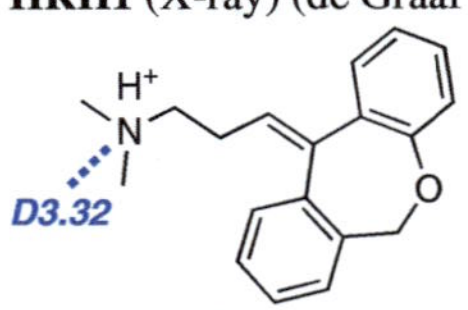

22 (ref)
$pK_i = 9.7$ (ant)

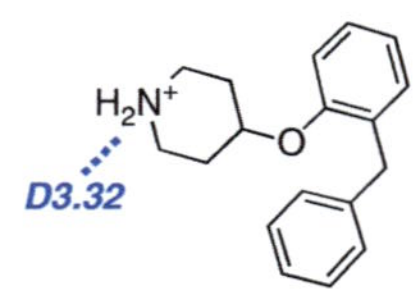

23
$pK_i = 8.2$ (ant)[b]

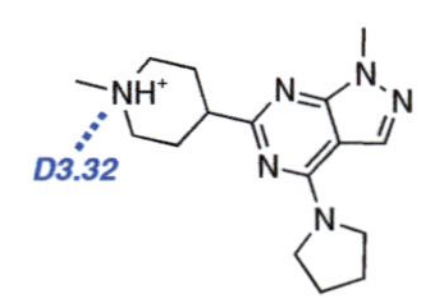

24
$pK_i = 7.2$ (ant)[b]

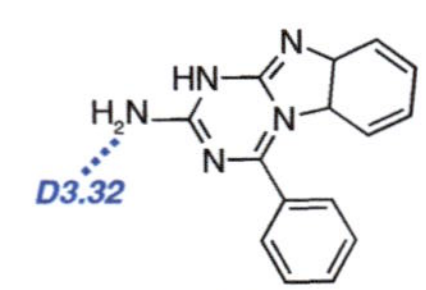

25
$K_i = 6.4$ (ant)[b]

HRH3 (model) (Sirci et al. 2012)

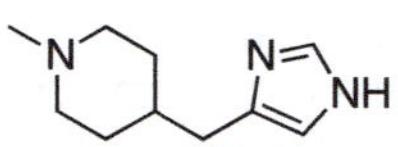

26 (ref)
$pK_i = 9.0$ (ago)

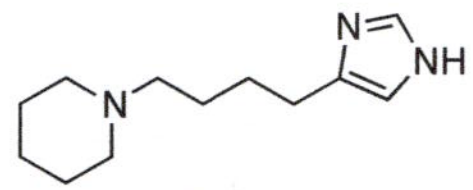

27 (ref)
$pK_i = 7.7$ (ago)

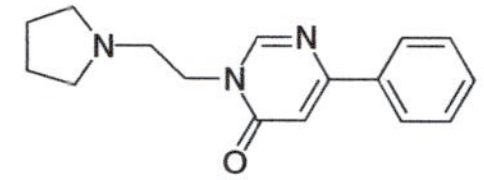

28
$pK_i = 5.9$

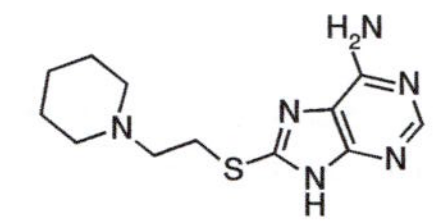

29
$pK_i = 5.9$

HRH4 (model) (Istyastono 2012)

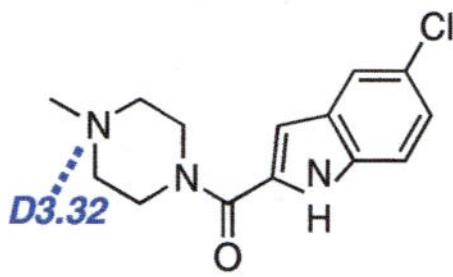

30 (ref)
$pK_i = 7.8$ (biased ago)

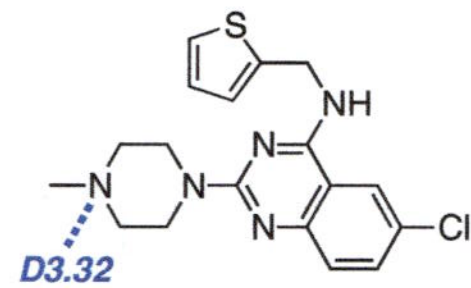

31 (ref)
$pK_i = 8.1$ (ant)

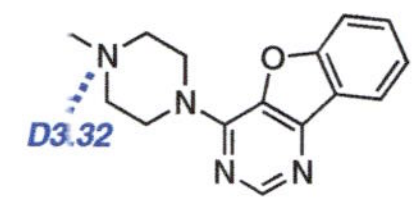

32
$pK_i = 6.9$

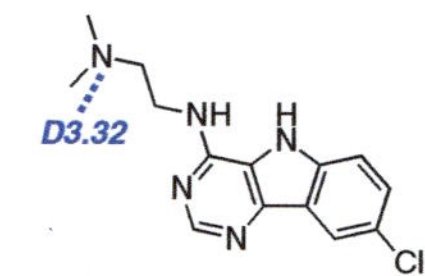

33
$pK_i = 5.2$ (ant)[b]

5HTR2A (model) (Lin et al. 2012)

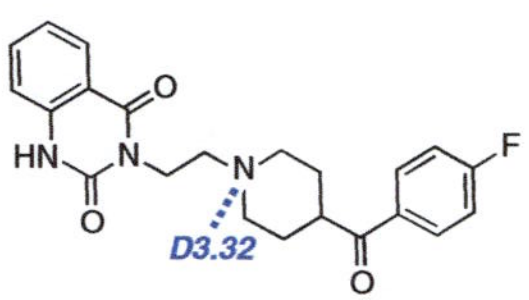

34 (ref)
$pK_i = 8.5$ (ant)

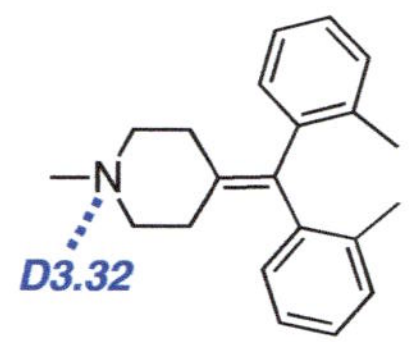

35 (ref)
$pK_i = 9.3$ (ant)[b]

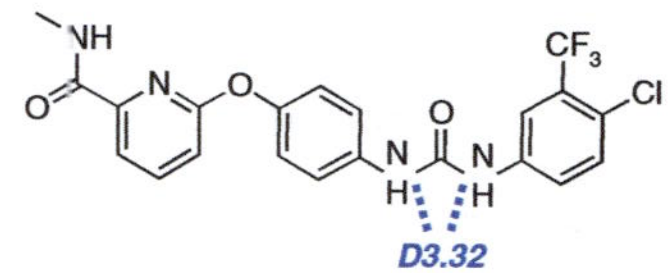

36
$pK_i = 7.3$

CXCR4 (model/X-ray)

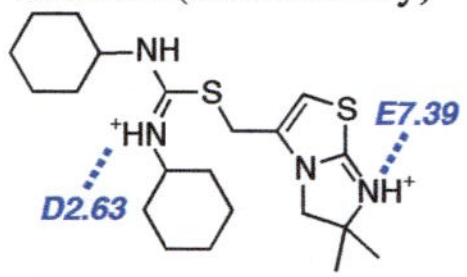

37 (ref) (Mysinger et al. 2012)
$pIC_{50} = 8.1$

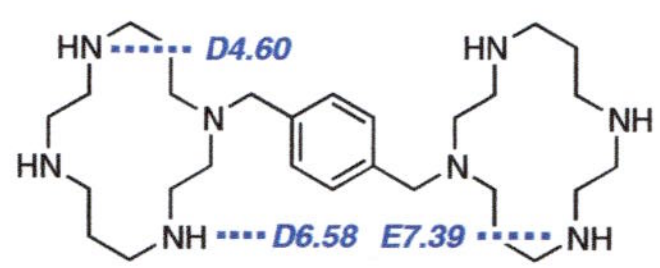

38 (ref) (Kim et al. 2012)
$pIC_{50} = 7.2$ (ant)

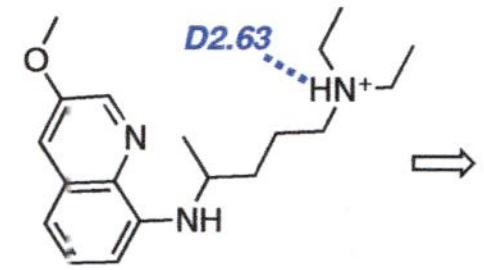

39a (Kim et al. 2012)
$pIC_{50} = 5.3$ (ant)

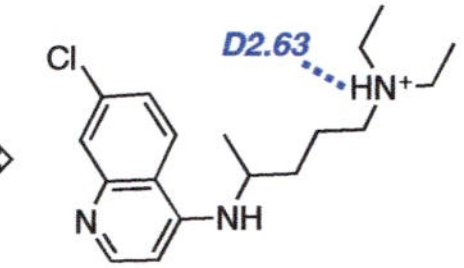

39b (analogue)[a] (Kim et al. 2012)
$IC_{50} = 5.2$ (ant)[b]

(continued)

Table 7.2 (continued)

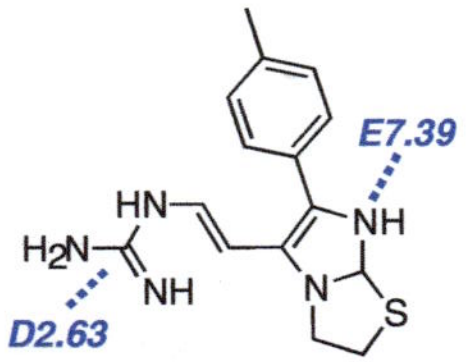

40 (Mysinger et al. 2012)
$pIC_{50} = 3.6$ (ant)

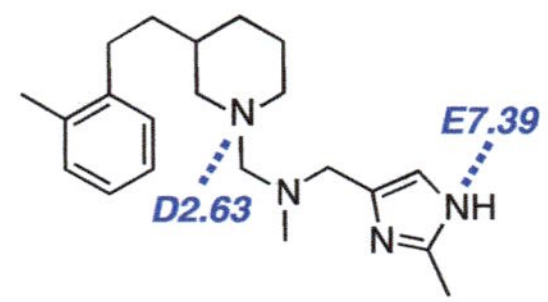

41 (Mysinger et al. 2012)
$pIC_{50} = 6.5$ (ant)[b]

OXER1 (model) (Blättermann et al. 2012)

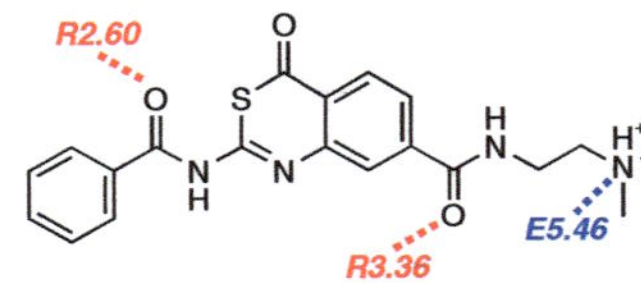

42 (ref)
- (biased ago)

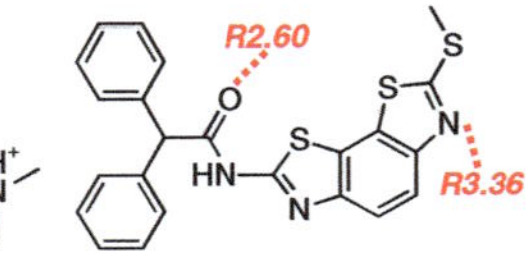

43
- (biased ant)

CA3R1 (model) (Klabunde et al. 2009)

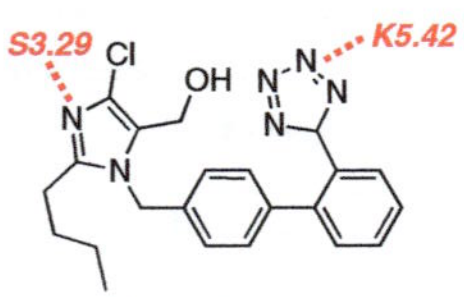

44 (ref)
pEC_{50} (AT1) = 8.1 (ago)

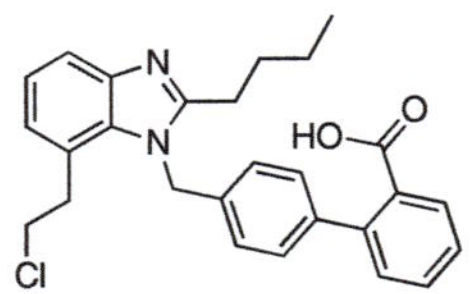

45
$pEC_{50} = 6.5$ (ago)

CNR2 (model) (Yrjola et al. 2013)

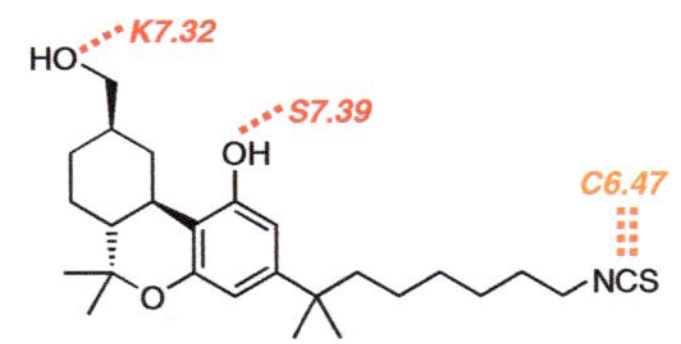

46 (ref)
$pEC_{50} = 9.0$ (ago)

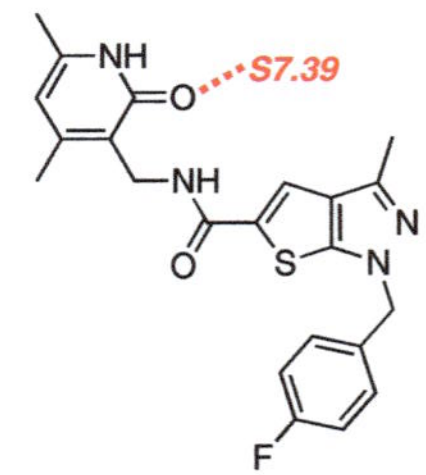

47
$pEC_{50} = 8.5$ (ago)

P2RY1 (model) (Costanzi et al. 2012)

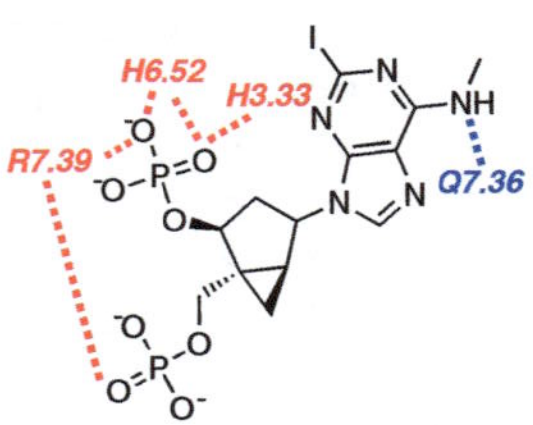

48 (ref)
$pK_i = 9.0$ (ant)

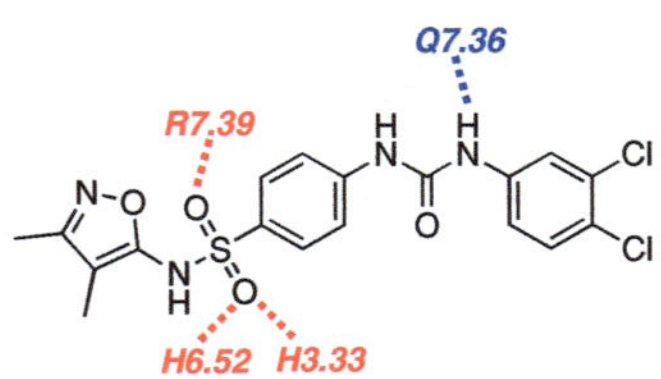

49
$pK_i = 4.9$ (ant)

GLR (model) (de Graaf et al. 2011a)

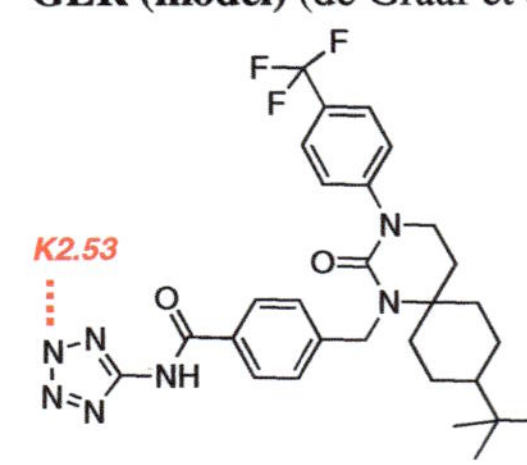

50 (ref)
$pIC_{50} = 7.5$ (ant)

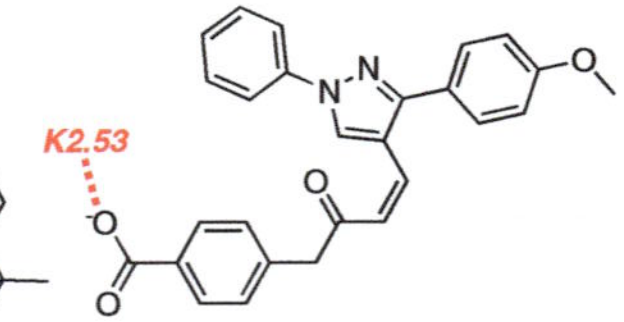

51
$pIC_{50} = 4.6$ (ant)[b]

H-bond donor/positive ionizable (blue), H-bond acceptor/negatively ionizable (red) or both (magenta) pharmacophore features (assigned to functional groups in the ligands) and docking constraints/post-processing interaction filters (indicated by dotted lines to complementary residue(s) in the receptor) are derived from reference ligands/binding modes and used to select hits along the virtual screening work flow (Fig. 7.2, Table 7.1). The experimentally determined affinity/activity parameters and function (full/partial agonist (ago) and/or inverse agonist/antagonist (ant)) of reference and discovered ligands are indicated. Homology model-based and crystal structure-based virtual screening reference compounds as well as hits are highlighted in blue and red respectively

[a]Analogue of SBVS hit: known antimalarial drug chloroquine

[b]Novelty of the hit was explicitly assessed

and SBVS studies against class C GPCRs have so far only focused on the orthosteric N-terminal binding site (Venus Fly trap (Triballeau et al. 2008)). The final sections will discuss two of the important challenges in SBVS for GPCR ligands: *structure-based optimization* of virtual screening hits (Sect. 7.4), and *selective* virtual screening for specific ligand types (Sect. 7.5).

SBVS campaigns against the first crystal structures of druggable GPCRs (ADRB2, AA2AR, DRDR3, and HRH1) have resulted in relatively high hit rates (Fig. 7.3, up to 73 % hit rate (de Graaf et al. 2011b)). It should however be noticed that high hit rates have not only been obtained based on docking studies against GPCR crystal structures, but also successful SBVS studies with high hit rates (>20 %) have been reported based on GPCR homology models (Fig. 7.3) (Varady et al. 2003; Evers et al. 2005). Moreover, in a recent comparative virtual screening comparable high hit rates were obtained for prospective virtual screening runs against the recently solved dopamine D_3 receptor (DRD3) crystal structure and a previously constructed DRD3 homology model (Carlsson et al. 2011). A different result was obtained in a comparative CXCR4 SBVS study in which the screen against the CXCR4 crystal structure was more successful. SBVS studies using homology models are highly dependent on the applied method as well as the homology model used as recently demonstrated by Kolb et al. (2012) who created four AA1R homology models with highly varying hit-rates.

The first crystal structure-based virtual screening study for a druggable GPCR was reported for ADRB2 (Cherezov et al. 2007). Kolb et al. (2009) performed docking simulations of 972,608 lead-like compounds against the ADRB2 crystal structure. They selected 25 compounds for experimental testing from the 500 top-ranking molecules based on chemical clustering, visual inspection, and favorable interaction energies with $D^{3.32}$ (Tables 7.1 and 7.2). 6 of the 25 hits had detectable binding affinity for ADRB2 (K_i values ranging from 4 to 0.009 μM) and were characterized as inverse agonists. Interestingly, the predicted binding mode of the highest affinity hit **19** (Table 7.2), was later corroborated by crystallization studies. Although this is not unexpected given its chemical similarity to the reference ligand used (carazolol, **18**), this demonstrates that GPCR structure-based virtual screening can yield not only new ligands, but also suitable starting points for structure-based hit optimization. Also several chemically novel ADRB2 ligands were reported (e.g., **21**), which is particularly challenging for ADRB2, a receptor with a relatively low ligand diversity (van der Horst et al. 2009). In fact, in 2008 Topiol et al. discovered several submicromolar affinity ligands based on the ADRB2 crystal structure that were chemically very close to known ADRB2 ligands (Topiol and Sabio 2008; Sabio et al. 2008).

The ADRB2 crystal structure has been used in several studies as a template for the creation of homology models. Istyastono et al., for example, used the ADRB2 structure to generate JNJ7777120 (cpd. **30**) and VUF10497 (cpd. **31**) bound HRH4 homology models (Istyastono 2012). These models were subsequently used for VS purposes. 23,112 fragment-like molecules were obtained from the ZINC database after filtering out the compounds containing reactive groups. After docking all compounds, the highest-ranking (according to the IFP score) and novel (according to their low 2D similarity to known HRH4 ligands from the ChEMBL) compounds were selected and clustered using a 2D ligand-based fingerprint. From each cluster the top compound was selected and in total 164 compounds were remaining for visual inspection. During the visual inspection compounds with an ionic interaction with $D^{3.32}$ and a polar moiety near $E^{5.46}$ were prioritized resulting in the purchase of 23 compounds. Six of the 23 compounds (including hits **32** and **33**) were shown to have affinities in the range of 0.14–6.3 μM (Istyastono 2012).

Two successful docking-based virtual screening studies against the AA2AR crystal structure have been performed (Carlsson et al. 2010; Katritch et al. 2010b), yielding diverse sets of novel ligands (**6–8**, **9a**, **10a**, Table 7.2). Katritch et al. selected 56 high ranking compounds from docking based virtual screening runs of four million compounds against the AA2AR crystal structure and discovered 23 new AA2AR ligands with K_i values between 0.032 and 10 μM (incl. **6**, **7**) (Katritch et al. 2010b). Interestingly, specific water molecules in the AA2AR binding site were included in the docking simulations, because retrospective virtual screening evaluations gave better results with water than without consideration of water (Katritch et al. 2010b). Carlsson et al. emphasized favorable H-bond interactions with $N^{6.55}$, an important AA2AR ligand binding residue (Kim et al. 1995; Jaakola et al. 2010; Zhukov et al. 2011), in their scoring protocol to rank the docking poses of 1.4 million cpds in the AA2AR crystal structure by increasing the dipole moment of the side chain amide group but without taking water into account (Carlsson et al. 2010). Using this experimentally guided SBVS approach, 7 of the 20 selected hits were experimentally confirmed as novel AA2AR ligands, with K_i values ranging from 0.2 to 9 μM (incl. **8**) (Carlsson et al. 2010). Recently an ADRB1-based AA2AR homology model (Zhukov et al. 2011) (that correctly predicted the binding mode of ZM241385 prior to publication of the AA2AR-ZM243185 crystal structure (Jaakola et al. 2008)) was successfully used to discover new AA2AR ligands (Langmead et al. 2012). The overall hit rate of this *in silico* screening campaign was somewhat lower than the AA2AR crystal structure-based screening runs (Fig. 7.3), but yielded a diverse set of chemically novel ligands (e.g., **9a**, **10a**) (Langmead et al. 2012), that were subsequently optimized by structure-based design to improve affinity and adenosine receptor selectivity (Sect. 7.4) (Langmead et al. 2012; Congreve et al. 2012). Both receptor model construction as ligand optimization was driven by experimental (mutagenesis/biophysical mapping (Zhukov et al. 2011) of receptor-ligand interaction hotspots) and computational analysis (identification of thermodynamically unstable water molecules that can be displaced by the ligand) (Langmead et al. 2012; Congreve et al. 2012; Andrews and Benjamin 2013).

With the antagonist bound AA2AR crystal structure as a template, Costanzi et al. (2012) created a homology model of the $P2Y_1$ receptor (P2RY1). The ligand and receptor based pharmacophore screen was performed on a set of 133,999 compounds that were selected from a set of 250,675 compounds using physicochemical filtering. The resulting 362 hit compounds were subsequently clustered and from each cluster one unique compound was selected. The resulting 110 compounds were subsequently validated *in vitro* for their binding affinity for P2RY1, yielding multiple hits (exact number was not reported). One of the hit compounds (with a reported binding affinity of 13 μM) was further explored through use of an analogue search and SAR studies (Sect. 7.4). This work resulted in new insights in the binding mode properties of this non-nucleotide P2RY1 antagonist series (Costanzi et al. 2012).

Carlsson et al. compared the SBVS performance of the dopamine D3 receptor (DRD3) crystal structure and an ADRB2-based homology model of DRD3 (that was constructed prior to the release of the DRD3 X-ray structure) (Carlsson et al. 2011). 26 and 25 of the highest ranking molecules with strong electrostatic interactions with $D^{3.32}$ (Tables 7.1 and 7.2) were selected for experimental testing from 3.3 million molecules docked in the DRD3 homology model and DRD3 crystal structure, respectively. Interestingly, both models performed equally well in terms of virtual screening hit rate (Fig. 7.3). 6 of the homology model based hits had affinities ranging from 0.2 to 3.1 μM (incl. hits **15**, **16**, **17a**, Table 7.2), while 5 of the X-ray structure based hits had affinities ranging from 0.3 to 3.0 μM (incl. hits **13**, **14**). Ligand **17** from the homology model screen was optimized for affinity to 81 nM (Sect. 7.4) (Carlsson et al. 2011).

A customized structure-based virtual screening protocol against the histamine H_1 receptor (HRH1) crystal structure was used to dock and score a database of 108,790 fragment-like compounds (heavy atoms ≤ 22) containing a basic moiety (de Graaf et al. 2011b). The method combined molecular docking simulations with a protein-ligand interaction fingerprint (IFP) scoring method (Fig. 7.4). The optimized *in silico* screening approach was successfully applied to identify a chemically diverse set of novel fragment-like ($\leq$22 heavy atoms) HRH1 ligands with an exceptionally high hit rate of 73 %. Of the 26 tested fragments, 19 compounds had affinities ranging from 10 μM to 6 nM (incl. hits **23–25**) This study shows the potential of *in silico* screening against GPCR crystal structures to explore novel, *fragment-like* GPCR ligand space (de Graaf et al. 2011b).

The HRH1 crystal structure also allowed Sirci et al. to create high-resolution homology models of HRH3 (Sirci et al. 2012). Multiple MD snapshots of the homology models were subjected to retrospective validation using the experimental data from a VU-MedChem fragment library screen against HRH3. The snapshot with the highest early enrichment was selected from both the methimepip-based (cpd. **26**) and the VUF5228-based homology model. After a retrospective comparison of both models using docking and FLAP-linear discriminant analysis (FLAP-LDA) the latter in combination with the methimepip-based homology model was shown to be superior in the retrieval of active fragment-like compounds. Multiple different ligand-based FLAP models were build (based on amongst others reference compound **27**) in parallel as well and were also retrospectively validated. Subsequently a prospective screening of 156,090 fragment-like compounds against both the best structure-based and ligand-based FLAP-LDA models was performed. From the highest scoring compounds for both the ligand *and* structure-based model 21 compounds were purchased as well as 8 from the highest scoring compounds from only the ligand-based model. Experimental validation pointed out a remarkably high hit rate of 62 % as 18 of the 29 selected compounds were found to be active. This study showed that the combined use of ligand-based and structure-based models with a thorough retrospective validation can be the key to a successful VS (Sirci et al. 2012).

Mysinger et al. used a chemokine receptor CXCR4 crystal structure and a CXCR4 homology model (constructed before the release of the crystal structure) in a comparative virtual screening study (Mysinger et al. 2012). 3.3 and 4.2 million lead-like compounds were docked against the homology model and crystal structure respectively. The binding modes of the top 500 compounds from each screen were visually inspected. Based on availability, internal energy, unsatisfied polar interactions, correctness of the protonation state and hit diversity 24 and 23 compounds were selected for the crystal structure and homology model screening respectively. Experimental validation yielded 1 and 4 hit compounds (a hit rate of 4 and 17 %) respectively and showed affinities ranging from 0.31 μM (hit **41**) to 225 μM (hit **40**). This study underlined the impact of the available templates and experimental knowledge for the creation of homology models (Mysinger et al. 2012).

Kim et al. (2012) constructed a CXCR4 homology model based on binding mode of AMD3100 (cpd. **38**), and validated this model using published site-directed mutagenesis data. 350,000 compounds from the open NCI database were docked in three conformations of the homology model. From the docked compounds the neutral molecules with a buried surface area of at least 80 % in close proximity to the acidic residues $D^{4.60}$, $D^{6.58}$ and $E^{7.39}$ were selected, excluding 90 % of the small molecules. For the remaining molecules the binding energies were calculated and the top 200 compounds in each receptor conformation were visually inspected, resulting in a subset of 50 compounds of which 32 were procured and experimentally validated. One hit was obtained (cpd. **39a**) which was further investigated because of its close similarity to known anti-malarial drugs (Sect. 7.4) (Kim et al. 2012).

Blättermann et al. (2012) used the CXCR4 crystal structure (Wu et al. 2010) to construct a homology model of an oxoeicosanoid receptor 1 (OXER1) homology model. Commercially available compounds from the ZINC database were filtered based on physicochemical properties, resulting in a subset of 1,047 compounds. This subset was docked into the OXER1 homology model and from the top 100 (as ranked by AutoDock), 10 compounds were visually selected and tested for OXER1 mediated Ca^{2+} release-inhibition. One of the 10 compounds (hit **43**) showed significant inhibition and was further investigated. Ligand deconstruction of this

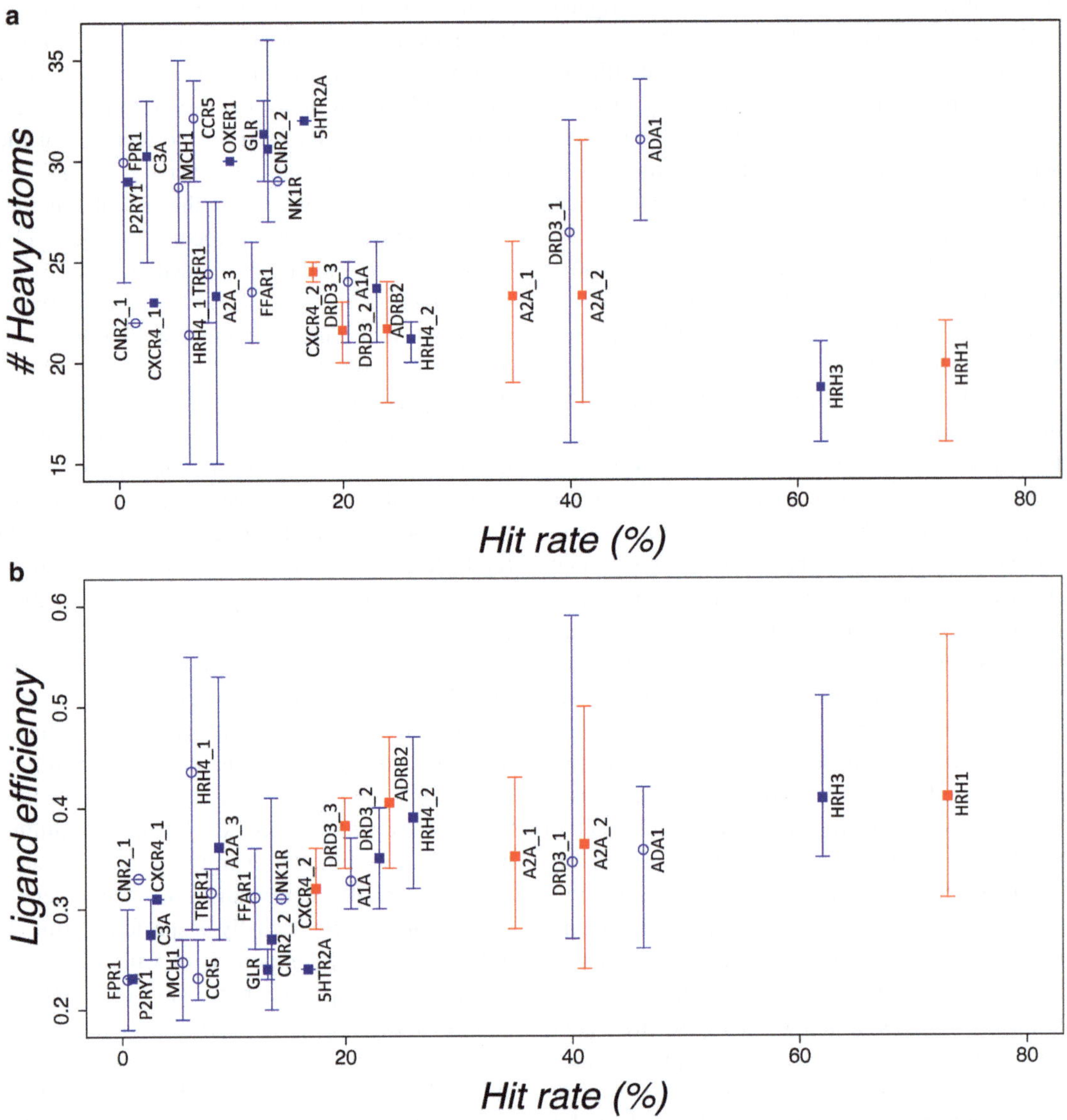

Fig. 7.3 Hit rate versus size (heavy atom count) (**a**) and hit rate versus ligand efficiency (**b**) of hits identified in prospective structure-based virtual screening studies against GPCR crystal structures (*red*) and homology models (*blue*). Open circles indicate a SBVS studies published before 2009, studies since 2009 are indicated with a closed square. The bars shown indicate the minimum and maximum heavy atom and ligand efficiency count for all hits of each SBVS respectively. The labels indicate the screening on the following receptors adenosine A_1 (A1A), adenosine A_2A receptor (A2A_1 (Carlsson et al. 2010), A2A_2 (Katritch et al. 2010b), A2A_3 (Langmead et al. 2012)), adrenergic alpha-1a receptor (ADA1 (Evers and Klabunde 2005)), adrenergic beta-2 receptor (ADRB2 (Kolb et al. 2009)), complement component 3a receptor 1 (C3A (Klabunde et al. 2009)), C-C chemokine receptor type 5 (CCR5 (Kellenberger et al. 2007)), cannabinoid receptor 2 (CNR2_1 (Salo et al. 2005), CNR2_2 (Renault et al. 2012)), chemokine receptor CXCR4(CXCR4_1 (Kim et al. 2012), CXCR4_2 (Mysinger et al. 2012)), dopamine receptor D3 (DRD3_1 (Varady et al. 2003) and DRD3_2 (Carlsson et al. 2011) and DRD3_3 (Carlsson et al. 2011)), free fatty acid receptor 1 (FFAR1 (Tikhonova et al. 2008)), formyl peptide receptor 1 (FPR1 (Edwards et al. 2005)), glucagon receptor (GLR (de Graaf et al. 2011a)), histamine receptor H1 (HRH1 (de Graaf et al. 2011b)), histamine receptor H3 (HRH3 (Sirci et al. 2012)), histamine receptor H4 (HRH4_1 (Kiss et al. 2008), HRH4_2 (Istyastono 2012)), 5-hydroxytryptamine receptor 2A (5HTR2A (Lin et al. 2012)), melanin-concentrating hormone receptor 1 (MCH1 (Cavasotto et al. 2008)), neurokinin 1 receptor

hit compound yielded no substructures with any binding affinity for OXER1 implying that the complete molecule, as discovered, was needed in order to successfully bind to OXER1 and inhibit the CA^{2+} flux. Further functional assays highlighted that this compound was biased and inhibited Gβγ but not Gαi signaling (Blättermann et al. 2012).

The work by Kolb et al. (2012) discussed earlier pushed the boundaries of virtual screening as they tried to screen for subtype selective ligands and evaluate the impact of homology model inaccuracies/differences. Four different AA1R homology models based were created based on the antagonist bound AA2AR crystal structure. These models were subjected to docking simulations of 2.2 million lead-like compounds from the ZINC database. Within this docking simulation the conserved interaction with $N254^{6.55}$ was accentuated by amplifying this term. The top 500 of the docking results from each of the models was inspected to filter out molecules with unsatisfied hydrogen bond donors and acceptors, incorrect protonation states, unlikely binding modes or highly strained conformation which resulted in the purchase of 39 compounds in total. All 39 compounds were tested for their AA1R, AA2AR and AA3R affinity in order to assess their selectivity. 8 of the compounds showed activity for the AA1R receptor (incl. hits **3** and **4**), but surprisingly, 15 and 14 compounds were active on the AA2AR and AA3R respectively. In total 20 of the 39 compounds were active on one (or more) of these three receptors. Interestingly, not all homology models performed equally well; one of the models did not yield any hits and another model accounted for half of the active compounds (but only one of the AA1R active compounds, namely hit **3**). This study highlighted once more the effect of small structural differences in the receptor model or crystal structure on virtual screening, and the challenges in virtual screening for receptor subtype selective ligands.

Virtual screening studies can also highlight surprising cross-pharmacology as was demonstrated in the VS study by Lin et al. (2012) A 5HTR2A homology model was created based on the timolol bound ADRB2 crystal structure (Hanson et al. 2008). Using MD, ten induced-fit models for both ketanserin and cyproheptadine bound 5HTR2A were generated. These 20 snapshots were subsequently used to include protein flexibility and afterwards subjected to a retrospective validation. The ketanserin (cpd. **34**) and cyproheptadine (cpd. **35**) bound models with the best early enrichment were selected for a prospective screening study. A filtered ($100 \leq MW \leq 600$) compound library, comprising 1,430 FDA approved drugs, was screened against the two models using docking and a MM-GB/SA refinement and rescoring procedure. The top 200 hits were filtered based on an essential hydrogen bond filter with $D^{3.32}$ and chemical novelty assessment based on comparisons with known 5HTR2A ligands in ChEMBL (Gaulton et al. 2012) and DrugBank (Knox et al. 2011) and SEA predictions (Keiser et al. 2007). Of the six compounds that were selected for experimental validation, the kinase inhibitor sorafenib (hit **36**) showed a binding affinity of 1,959 nM for 5HTR2A. This compound **36** was also tested on the other HT receptor subtypes and found to be active against all with an affinity ranging from 56 to 7,071 nM. It should be noted that although sorafenib is not able to form the conserved salt bridge with $D^{3.32}$, the urea moiety of sorafenib is seemingly able to provide a strong hydrogen bond to $D^{3.32}$ thereby replacing the necessity for a basic moiety.

Renault et al. (2012) used a combined approach of a trained ligand-based filter and an automated docking approach for the

Fig. 7.3 (continued) (NK1R (Evers and Klebe 2004)), oxoeicosanoid receptor 1 (OXER1 (Blättermann et al. 2012)), P2Y purinoceptor 1 (P2RY1 (Costanzi et al. 2012)) and transferrin receptor 1 (TRFR1 (Engel et al. 2008)). The maximum heavy atom count of FPR1 (Lipinski et al. 2001) is not shown for clarity purposes. Only hits for which: (i) binding affinity Ki/IC50/EC50/KB ≤ 15 μM was experimentally determined; and (ii) for which a molecular structure was reported, are included in the analysis. The affinity of the OXER1 hit compound was not quantified (Blättermann et al. 2012) and is therefore only included in **a**

identification of novel ligands for the cannabinoid receptor 2 (CN2R). A database of 5,513,820 compounds was reduced using physicochemical filtering resulting to a subset of 3,495,595 compounds. This dataset was subjected to a trained 2D-based Bayesian classifier (based on 90 compounds with measured affinity for CN2R) that further reduced the compound set to 209,442 compounds. Docking in an active-state CN2R homology model (using carazolol-bound (**18**) inactivate ADRB2 (Cherezov et al. 2007) as a template) with an interaction filter on $S^{7.39}$ resulted in the selection of 1,385 compounds of which 150 were selected and 149 were available for purchase. Due to solvation issues only 97 of the compounds could be tested which resulted in the confirmation of 13 hits with affinities ranging from 2.3 nM (hit **47**) to 71 μM.

De Graaf et al. (2011a) constructed a homology model of the class B GPCR GLR based on a validated structural model of the CRFR1 receptor. For this representative class B GPCR numerous experimental ligand and receptor data were available to guide the modeling procedure and to validate the refined model with retrospective virtual screening studies (see Sect. 7.2). A database of 1.9 million commercially available drug-like compounds was screened for chemical similarity to existing GLR noncompetitive antagonists and docked to the transmembrane cavity of the GLR homology model. 23 compounds were selected based on binding mode similarity to the protein-ligand interaction fingerprints of the docking poses of two different reference ligands (known GLR ligands **65** and L-168,049) in the GLR homology model. Two of the 23 compounds inhibited the effect of glucagon in a dose-dependent manner (both from the hit list ranked according to IFP similarity to the reference binding mode of **50**). Interestingly, one *in silico* hit that was inactive at the GLR was shown to bind to GLP-1R and potentiate the response to the endogenous GLP-1 ligand. This illustrates the strength of using two alternative binding mode hypotheses in prospective SBVS studies. Although the potencies of the ligands are still modest, this study showed for the first time that structure-based approaches can indeed be used to identify novel class B non-competitive ligands (de Graaf et al. 2011a).

7.4 *In Silico Guidance*: Optimization of Virtual Screening Hits

The identification of active compounds for a specific target using VS is only the first step in the search for novel leads with the targeted affinity, selectivity and/or pharmacological properties and is often followed by SAR exploration, site-directed mutagenesis studies and structure-guided optimization. True rational prospective structure-based hit optimizations, however, remain relatively scarce (Congreve et al. 2011; Tautermann 2011). Nevertheless some of the new hits identified in GPCR structure-based virtual screening campaigns (Table 7.2) have been optimized to improve ligand affinity (Cavasotto et al. 2008; Liu et al. 2008), receptor selectivity (Langmead et al. 2012; Congreve et al. 2012; Becker et al. 2006), pharmacokinetic properties (Congreve et al. 2012), and to explore SAR to validate predicted binding modes (Tikhonova et al. 2008; Lin et al. 2012; Engel et al. 2008; Wacker et al. 2010). Some of these hit optimization studies have been driven by *structure*-based design (Langmead et al. 2012; Congreve et al. 2012; Lin et al. 2012; Becker et al. 2006; Wacker et al. 2010).

One of the most frequently observed steps in a VS hit exploration is the search for close analogues of one or more hit compounds (Carlsson et al. 2010, 2011; Langmead et al. 2012; Kim et al. 2012; Kiss et al. 2012) or to synthesize a small series around one or more hit compounds (Costanzi et al. 2012; Yrjola et al. 2013) (Table 7.2). The experimental validation of compound **8**, an AA2AR-selective VS hit from Carlsson et al. (2010), was followed by an analogue search which yielded 5 equally selective analogues but all with a lower affinity. Langmead et al. (2012) also performed an analogue search for one of their AA2AR structure-based virtual screening hits (**9a**). This search yielded three interesting analogues, amongst them was a selective (16-fold over AA1R) and highly potent ($pK_i = 8.5$) compound (**9b**) and an even more selective (>100-fold over AA1R) but slightly less potent ($pK_i = 7.8$) compound. Compound **10a**, the compound with the highest LE (LE of 0.53, see Fig. 7.3b), and a chromone-containing compound were also further investigated with a ligand deconstruction study and a hit-to-lead synthesis program.

The SAR obtained from the deconstruction of compound **10a** highlighted that the phenol in combination with the amino functionality was the core scaffold for high potency ligands. Subsequently the propenyl-thiophene was reduced to an isopropyl, which yielded a simplified compound with only 17 heavy atoms and a pKi of 7.9 (resulting in a LE of 0.64). Further substitutions yielded amongst others a highly potent (pKi = 9.0) and highly selective (59-fold over AA1R) compound (**10b**) by introducing a n-methylpiperazine moiety. Optimization of these triazine compounds (**10a** and **10b**) was continued in a rational structure-based optimization study (Congreve et al. 2012; Andrews and Benjamin 2013). Instead of 1,3,5-triazines derivatives, 1,2,4-triazine derivatives were further explored by combined SPR, SDM, *in silico* binding mode studies and *in silico* thermodynamic stability calculations of water molecules. Three of the designed compounds (**11a, 11b** and **11c**) were also successfully crystallized in AA2AR and corroborated the modeling approach (Congreve et al. 2012; Andrews and Benjamin 2013). Also the proposed binding mode of the ADRB2 antagonist **17**, discovered by ADRB2 crystal structure-based *in silico* screening (Kolb et al. 2009), was later experimentally validated by a co-crystal structure of ADRB2 and **19** (PDB-code: 3NY9) (Wacker et al. 2010).

The SBVS study performed by Carlsson et al. against both a homology model and a crystal structure yielded 11 new compounds. The compound with the most novel scaffold (compound **17a**) was further explored by the purchase of 20 analogues. Most of the analogues showed the submicromolar affinity for DRD3 and (sub)micromolar affinity for DRD2. The 2,5-dichlorophenyl derivatives appeared to have a slight selectivity towards DRD3 over DRD2 in contrast to the 3,5-dichlorophenyl derivatives as they showed comparable affinity towards both receptors. Reduction of the 2-methylpropane-1,3-diol tail into a 2-methylpropane-1-ol (**17b**) and a methyl (**17c**) yielded submicromolar affinity compounds (for DRD2 and DRD3) with high ligand efficiencies (LE of 0.38, 0.47 and 0.57 for **17a**, **17b** and **17c** respectively).

Lin et al. discovered that the kinase-inhibitor sorafenib (**36**) is also a 5HTR antagonist based on a SBVS study against a 5HTR2A homology model. Based on the predicted binding mode of sorafenib in 5HTR2A multiple analogues were designed and two of them were synthesized to further validate this surprising discovery. The replacement of the aromatic nitrogen with a carbon slightly improved the binding affinity indicating that this atom was not engaged in a polar interaction with the receptor. However, the subsequent addition of a methyl group on the nitrogen atom of the amide resulted in a more than ninefold loss of affinity which again indicated that the hypothesized binding mode was correct in which this nitrogen was engaged in hydrogen bonding with $L^{7.35}$.

Kim et al. (2012) obtained a comparably surprising result in their SBVS study against a CXCR4 homology model in which one hit (**39a**) was identified that was very closely related to known anti-malarial drugs. The obvious next step was to also test these compounds (chloroquine (**39b**), hydroxychloroquine and quinacrine) and they were all found to be active against CXCR4. Apparently the replacement of the methoxy moiety with a chlorine atom (the sole difference between hit compound **39a** and anti-malarial drug chloroquine **39b**) did not have a significant impact of the affinity and cell migration, thereby opening up new repurposing possibilities for anti-malarial drugs.

The most potent of the three hits (compound **49**) from the structure-based pharmacophore screen against P2RY1 was explored by both the purchase of commercially available analogues as well as the synthesis of multiple analogues. Synthesis of the analogues focused on the redecoration of the phenyl ring (from the 3,4-

dichlorophenyl moiety in **49**) with different substituents on all positions. However, all synthesized analogues showed a lower binding affinity as well as lower inhibition of the Ca^{2+} efflux than hit **49**. Moreover, the replacement of the chlorine atoms with fluorine atoms or a methoxy moiety resulted in the complete loss of binding affinity for P2RY1. Two analogues that were purchased contained different substituents for the 3,4-dimethylisoxazole moiety of hit **49**, and had a slightly higher binding affinity ($K_i = 6–10\ \mu M$) than the initial hit (13 μM) (Costanzi et al. 2012).

Tosh et al. (2012) optimized adenosine 5′-carboxamide derivatives for AA2AR by combining binding mode predictions with *in silico* fragment replacement and redocking techniques. The initial binding modes of some adenosine 5′-carboxamide derivatives suggested optimization of these compounds towards $W^{6.48}$. By manual fragment replacement several optimizations were proposed and an automated fragment replacement resulted in 2,000 proposals. The proposed compounds were energy optimized in the binding pocket from which the top 100 compounds were selected for redocking and rescoring. 23 compounds were subsequently selected and synthesized for further testing. This resulted in 2 inactive compounds and 22 compounds with highly differing affinity and selectivity profiles, but most of them had submicromolar affinity for AA2AR and were full or partial agonists at AA1R.

The overview in Fig. 7.3a shows that several of the GPCR ligands identified in SBVS campaigns are relatively small ($\leq$22 heavy atoms), this offers new possibilities for a fragment-based drug design approach. In fact, in the first SBVS study against the HRH1 (de Graaf et al. 2011b) crystal structure successfully retrieved many novel ligands with good ligand efficiency (LE) (Hopkins et al. 2004) from a library of fragment-like molecules (incl. **23** (LE = 0.57), **24** (LE = 0.45) and **25** (LE = 0.44)) (de Graaf et al. 2011b). Also SBVS runs against, for example, AA2AR have resulted in fragment-like hits with good ligand efficiency (9 hits with LE between 0.3 and 0.5 incl. **7** (LE = 0.34)) (Katritch et al. 2010b). It should be noted, however, that structure-based optimization of fragments can be challenging because small fragments can adopt a larger variety of binding modes in different protein (sub)pockets (Leach and Hann 2011; Loving et al. 2010). Furthermore scoring functions used to estimate the binding affinity and determine binding modes in molecular docking simulations are not trained for ranking (the poses of) small fragment-like compounds (Loving et al. 2010; Deng et al. 2004; Marcou and Rognan 2007). Virtual fragment screening with a molecular interaction fingerprint (IFP (Deng et al. 2004; Marcou and Rognan 2007)) is a suitable alternative approach to select ligands that can make specific interactions in specific binding pockets (de Graaf et al. 2011a, b; de Graaf and Rognan 2008). Validated fragments can then efficiently be used as starting points for ligand optimization strategies by fragment growing, linking, or merging (Loving et al. 2010; de Kloe et al. 2009) of fragments originating from different IFP ranking lists.

7.5 *Fitting Functional Selectivity*: Selective Screening for Ligands with Specific Functional Effects

The growing number of available GPCR crystal structures has not only provided insights in receptor-ligand binding modes, but has also increased our understanding in the activation and signaling mechanism of GPCRs. For the bRho, AA2AR, ADRB1, and ADRB2 multiple GPCR crystal structures are available comprising inverse agonist, antagonist and (biased) agonist bound states as well as agonist bound *and* G-protein (mimic) coupled states (Katritch et al. 2013). Although many VS studies are based on the inactive (inverse agonist/antagonist bound) state models with which many new antagonist were found (de Graaf and Rognan 2009; Kooistra et al. 2013), these inactive state models have also been shown to be useful in the search for agonists (de Graaf et al. 2011a; Varady et al. 2003; Kellenberger et al. 2007; Tikhonova et al. 2008; Salo et al. 2005; Kiss et al. 2008). In some of these studies (Varady et al. 2003; Tikhonova et al. 2008; Salo et al. 2005), the initial ground state model was refined by true agonists, but in other cases (de Graaf et al. 2011a; Kellenberger et al. 2007) the inactive model was even refined by antagonists. The other way around, *agonist*-biased models have also been successfully applied to find antagonists (Kiss et al. 2008).

Thereare currently 23 crystal structures deposited in the PDB for the beta-adrenergic receptors (12 ADRB1 and 11 ADRB2 structures) (Warne et al. 2008, 2011, 2012; Cherezov et al. 2007; Wacker et al. 2010; Moukhametzianov et al. 2011; Rasmussen et al. 2007, 2011a, b; Rosenbaum et al. 2007, 2011). Most of them (15) are in the inactive antagonist-bound (or inverse agonist) state, six in the inactive (biased) agonist-bound state and two in the agonist *and* G-protein (mimic) coupled states and all show a well-conserved mode of binding. Despite this similar binding mode co-crystallized antagonists, inverse agonists and full/partial agonists show specific interactions with the receptor binding site (Tang et al. 2012). IFP analysis for these crystal structures highlights for example the agonist specific hydrogen-bond interaction with $S^{5.46}$ (Costanzi and Vilar 2012). In 2008 de Graaf and Rognan showed in a retrospective study that by adjusting the rotameric states of $S^{5.43}$ and $S^{5.46}$ the inactive state ADRB2 structure bound by the inverse agonist carazolol (PDB-code: 2RH1 (Cherezov et al. 2007)) could be used for the selective retrieval of partial and full agonists (de Graaf and Rognan 2008). The more recent agonist-bound ADRB2 crystal structures have validated these rotameric changes, e.g. the BI-167107 (and nanobody) bound ADRB2 structure (PDB-code: 3P0G, Fig. 7.4a) (Rasmussen et al. 2011a) versus the carazolol bound ADRB1 structure (PDB-code: 2YCW, Fig. 7.4b) (Moukhametzianov et al. 2011).

In a retrospective docking study against all ligand-bound (21) ADRB1 and ADRB2 crystal structures were tested for their selectivity in a SBVS towards antagonist/inverse agonists and agonists. A set of compounds comprising 24 agonists (Baker 2010), 25 antagonist/inverse agonists (Baker 2005) and 980 physicochemically similar decoys (de Graaf and Rognan 2008) were automatically docked in one chain of each of the crystal structures. The docking poses of all compounds were subsequently post-processed with IFP and the enrichment factor at 1 % false positive rate (EF1 %) was determined for both the agonists and the antagonists (or inverse agonists) versus the decoys (Fig. 7.5a). All agonist-bound crystal structures (indicated by red dots in Fig. 7.5a) showed a preference towards agonists over antagonists (the dots are located above the black dashed line) except for the biased agonist bucindolol-bound structure (PDB-code: 4AMI) and vice versa for the antagonist bound crystal structures (indicated by blue dots in Fig. 7.5a). The full ROC curves of the crystal structures showing the highest selectivity (EF1% for agonist versus EF1 % for antagonists) towards antagonists and agonist are shown in Fig. 7.5b, c respectively.

The retrospective virtual screening results presented in Fig. 7.5a show that the unique combination of the IFP of the bound ligand and the orientation of the residues in the pocket determines the selectivity as can be seen from the difference in the retrieval rate of e.g. both carazolol-bound structures (PBB-codes: 2YCW and 2RH1). It should furthermore be noted that both the isoprenaline (PDB-code: 2Y03) and BI-167107 (PDB-code: 3P0G) bound structures are highly selective towards agonists. These ligands are different in size but share the same key interactions with the ADRB2 binding site (Fig. 7.4c), including the hydrogen bond with $S^{5.46}$.

In summary, SBVS campaigns can in principle be biased towards ligands with a specific pharmacological effect by carefully selecting a crystal structure and reference ligand (or carefully constructing a homology model) for molecular docking simulations and combining it with selective interaction filters. IFP analyses can be used to identify and emphasize key interactions that are linked to specific pharmacological effects. This information can ultimately be used to rescore docking poses and identify novel chemotypes with similar interaction profiles and desired functional effects.

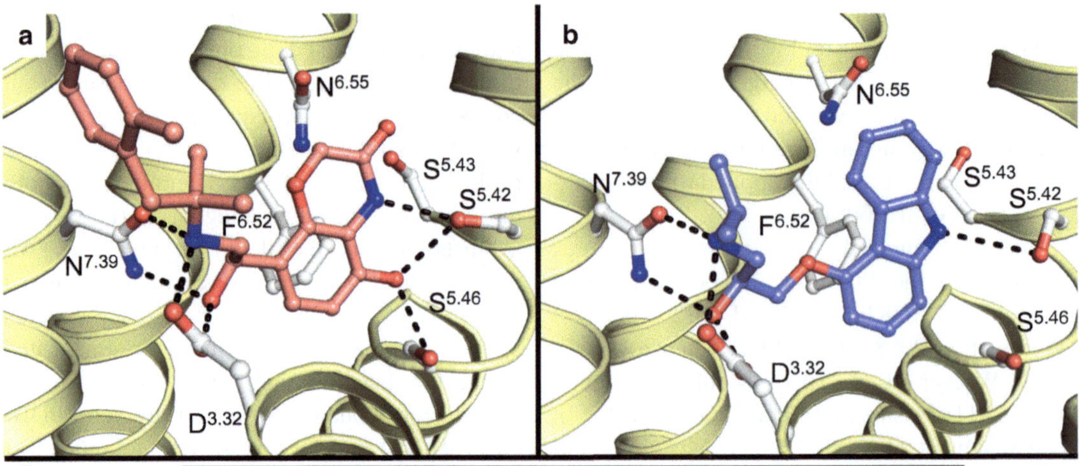

c

	D3.32	S5.42	S5.43	S5.46	F6.52	N7.39
Carazolol (2YCW)	1 0 0 0 1 0 1	1 0 0 0 1 0 0	0 0 0 0 0 0 0	1 0 0 0 0 0 0	1 0 1 0 0 0 0	1 0 0 1 1 0 0
Iodocyanopindolol (2YCZ)	1 0 0 0 1 0 1	1 0 0 0 1 0 0	0 0 0 0 0 0 0	1 0 0 0 0 0 0	1 0 1 0 0 0 0	1 0 0 1 1 0 0
BI-167107 (3P0G)	1 0 0 0 1 0 1	0 0 0 1 1 0 0	0 0 0 0 0 0 0	1 0 0 0 1 0 0	1 0 1 0 0 0 0	1 0 0 1 1 0 0
Isoprenaline (2Y04)	1 0 0 0 1 0 1	0 0 0 0 1 0 0	0 0 0 0 0 0 0	0 0 0 0 1 0 0	1 0 1 0 0 0 0	1 0 0 1 1 0 0

Ionic (residue negative)
Ionic (residue positive)
Hbond (residue acceptor)
Hbond (residue donor)
Aromatic edge-to-face
Aromatic face-to-face
Apolar

Fig. 7.4 Binding mode of full agonist BI-167107 (**a**, PDB-code: 3P0G) (Rasmussen et al. 2011a) and inverse agonist carazolol (**b**, PDB-code: 2YCW) (Moukhametzianov et al. 2011) in the active agonist-bound of ADRB2 and inactive antagonist-bound state of ADRB1, respectively. In panel **c** the interaction fingerprints (IFP, only key residues are shown) of the binding modes of crystallized BI-167107 (**a**) and carazolol (**b**) are compared with the also highly discriminating (see Fig. 7.5) binding modes of co-crystallized partial inverse agonist iodocyanopindolol (PDB-code: 2YCZ) (Moukhametzianov et al. 2011) and full agonist isoprenaline (PDB-code: 2Y04) (Warne et al. 2011)

7.6 Exploring Novel GPCR-Ligand Interaction Space in More Detail

The presented overview shows that structure-based GPCR ligand discovery is growing in popularity and success. Many virtual screening studies have resulted in the discovery of novel ligands for diverse class A GPCRs and recently also the first class B SBVS has been reported (de Graaf et al. 2011a). As opposed to the more recent computer-aided ligand discovery and design studies, previous studies were mostly focused on the identification of larger ligands (Fig. 7.3a). However, the focus is currently gradually shifting towards smaller ligands (fragments) with high ligand efficiency (de Graaf et al. 2011b; Katritch et al. 2010b; Hopkins et al. 2004). This fragment-based drug discovery approach (de Graaf et al. 2013) is starting to have an impact on the targeted compounds in SBVS studies, as the average heavy atom count of the obtained ligands is decreasing (Fig 7.3a) and simultaneously their ligand efficiency is increasing (Fig. 7.3b). Several discussed studies have demonstrated that insights in GPCR-ligand interactions obtained from combined experimental and *in silico* modeling techniques can be used to enable fragment-based drug discovery and design against GPCRs (Sirci et al. 2012; de Graaf et al. 2011b; Istyastono 2012). Despite this progress, the sometimes low affinity of small fragments can be a problem as the difference in size of the endogenous ligands between GPCRs (e.g. neurotransmitters versus chemokines) influences the shape of the orthosteric binding pocket and thereby its recognition of fragments-like ligands (de Kloe et al. 2009). Therefore a thorough retrospective validation is *pivotal* for a successful SBVS for fragment-like molecules, which emphasizes the need for GPCR-target annotated fragment libraries with true actives *and* true inactives (de Graaf et al. 2013).

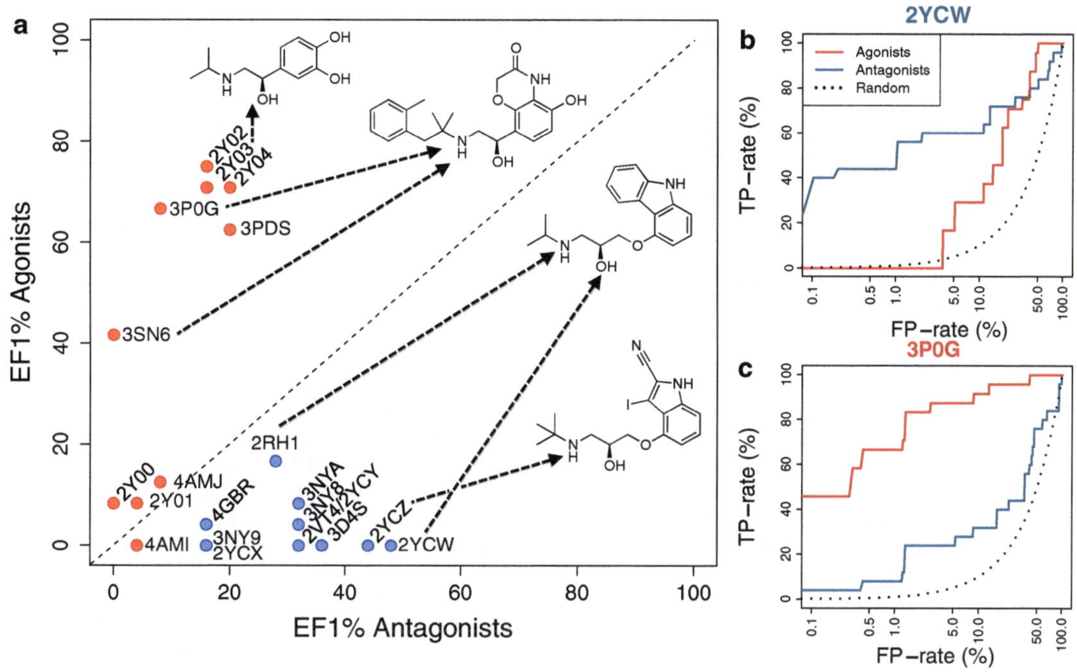

Fig. 7.5 Overall enrichment at 1 % false positive rate (FP-rate) for the retrieval of 25 partial/full agonists (de Graaf and Rognan 2008; Baker 2010) and 24 inverse agonists/antagonists of ADRB1/ADRB2 (de Graaf and Rognan 2008; Baker 2005) over a set of 980 decoy molecules (de Graaf and Rognan 2008) (**a**) using IFP scoring (see Fig. 7.4c). Full ROC curves visualizing the retrieval rate (TP-rate) of agonists (*red*) and antagonists (*blue*) in the best antagonist crystal structure (**b**) and best agonist crystal structure (**c**, legend shown in **b**)

The increasing number of GPCR crystal structures (Katritch et al. 2013) has brought a wealth of detailed structural insights into the molecular recognition mechanisms of (small) ligands. This has opened up new possibilities for (high resolution) homology modeling as they facilitate better templates and insights in regions with relatively high structural variability (e.g., extracellular loops) as well as the structural features of sequence-specific motifs (e.g. the helical kink induced by the TXP motif in CXCR4 (Wu et al. 2010)). The new GPCR crystal structures in combination with molecular dynamics simulations give insights into receptor flexibility and (potential) ligand access and exit channels (Kruse et al. 2012; Zhang et al. 2012; Dror et al. 2011). The association and dissociation pathways revealed by computational simulations in combination with experimental studies (e.g. mutagenesis data and biophysical measurements (Congreve et al. 2012)) can in time be used to relate ligand structure to kinetic properties, thereby changing the focus from solely affinity-based optimizations to optimization of kinetic properties (Tresadern et al. 2011; Miller et al. 2012). As discussed in Sect. 7.5 the (relative) plethora of crystal structures of the adrenergic receptors including the first (biased) agonist bound (active and inactive) GPCR crystal structures show how subtle differences in the binding pocket can accommodate agonist binding, but also the molecular mechanisms of G-protein binding and activation (Nygaard et al. 2013). A better understanding of the ligand binding modes (Katritch and Abagyan 2011), binding pockets (Rosenkilde et al. 2010), and conformational changes (Liu et al. 2012; Bokoch et al. 2010) associated with (specific) signaling pathways will allow the development of selective SBVS strategies for agonists over antagonists (or *vice versa*) or ultimately even for *biased* ligands.

References

Andrews SP, Benjamin T (2013) Stabilised G protein-coupled receptors in structure-based drug design: a case study with adenosine A2A receptor. MedChemComm 4(1):52–67. doi:10.1039/c2md20164j

Baell JB, Holloway GA (2010) New substructure filters for removal of pan assay interference compounds (PAINS) from screening libraries and for their exclusion in bioassays. J Med Chem 53(7):2719–2740. doi:10.1021/jm901137j

Baker JG (2005) The selectivity of beta-adrenoceptor antagonists at the human beta1, beta2 and beta3 adrenoceptors. Br J Pharmacol 144(3):317–322. doi:10.1038/sj.bjp.0706048

Baker JG (2010) The selectivity of beta-adrenoceptor agonists at human beta1-, beta2- and beta3-adrenoceptors. Br J Pharmacol 160(5):1048–1061. doi:10.1111/j.1476-5381.2010.00754.x

Ballesteros JA, Weinstein H (1995) Integrated methods for the construction of three-dimensional models and computational probing of structure-function relations of G protein-coupled receptors. Methods Neurosci 25:366–428. doi:10.1016/S1043-9471(05)80049-7

Barillari C, Marcou G, Rognan D (2008) Hot-spots-guided receptor-based pharmacophores (HS-Pharm): a knowledge-based approach to identify ligand-anchoring atoms in protein cavities and prioritize structure-based pharmacophores. J Chem Inf Model 48(7):1396–1410. doi:10.1021/ci800064z

Becker OM, Marantz Y, Shacham S, Inbal B, Heifetz A, Kalid O, Bar-Haim S, Warshaviak D, Fichman M, Noiman S (2004) G protein-coupled receptors: in silico drug discovery in 3D. Proc Natl Acad Sci U S A 101(31):11304–11309. doi:10.1073/pnas.0401862101

Becker OM, Dhanoa DS, Marantz Y, Chen D, Shacham S, Cheruku S, Heifetz A, Mohanty P, Fichman M, Sharadendu A, Nudelman R, Kauffman M, Noiman S (2006) An integrated in silico 3D model-driven discovery of a novel, potent, and selective amidosulfonamide 5-HT1A agonist (PRX-00023) for the treatment of anxiety and depression. J Med Chem 49(11):3116–3135. doi:10.1021/jm0508641

Bissantz C, Bernard P, Hibert M, Rognan D (2003) Protein-based virtual screening of chemical databases. II. Are homology models of G-protein coupled receptors suitable targets? Proteins 50(1):5–25. doi:10.1002/prot.10237

Blättermann S, Peters L, Ottersbach PA, Bock A, Konya V, Weaver CD, Gonzalez A, Schroder R, Tyagi R, Luschnig P, Gab J, Hennen S, Ulven T, Pardo L, Mohr K, Gutschow M, Heinemann A, Kostenis E (2012) A biased ligand for OXE-R uncouples G alpha and G beta gamma signaling within a heterotrimer. Nat Chem Biol 8(7):631–638. doi:10.1038/nchembio.962

Bokoch MP, Zou Y, Rasmussen SG, Liu CW, Nygaard R, Rosenbaum DM, Fung JJ, Choi HJ, Thian FS, Kobilka TS, Puglisi JD, Weis WI, Pardo L, Prosser RS, Mueller L, Kobilka BK (2010) Ligand-specific regulation of the extracellular surface of a G-protein-coupled receptor. Nature 463(7277):108–112. doi:10.1038/nature08650

Carlsson J, Yoo L, Gao ZG, Irwin JJ, Shoichet BK, Jacobson KA (2010) Structure-based discovery of A2A adenosine receptor ligands. J Med Chem 53(9):3748–3755. doi:10.1021/jm100240h

Carlsson J, Coleman RG, Setola V, Irwin JJ, Fan H, Schlessinger A, Sali A, Roth BL, Shoichet BK (2011) Ligand discovery from a dopamine D3 receptor homology model and crystal structure. Nat Chem Biol 7(11):769–778. doi:10.1038/nchembio.662

Cavasotto CN (2011) Homology models in docking and high-throughput docking. Curr Top Med Chem 11(12):1528–1534. doi:10.2174/156802611795860951

Cavasotto CN, Orry AJ, Murgolo NJ, Czarniecki MF, Kocsi SA, Hawes BE, O'Neill KA, Hine H, Burton MS, Voigt JH, Abagyan RA, Bayne ML, Monsma FJ Jr (2008) Discovery of novel chemotypes to a G-protein-coupled receptor through ligand-steered homology modeling and structure-based virtual screening. J Med Chem 51(3):581–588. doi:10.1021/jm070759m

Chen JZ, Wang J, Xie XQ (2007) GPCR structure-based virtual screening approach for CB2 antagonist search. J Chem Inf Model 47(4):1626–1637. doi:10.1021/ci7000814

Cherezov V, Rosenbaum DM, Hanson MA, Rasmussen SG, Thian FS, Kobilka TS, Choi HJ, Kuhn P, Weis WI, Kobilka BK, Stevens RC (2007) High-resolution crystal structure of an engineered human beta2-adrenergic G protein-coupled receptor. Science 318(5854):1258–1265. doi:10.1126/science.1150577

Cherezov V, Abola E, Stevens RC (2010) Recent progress in the structure determination of GPCRs, a membrane protein family with high potential as pharmaceutical targets. Methods Mol Biol 654:141–168. doi:10.1007/978-1-60761-762-4_8

Chien EY, Liu W, Zhao Q, Katritch V, Han GW, Hanson MA, Shi L, Newman AH, Javitch JA, Cherezov V, Stevens RC (2010) Structure of the human dopamine D3 receptor in complex with a D2/D3 selective antagonist. Science 330(6007):1091–1095. doi:10.1126/science.1197410

Congreve M, Langmead CJ, Mason JS, Marshall FH (2011) Progress in structure based drug design for G protein-coupled receptors. J Med Chem 54(13):4283–4311. doi:10.1021/jm200371q

Congreve M, Andrews SP, Dore AS, Hollenstein K, Hurrell E, Langmead CJ, Mason JS, Ng IW, Tehan B, Zhukov A, Weir M, Marshall FH (2012) Discovery of 1,2,4-triazine derivatives as adenosine A(2A) antagonists using structure based drug design. J Med Chem 55(5):1898–1903. doi:10.1021/jm201376w

Costanzi S, Vilar S (2012) In silico screening for agonists and blockers of the beta(2) adrenergic receptor: implications of inactive and activated state structures. J Comput Chem 33(5):561–572. doi:10.1002/jcc.22893

Costanzi S, Santhosh Kumar T, Balasubramanian R, Kendall Harden T, Jacobson KA (2012) Virtual screening leads to the discovery of novel non-nucleotide P2Y(1) receptor antagonists. Bioorg Med Chem 20(17):5254–5261. doi:10.1016/j.bmc.2012.06.044

de Graaf C, Rognan D (2008) Selective structure-based virtual screening for full and partial agonists of the beta2 adrenergic receptor. J Med Chem 51(16):4978–4985. doi:10.1021/jm800710x

de Graaf C, Rognan D (2009) Customizing G protein-coupled receptor models for structure-based virtual screening. Curr Pharm Des 15(35):4026–4048. doi:10.2174/138161209789824786

de Graaf C, Foata N, Engkvist O, Rognan D (2008) Molecular modeling of the second extracellular loop of G-protein coupled receptors and its implication on structure-based virtual screening. Proteins 71(2):599–620. doi:10.1002/prot.21724

de Graaf C, Rein C, Piwnica D, Giordanetto F, Rognan D (2011a) Structure-based discovery of allosteric modulators of two related class B G-protein-coupled receptors. ChemMedChem 6(12):2159–2169. doi:10.1002/cmdc.201100317

de Graaf C, Kooistra AJ, Vischer HF, Katritch V, Kuijer M, Shiroishi M, Iwata S, Shimamura T, Stevens RC, de Esch IJ, Leurs R (2011b) Crystal structure-based virtual screening for fragment-like ligands of the human histamine H(1) receptor. J Med Chem 54(23):8195–8206. doi:10.1021/jm2011589

de Graaf C, Vischer HF, de Kloe GE, Kooistra AJ, Nijmeijer S, Kuijer M, Verheij MH, England PJ, van Muijlwijk-Koezen JE, Leurs R, de Esch IJ (2013) Small and colorful stones make beautiful mosaics: fragment-based chemogenomics. Drug Discov Today 18(7–8):323–330. doi:10.1016/j.drudis.2012.12.003

de Kloe GE, Bailey D, Leurs R, de Esch IJ (2009) Transforming fragments into candidates: small becomes big in medicinal chemistry. Drug Discov Today 14(13–14):630–646. doi:10.1016/j.drudis.2009.03.009

Deng Z, Chuaqui C, Singh J (2004) Structural interaction fingerprint (SIFt): a novel method for analyzing three-dimensional protein-ligand binding interactions. J Med Chem 47(2):337–344. doi:10.1021/jm030331x

Dror RO, Pan AC, Arlow DH, Borhani DW, Maragakis P, Shan Y, Xu H, Shaw DE (2011) Pathway and mechanism of drug binding to G-protein-coupled receptors. Proc Natl Acad Sci U S A 108(32):13118–13123. doi:10.1073/pnas.1104614108

Edwards BS, Bologa C, Young SM, Balakin KV, Prossnitz ER, Savchuck NP, Sklar LA, Oprea TI (2005) Integration of virtual screening with high-throughput flow cytometry to identify novel small molecule formylpeptide receptor antagonists. Mol Pharmacol 68(5):1301–1310. doi:10.1124/mol.105.014068

Engel S, Skoumbourdis AP, Childress J, Neumann S, Deschamps JR, Thomas CJ, Colson AO, Costanzi S, Gershengorn MC (2008) A virtual screen for diverse ligands: discovery of selective G protein-coupled receptor antagonists. J Am Chem Soc 130(15):5115–5123. doi:10.1021/ja077620l

Evers A, Klabunde T (2005) Structure-based drug discovery using GPCR homology modeling: successful virtual screening for antagonists of the alpha1A adrenergic receptor. J Med Chem 48(4):1088–1097. doi:10.1021/jm0491804

Evers A, Klebe G (2004) Successful virtual screening for a submicromolar antagonist of the neurokinin-1 receptor based on a ligand-supported homology model. J Med Chem 47(22):5381–5392. doi:10.1021/jm0311487

Evers A, Hessler G, Matter H, Klabunde T (2005) Virtual screening of biogenic amine-binding G-protein coupled receptors: comparative evaluation of protein- and ligand-based virtual screening protocols. J Med Chem 48(17):5448–5465. doi:10.1021/jm050090o

Gaulton A, Bellis LJ, Bento AP, Chambers J, Davies M, Hersey A, Light Y, McGlinchey S, Michalovich D, Al-Lazikani B, Overington JP (2012) ChEMBL: a large-scale bioactivity database for drug discovery. Nucleic Acids Res 40(Database issue):D1100–D1107. doi:10.1093/nar/gkr777

Gloriam DE, Wellendorph P, Johansen LD, Thomsen AR, Phonekeo K, Pedersen DS, Brauner-Osborne H (2011) Chemogenomic discovery of allosteric antagonists at the GPRC6A receptor. Chem Biol 18(11):1489–1498. doi:10.1016/j.chembiol.2011.09.012

Granier S, Manglik A, Kruse AC, Kobilka TS, Thian FS, Weis WI, Kobilka BK (2012) Structure of the delta-opioid receptor bound to naltrindole. Nature 485(7398):400–404. doi:10.1038/nature11111

Haga K, Kruse AC, Asada H, Yurugi-Kobayashi T, Shiroishi M, Zhang C, Weis WI, Okada T, Kobilka BK, Haga T, Kobayashi T (2012) Structure of the human M2 muscarinic acetylcholine receptor bound to an antagonist. Nature 482:547–551. doi:10.1038/nature10753

Hanson MA, Cherezov V, Griffith MT, Roth CB, Jaakola VP, Chien EY, Velasquez J, Kuhn P, Stevens RC (2008) A specific cholesterol binding site is established by the 2.8 A structure of the human beta2-adrenergic receptor. Structure 16(6):897–905. doi:10.1016/j.str.2008.05.001

Hanson MA, Roth CB, Jo E, Griffith MT, Scott FL, Reinhart G, Desale H, Clemons B, Cahalan SM, Schuerer SC, Sanna MG, Han GW, Kuhn P, Rosen H, Stevens RC (2012) Crystal structure of a lipid G protein-coupled receptor. Science 335(6070):851–855. doi:10.1126/science.1215904

Hopkins AL, Groom CR, Alex A (2004) Ligand efficiency: a useful metric for lead selection. Drug Discov Today 9(10):430–431. doi:10.1016/S1359-6446(04)03069-7

Istyastono EP (2012) Computational studies of histamine H4 receptor-ligand interactions. VU University Amsterdam, Amsterdam. ISBN 978-90-8570-994-7

Jaakola VP, Griffith MT, Hanson MA, Cherezov V, Chien EY, Lane JR, Ijzerman AP, Stevens RC (2008) The 2.6 angstrom crystal structure of a human A2A adenosine receptor bound to an antagonist. Science 322(5905):1211–1217. doi:10.1126/science.1164772

Jaakola VP, Lane JR, Lin JY, Katritch V, Ijzerman AP, Stevens RC (2010) Ligand binding and subtype selectivity of the human A(2A) adenosine receptor:

identification and characterization of essential amino acid residues. J Biol Chem 285(17):13032–13044. doi:10.1074/jbc.M109.096974

Katritch V, Abagyan R (2011) GPCR agonist binding revealed by modeling and crystallography. Trends Pharmacol Sci 32(11):637–643. doi:10.1016/j.tips.2011.08.001

Katritch V, Rueda M, Lam PC, Yeager M, Abagyan R (2010a) GPCR 3D homology models for ligand screening: lessons learned from blind predictions of adenosine A2a receptor complex. Proteins 78(1):197–211. doi:10.1002/prot.22507

Katritch V, Jaakola VP, Lane JR, Lin J, Ijzerman AP, Yeager M, Kufareva I, Stevens RC, Abagyan R (2010b) Structure-based discovery of novel chemotypes for adenosine A(2A) receptor antagonists. J Med Chem 53(4):1799–1809. doi:10.1021/jm901647p

Katritch V, Cherezov V, Stevens RC (2012) Diversity and modularity of G protein-coupled receptor structures. Trends Pharmacol Sci 33(1):17–27. doi:10.1016/j.tips.2011.09.003

Katritch V, Cherezov V, Stevens RC (2013) Structure-function of the g protein-coupled receptor superfamily. Annu Rev Pharmacol Toxicol 53:531–556. doi:10.1146/annurev-pharmtox-032112-135923

Keiser MJ, Roth BL, Armbruster BN, Ernsberger P, Irwin JJ, Shoichet BK (2007) Relating protein pharmacology by ligand chemistry. Nat Biotechnol 25(2):197–206. doi:10.1038/nbt1284

Kellenberger E, Springael JY, Parmentier M, Hachet-Haas M, Galzi JL, Rognan D (2007) Identification of nonpeptide CCR5 receptor agonists by structure-based virtual screening. J Med Chem 50(6):1294–1303. doi:10.1021/jm061389p

Kim J, Wess J, van Rhee AM, Schoneberg T, Jacobson KA (1995) Site-directed mutagenesis identifies residues involved in ligand recognition in the human A2a adenosine receptor. J Biol Chem 270(23):13987–13997. doi:10.1074/jbc.270.23.13987

Kim J, Yip ML, Shen X, Li H, Hsin LY, Labarge S, Heinrich EL, Lee W, Lu J, Vaidehi N (2012) Identification of anti-malarial compounds as novel antagonists to chemokine receptor CXCR4 in pancreatic cancer cells. PLoS One 7(2):e31004. doi:10.1371/journal.pone.0031004

Kiss R, Kiss B, Konczol A, Szalai F, Jelinek I, Laszlo V, Noszal B, Falus A, Keseru GM (2008) Discovery of novel human histamine H4 receptor ligands by large-scale structure-based virtual screening. J Med Chem 51(11):3145–3153. doi:10.1021/jm7014777

Kiss GN, Fells JI, Gupte R, Lee SC, Liu J, Nusser N, Lim KG, Ray RM, Lin FT, Parrill AL, Sumegi B, Miller DD, Tigyi G (2012) Virtual screening for LPA2-specific agonists identifies a nonlipid compound with antiapoptotic actions. Mol Pharmacol 82(6):1162–1173. doi:10.1124/mol.112.079699

Klabunde T, Giegerich C, Evers A (2009) Sequence-derived three-dimensional pharmacophore models for G-protein-coupled receptors and their application in virtual screening. J Med Chem 52(9):2923–2932. doi:10.1021/jm9001346

Knox C, Law V, Jewison T, Liu P, Ly S, Frolkis A, Pon A, Banco K, Mak C, Neveu V, Djoumbou Y, Eisner R, Guo AC, Wishart DS (2011) DrugBank 3.0: a comprehensive resource for 'omics' research on drugs. Nucleic Acids Res 39(Database issue):D1035–D1041. doi:10.1093/nar/gkq1126

Kolb P, Rosenbaum DM, Irwin JJ, Fung JJ, Kobilka BK, Shoichet BK (2009) Structure-based discovery of beta2-adrenergic receptor ligands. Proc Natl Acad Sci U S A 106(16):6843–6848. doi:10.1073/pnas.0812657106

Kolb P, Phan K, Gao ZG, Marko AC, Sali A, Jacobson KA (2012) Limits of ligand selectivity from docking to models: in silico screening for A(1) adenosine receptor antagonists. PLoS One 7(11):e49910. doi:10.1371/journal.pone.0049910

Kooistra AJ, Roumen L, Leurs R, de Esch IJP, de Graaf C (2013) From heptahelical bundle to hits from the Haystack: structure-based virtual screening for GPCR ligands. Methods Enzymol 522:279–336. doi:10.1016/B978-0-12-407865-9.00015-7

Kruse AC, Hu J, Pan AC, Arlow DH, Rosenbaum DM, Rosemond E, Green HF, Liu T, Chae PS, Dror RO, Shaw DE, Weis WI, Wess J, Kobilka BK (2012) Structure and dynamics of the M3 muscarinic acetylcholine receptor. Nature 482:552–556. doi:10.1038/nature10753

Kufareva I, Rueda M, Katritch V, Stevens RC, Abagyan R (2011) Participants, G.D.: status of GPCR modeling and docking as reflected by community-wide GPCR dock 2010 assessment. Structure 19(8):1108–1126. doi:10.1016/j.str.2011.05.012

Lagerstrom MC, Schioth HB (2008) Structural diversity of G protein-coupled receptors and significance for drug discovery. Nat Rev Drug Discov 7(4):339–357. doi:10.1038/nrd2518

Langmead CJ, Andrews SP, Congreve M, Errey JC, Hurrell E, Marshall FH, Mason JS, Richardson CM, Robertson N, Zhukov A, Weir M (2012) Identification of novel adenosine A2A receptor antagonists by virtual screening. J Med Chem 55(5):1904–1909. doi:10.1021/jm201455y

Leach AR, Hann MM (2011) Molecular complexity and fragment-based drug discovery: ten years on. Curr Opin Chem Biol 15(4):489–496. doi:10.1016/j.cbpa.2011.05.008

Lin X, Huang XP, Chen G, Whaley R, Peng S, Wang Y, Zhang G, Wang SX, Wang S, Roth BL, Huang N (2012) Life beyond kinases: structure-based discovery of sorafenib as nanomolar antagonist of 5-HT receptors. J Med Chem 55(12):5749–5759. doi:10.1021/jm300338m

Lipinski CA, Lombardo F, Dominy BW, Feeney PJ (2001) Experimental and computational approaches to estimate solubility and permeability in drug discovery and development settings. Adv Drug Deliv Rev 46(1–3):3–26. doi:10.1016/S0169-409X(00)00129-0

Liu T, Lin Y, Wen X, Jorissen RN, Gilson MK (2007) BindingDB: a web-accessible database of experimentally determined protein-ligand binding affinities. Nucleic Acids Res 35(Database issue):D198–D201. doi:10.1093/nar/gkl999

Liu Y, Zhou E, Yu K, Zhu J, Zhang Y, Xie X, Li J, Jiang H (2008) Discovery of a novel CCR5 antagonist lead compound through fragment assembly. Molecules 13(10):2426–2441. doi:10.3390/molecules13102426

Liu JJ, Horst R, Katritch V, Stevens RC, Wuthrich K (2012) Biased signaling pathways in beta2-adrenergic receptor characterized by 19F-NMR. Science 335(6072):1106–1110. doi:10.1126/science.1215802

Loving K, Alberts I, Sherman W (2010) Computational approaches for fragment-based and de novo design. Curr Top Med Chem 10(1):14–32. doi:10.2174/156802610790232305

Malherbe P, Kratochwil N, Muhlemann A, Zenner MT, Fischer C, Stahl M, Gerber PR, Jaeschke G, Porter RH (2006) Comparison of the binding pockets of two chemically unrelated allosteric antagonists of the mGlu5 receptor and identification of crucial residues involved in the inverse agonism of MPEP. J Neurochem 98(2):601–615. doi:10.1111/j.1471-4159.2006.03886.x

Manglik A, Kruse AC, Kobilka TS, Thian FS, Mathiesen JM, Sunahara RK, Pardo L, Weis WI, Kobilka BK, Granier S (2012) Crystal structure of the micro-opioid receptor bound to a morphinan antagonist. Nature 485(7398):321–326. doi:10.1038/nature10954

Marcou G, Rognan D (2007) Optimizing fragment and scaffold docking by use of molecular interaction fingerprints. J Chem Inf Model 47(1):195–207. doi:10.1021/ci600342e

Michino M, Abola E, Participants GD, Brooks CL 3rd, Dixon JS, Moult J, Stevens RC (2009) Community-wide assessment of GPCR structure modelling and ligand docking: GPCR dock 2008. Nat Rev Drug Discov 8(6):455–463. doi:10.1038/nrd2877

Miller LJ, Chen Q, Lam PC, Pinon DI, Sexton PM, Abagyan R, Dong M (2011) Refinement of glucagon-like peptide 1 docking to its intact receptor using mid-region photolabile probes and molecular modeling. J Biol Chem 286(18):15895–15907. doi:10.1074/jbc.M110.217901

Miller DC, Lunn G, Jones P, Sabnis Y, Davies NL, Driscoll P (2012) Investigation of the effect of molecular properties on the binding kinetics of a ligand to its biological target. MedChemComm 3:449–452. doi:10.1021/ci200088d

Moitessier N, Englebienne P, Lee D, Lawandi J, Corbeil CR (2008) Towards the development of universal, fast and highly accurate docking/scoring methods: a long way to go. Br J Pharmacol 153(Suppl 1):S7–S26. doi:10.1038/sj.bjp.0707515

Moukhametzianov R, Warne T, Edwards PC, Serrano-Vega MJ, Leslie AG, Tate CG, Schertler GF (2011) Two distinct conformations of helix 6 observed in antagonist-bound structures of a beta1-adrenergic receptor. Proc Natl Acad Sci U S A 108(20):8228–8232. doi:10.1073/pnas.1100185108

Moura Barbosa AJ, Del Rio A (2012) Freely accessible databases of commercial compounds for high-throughput virtual screenings. Curr Top Med Chem 12(8):866–877. doi:10.2174/156802612800166710

Muegge I (2003) Selection criteria for drug-like compounds. Med Res Rev 23(3):302–321. doi:10.1002/med.10041

Mysinger MM, Weiss DR, Ziarek JJ, Gravel S, Doak AK, Karpiak J, Heveker N, Shoichet BK, Volkman BF (2012) Structure-based ligand discovery for the protein-protein interface of chemokine receptor CXCR4. Proc Natl Acad Sci U S A 109(14):5517–5522. doi:10.1073/pnas.1120431109

Nygaard R, Zou Y, Dror RO, Mildorf TJ, Arlow DH, Manglik A, Pan AC, Liu CW, Fung JJ, Bokoch MP, Thian FS, Kobilka TS, Shaw DE, Mueller L, Prosser RS, Kobilka BK (2013) The dynamic process of beta(2)-adrenergic receptor activation. Cell 152(3):532–542. doi:10.1016/j.cell.2013.01.008

Olah H, Rad R, Ostopovici L, Bora A, Hadaruga N, Hadaruga D, Moldovan R, Fulias A, Mracec M, Oprea TI (2007) WOMBAT and WOMBAT-PK: bioactivity databases for lead and drug discovery. In: Schreiber SL, Kapoor T, Wess G (eds) Chemical biology: from small molecules to systems biology and drug design. Wiley-VCH, New York, pp 760–786. doi:10.1002/9783527619375.ch13b

Palczewski K, Kumasaka T, Hori T, Behnke CA, Motoshima H, Fox BA, Le Trong I, Teller DC, Okada T, Stenkamp RE, Yamamoto M, Miyano M (2000) Crystal structure of rhodopsin: a G protein-coupled receptor. Science 289(5480):739–745. doi:10.1126/science.289.5480.739

Petrel C, Kessler A, Dauban P, Dodd RH, Rognan D, Ruat M (2004) Positive and negative allosteric modulators of the Ca2+-sensing receptor interact within overlapping but not identical binding sites in the transmembrane domain. J Biol Chem 279(18):18990–18997. doi:10.1074/jbc.M400724200

Rasmussen SG, Choi HJ, Rosenbaum DM, Kobilka TS, Thian FS, Edwards PC, Burghammer M, Ratnala VR, Sanishvili R, Fischetti RF, Schertler GF, Weis WI, Kobilka BK (2007) Crystal structure of the human beta2 adrenergic G-protein-coupled receptor. Nature 450(7168):383–387. doi:10.1038/nature06325

Rasmussen SG, Choi HJ, Fung JJ, Pardon E, Casarosa P, Chae PS, Devree BT, Rosenbaum DM, Thian FS, Kobilka TS, Schnapp A, Konetzki I, Sunahara RK, Gellman SH, Pautsch A, Steyaert J, Weis WI, Kobilka BK (2011a) Structure of a nanobody-stabilized active state of the beta(2) adrenoceptor. Nature 469(7329):175–180. doi:10.1038/nature09648

Rasmussen SG, DeVree BT, Zou Y, Kruse AC, Chung KY, Kobilka TS, Thian FS, Chae PS, Pardon E, Calinski D, Mathiesen JM, Shah ST, Lyons JA, Caffrey M, Gellman SH, Steyaert J, Skiniotis G, Weis WI, Sunahara RK, Kobilka BK (2011b) Crystal structure of the beta2 adrenergic receptor-Gs protein complex. Nature 477(7366):549–555. doi:10.1038/nature10361

Renault N, Laurent X, Farce A, El Bakali J, Mansouri R, Gervois P, Millet R, Desreumaux P, Furman C, Chavatte P (2012) Virtual screening of CB(2) receptor agonists from Bayesian network and high-throughput docking: structural insights into agonist-modulated GPCR features. Chem Biol Drug Des. doi:10.1111/cbdd.12095

Rodriguez D, Gutierrez-de-Teran H (2013) Computational approaches for ligand discovery and design in class-A G protein-coupled receptors. Curr Pharm Des 19(12):2216–2236. doi:10.2174/1381612811319120009

Rosenbaum DM, Cherezov V, Hanson MA, Rasmussen SG, Thian FS, Kobilka TS, Choi HJ, Yao XJ, Weis WI, Stevens RC, Kobilka BK (2007) GPCR engineering yields high-resolution structural insights into beta2-adrenergic receptor function. Science 318(5854):1266–1273. doi:10.1126/science.1150609

Rosenbaum DM, Zhang C, Lyons JA, Holl R, Aragao D, Arlow DH, Rasmussen SG, Choi HJ, Devree BT, Sunahara RK, Chae PS, Gellman SH, Dror RO, Shaw DE, Weis WI, Caffrey M, Gmeiner P, Kobilka BK (2011) Structure and function of an irreversible agonist-beta(2) adrenoceptor complex. Nature 469(7329):236–240. doi:10.1038/nature09665

Rosenkilde MM, Benned-Jensen T, Frimurer TM, Schwartz TW (2010) The minor binding pocket: a major player in 7TM receptor activation. Trends Pharmacol Sci 31(12):567–574. doi:10.1016/j.tips.2010.08.006

Roumen L, Sanders MP, Vroling B, de Esch IJ, de Vlieg J, Leurs R, Klomp JP, Nabuurs SB, de Graaf C (2011) In silico veritas: the pitfalls and challenges of predicting GPCR-ligand interactions. Pharmaceuticals 4(9):1196–1215. doi:10.1021/ci200088d

Sabio M, Jones K, Topiol S (2008) Use of the X-ray structure of the beta2-adrenergic receptor for drug discovery. Part 2: identification of active compounds. Bioorg Med Chem Lett 18(20):5391–5395. doi:10.1016/j.bmcl.2008.09.046

Salo OM, Raitio KH, Savinainen JR, Nevalainen T, Lahtela-Kakkonen M, Laitinen JT, Jarvinen T, Poso A (2005) Virtual screening of novel CB2 ligands using a comparative model of the human cannabinoid CB2 receptor. J Med Chem 48(23):7166–7171. doi:10.1021/jm050565b

Salon JA, Lodowski DT, Palczewski K (2011) The significance of G protein-coupled receptor crystallography for drug discovery. Pharmacol Rev 63(4):901–937. doi:10.1124/pr.110.003350

Sanders MP, Verhoeven S, de Graaf C, Roumen L, Vroling B, Nabuurs SB, de Vlieg J, Klomp JP (2011) Snooker: a structure-based pharmacophore generation tool applied to class A GPCRs. J Chem Inf Model 51(9):2277–2292. doi:10.1021/ci200088d

Sanders MP, McGuire R, Roumen L, de Esch IJ, de Vlieg J, Klomp JP, de Graaf C (2012) From the protein's perspective: the benefits and challenges of protein structure-based pharmacophore modeling. Med Chem Comm 3(1):28–38. doi:10.1021/ci200088d

Shimamura T, Shiroishi M, Weyand S, Tsujimoto H, Winter G, Katritch V, Abagyan R, Cherezov V, Liu W, Han GW, Kobayashi T, Stevens RC, Iwata S (2011) Structure of the human histamine H1 receptor complex with doxepin. Nature 475(7354):65–70. doi:10.1038/nature10236

Sirci F, Istyastono EP, Vischer HF, Kooistra AJ, Nijmeijer S, Kuijer M, Wijtmans M, Mannhold R, Leurs R, de Esch IJ, de Graaf C (2012) Virtual fragment screening: discovery of histamine h(3) receptor ligands using ligand-based and protein-based molecular fingerprints. J Chem Inf Model 52(12):3308–3324. doi:10.1021/ci3004094

Stevens RC, Cherezov V, Katritch V, Abagyan R, Kuhn P, Rosen H, Wuthrich K (2012) The GPCR network: a large-scale collaboration to determine human GPCR structure and function. Nat Rev Drug Discov 12(1):25–34. doi:10.1038/nrd3859

Sum CS, Tikhonova IG, Neumann S, Engel S, Raaka BM, Costanzi S, Gershengorn MC (2007) Identification of residues important for agonist recognition and activation in GPR40. J Biol Chem 282(40):29248–29255. doi:10.1074/jbc.M705077200

Surgand JS, Rodrigo J, Kellenberger E, Rognan D (2006) A chemogenomic analysis of the transmembrane binding cavity of human G-protein-coupled receptors. Proteins 62(2):509–538. doi:10.1002/prot.20768

Tang H, Wang XS, Hsieh JH, Tropsha A (2012) Do crystal structures obviate the need for theoretical models of GPCRs for structure based virtual screening. Proteins. doi:10.1002/prot.24035

Tautermann CS (2011) The use of G-protein coupled receptor models in lead optimization. Future Med Chem 3(6):709–721. doi:10.4155/fmc.11.24

Thompson AA, Liu W, Chun E, Katritch V, Wu H, Vardy E, Huang XP, Trapella C, Guerrini R, Calo G, Roth BL, Cherezov V, Stevens RC (2012) Structure of the nociceptin/orphanin FQ receptor in complex with a peptide mimetic. Nature 485(7398):395–399. doi:10.1038/nature11085

Tikhonova IG, Sum CS, Neumann S, Engel S, Raaka BM, Costanzi S, Gershengorn MC (2008) Discovery of novel agonists and antagonists of the free fatty acid receptor 1 (FFAR1) using virtual screening. J Med Chem 51(3):625–633. doi:10.1021/jm7012425

Topiol S, Sabio M (2008) Use of the X-ray structure of the Beta2-adrenergic receptor for drug discovery. Bioorg Med Chem Lett 18(5):1598–1602. doi:10.1016/j.bmcl.2008.01.063

Tosh DK, Phan K, Gao ZG, Gakh AA, Xu F, Deflorian F, Abagyan R, Stevens RC, Jacobson KA, Katritch V (2012) Optimization of adenosine 5′-carboxamide derivatives as adenosine receptor agonists using structure-based ligand design and fragment screening. J Med Chem 55(9):4297–4308. doi:10.1021/jm300095s

Tresadern G, Bartolome JM, Macdonald GJ, Langlois X (2011) Molecular properties affecting fast dissociation from the D2 receptor. Bioorg Med Chem 19(7):2231–2241. doi:10.1016/j.bmc.2011.02.033

Triballeau N, Van Name E, Laslier G, Cai D, Paillard G, Sorensen PW, Hoffmann R, Bertrand HO, Ngai J, Acher FC (2008) High-potency olfactory receptor agonists discovered by virtual high-throughput screening: molecular probes for receptor structure and olfactory function. Neuron 60(5):767–774. doi:10.1016/j.neuron.2008.11.014

van der Horst E, Okuno Y, Bender A, IJzerman AP (2009) Substructure mining of GPCR ligands reveals activity-class specific functional groups in an unbiased manner. J Chem Inf Model 49(2):348–360. doi:10.1021/ci8003896

Varady J, Wu X, Fang X, Min J, Hu Z, Levant B, Wang S (2003) Molecular modeling of the three-dimensional structure of dopamine 3 (D3) subtype receptor: discovery of novel and potent D3 ligands through a hybrid pharmacophore- and structure-based database searching approach. J Med Chem 46(21):4377–4392. doi:10.1021/jm030085p

Vohra S, Taddese B, Conner AC, Poyner DR, Hay DL, Barwell J, Reeves PJ, Upton GJ, Reynolds CA (2013) Similarity between class A and class B G-protein-coupled receptors exemplified through calcitonin gene-related peptide receptor modelling and mutagenesis studies. J R Soc Interface 10(79):20120846. doi:10.1098/rsif.2012.0846

Wacker D, Fenalti G, Brown MA, Katritch V, Abagyan R, Cherezov V, Stevens RC (2010) Conserved binding mode of human beta2 adrenergic receptor inverse agonists and antagonist revealed by X-ray crystallography. J Am Chem Soc 132(33):11443–11445. doi:10.1021/ja105108q

Warne T, Serrano-Vega MJ, Baker JG, Moukhametzianov R, Edwards PC, Henderson R, Leslie AG, Tate CG, Schertler GF (2008) Structure of a beta1-adrenergic G-protein-coupled receptor. Nature 454(7203):486–491. doi:10.1038/nature07101

Warne T, Moukhametzianov R, Baker JG, Nehme R, Edwards PC, Leslie AG, Schertler GF, Tate CG (2011) The structural basis for agonist and partial agonist action on a beta(1)-adrenergic receptor. Nature 469(7329):241–244. doi:10.1038/nature09746

Warne T, Edwards PC, Leslie AG, Tate CG (2012) Crystal structures of a stabilized beta1-adrenoceptor bound to the biased agonists bucindolol and carvedilol. Structure 20(5):841–849. doi:10.1016/j.str.2012.03.014

White JF, Noinaj N, Shibata Y, Love J, Kloss B, Xu F, Gvozdenovic-Jeremic J, Shah P, Shiloach J, Tate CG, Grisshammer R (2012) Structure of the agonist-bound neurotensin receptor. Nature 490(7421):508–513. doi:10.1038/nature11558

Wu B, Chien EY, Mol CD, Fenalti G, Liu W, Katritch V, Abagyan R, Brooun A, Wells P, Bi FC, Hamel DJ, Kuhn P, Handel TM, Cherezov V, Stevens RC (2010) Structures of the CXCR4 chemokine GPCR with small-molecule and cyclic peptide antagonists. Science 330(6007):1066–1071. doi:10.1126/science.1194396

Wu H, Wacker D, Mileni M, Katritch V, Han GW, Vardy E, Liu W, Thompson AA, Huang XP, Carroll FI, Mascarella SW, Westkaemper RB, Mosier PD, Roth BL, Cherezov V, Stevens RC (2012) Structure of the human kappa-opioid receptor in complex with JDTic. Nature 485(7398):327–332. doi:10.1038/nature10939

Yrjola S, Kalliokoski T, Laitinen T, Poso A, Parkkari T, Nevalainen T (2013) Discovery of novel cannabinoid receptor ligands by a virtual screening approach: further development of 2,4,6-trisubstituted 1,3,5-triazines as CB2 agonists. Eur J Pharm Sci 48(1–2):9–20. doi:10.1016/j.ejps.2012.10.020

Zhang C, Srinivasan Y, Arlow DH, Fung JJ, Palmer D, Zheng Y, Green HF, Pandey A, Dror RO, Shaw DE, Weis WI, Coughlin SR, Kobilka BK (2012) High-resolution crystal structure of human protease-activated receptor 1. Nature 492(7429):387–392. doi:10.1038/nature11701

Zhukov A, Andrews SP, Errey JC, Robertson N, Tehan B, Mason JS, Marshall FH, Weir M, Congreve M (2011) Biophysical mapping of the adenosine A2A receptor. J Med Chem 54(13):4312–4323. doi:10.1021/jm2003798

Mathematical Modeling of G Protein-Coupled Receptor Function: What Can We Learn from Empirical and Mechanistic Models?

8

David Roche, Debora Gil, and Jesús Giraldo

Abstract

Empirical and mechanistic models differ in their approaches to the analysis of pharmacological effect. Whereas the parameters of the former are not physical constants those of the latter embody the nature, often complex, of biology. Empirical models are exclusively used for curve fitting, merely to characterize the shape of the E/[A] curves. Mechanistic models, on the contrary, enable the examination of mechanistic hypotheses by parameter simulation. Regretfully, the many parameters that mechanistic models may include can represent a great difficulty for curve fitting, representing, thus, a challenge for computational method development. In the present study some empirical and mechanistic models are shown and the connections, which may appear in a number of cases between them, are analyzed from the curves they yield. It may be concluded that systematic and careful curve shape analysis can be extremely useful for the understanding of receptor function, ligand classification and drug discovery, thus providing a common language for the communication between pharmacologists and medicinal chemists.

Keywords

β-arrestin • Biased agonism • Curve fitting • Empirical modeling • Evolutionary algorithm • Functional selectivity • G protein • GPCR • Hill coefficient • Intrinsic efficacy • Inverse agonism • Mathematical modeling • Mechanistic modeling • Operational model • Parameter optimization • Receptor dimer • Receptor oligomerization • Receptor constitutive activity • Signal transduction • Two-state model

D. Roche • J. Giraldo (✉)
Laboratory of Systems Pharmacology and Bioinformatics, Institut de Neurociències and Unitat de Bioestadística, Universitat Autònoma de Barcelona, 08193 Bellaterra, Spain
e-mail: Jesus.Giraldo@uab.es

D. Gil
Computer Vision Center, Computer Science Department, Universitat Autònoma de Barcelona, 08193 Bellaterra, Spain

M. Filizola (ed.), *G Protein-Coupled Receptors - Modeling and Simulation*, Advances in Experimental Medicine and Biology 796, DOI 10.1007/978-94-007-7423-0_8,

Abbreviations

AS3D asymmetric/symmetric three-state dimer receptor model
EA evolutionary algorithm
GPCR G protein-coupled receptor

8.1 Biological Reality and the Model: Some Introductory Words

Knowledge of biological systems can be gained by examining the physiological effects that they produce under particular experimental conditions (Fig. 8.1). These effects are the result of the transduction by the system of the signal embodied in the molecular structure of a drug, where for drug we mean, in a general sense, any substance able to perturb a biological system in a concentration-dependent fashion. Because the true transduction function is generally unknown an estimate must be obtained. These functional estimates, expressed as mathematical models, are useful for the characterization of the biological system and for the classification and discovery of new drugs.

Yet, what do we understand by a mathematical model in pharmacology? A pharmacological effect is mathematically described by an equation, $E = f([A])$, relating the pharmacological effect, E, to drug concentration, [A], through an analytic function, f, which depends on some parameters. The definition of these model parameters determines the approach we are using for modeling the experimental data. In this context, there are two types of approaches: empirical and mechanistic. Empirical approaches aim at describing the function profile and, thus, model parameters are set to empirically fit experimental data. Although very sensible from a mathematical point of view, a main concern is that empirical parameters lack of any physical interpretation. An alternative approach is to identify parameters conveying part of the biological information the function is intended to represent by using mechanistic models (Giraldo et al. 2002).

Although this chapter is mostly devoted to mechanistic models, a discussion on empirical ones will be included to show that, apart from the widely used Hill equation, other models can be of interest. Our discussion focuses on a comparative analysis between empirical and mechanistic models in order to identify parallelisms

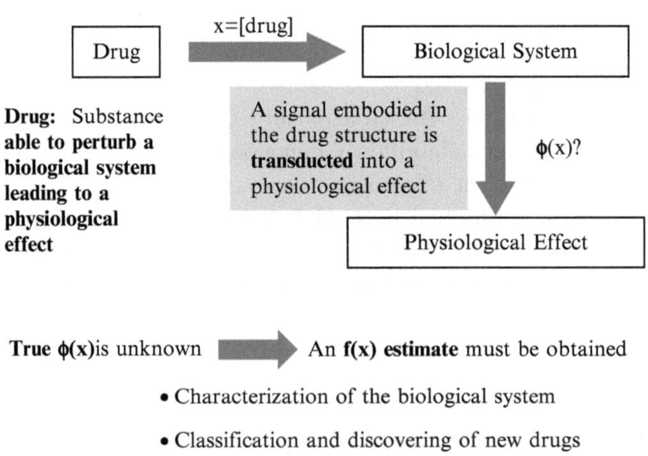

Fig. 8.1 Schematic representation of the drug action process. The perturbation that the drug exerts on the biological system can, in theory, be described by a "true" theoretical function $\phi(x)$ of all the parameters present in the system. Because this function is unknown an $f(x)$ estimate must be obtained from E/[A] experimental data. The $f(x)$ estimate can be used for both characterization of the biological system and classification and design of new drugs

and coincidences that might lead to new ideas for debate and analysis.

8.2 What Can We Learn from Empirical Models?

Customarily, drugs and receptors are compared and classified by monitoring the pharmacological effect the receptor produces with increasing [A]. The E/[A] data are commonly depicted in a semi-logarithmic scale, E/x with $x = \log[A]$, typically leading to sigmoid-shaped plots. To properly compare the experimental scatter plots, $E = f(x)$ functions are fitted to the data points and their parameters estimated. The parameter estimates allow the quantification of the geometric characteristics of the E/x curves. If we assume that basal response is absent ($E = 0$ for $[A] = 0$), then E/x curves can be geometrically described by four quantities: the upper asymptote (maximum response), the mid-point (curve location), the mid-point slope (steepness) and the inflection point (symmetry of the curve) (Fig. 8.2). Each of these quantities can be mathematically defined and have a geometric meaning but also a pharmacological interpretation.

The upper asymptote, top, reflects the efficacy of the agonist-receptor system and is defined as the value towards the effect tends as [A] increases, $\text{Top} = \lim_{x\to\infty} E$. The mid-point, x_{50}, measures the agonist potency and is defined as the x for half the top. The mid-point slope is the value of the slope of the E/x curve at the mid-point, $\left(\frac{dE}{dx}\right)_{x=x_{50}}$, and displays the sensitivity of the system to small changes in agonist concentration. Rectangular hyperbolic curves give a typical mid-point slope of 0.576 in the case they are normalized (the derivative is divided by top), while non-hyperbolic curves can be steep ($\left(\frac{dE}{dx}\right)_{x=x_{50}} > 0.576$) or flat ($\left(\frac{dE}{dx}\right)_{x=x_{50}} < 0.576$). The point of inflection, x_I, is a point on a curve at which the curvature changes from convex to concave or vice versa. For an E/x curve, this is a point at which the first derivative of the function is a maximum whereas the second derivative is equal to zero. An important remark is that the location of the point of inflection serves to assess the curve degree of symmetry. In particular, an E/x curve is symmetric if the point of inflection matches the mid-point, $x_I = x_{50}$, and asymmetric if it does not, $x_I \neq x_{50}$ (Fig. 8.2).

The above geometric features are obtained after fitting the mathematical equation $E = f([A])$ to the experimental E/x plots. Among existing $E = f(x)$ functions for data fitting, the Hill equation is, without doubt, the most widely used function in pharmacology.

8.2.1 The Hill Equation: A Model for Symmetric Concentration-Effect Curves

The Hill equation (Hill 1913) is the three-parameter Eq. 8.1.

$$E = \frac{a}{1 + 10^{m(x_b - x)}} \tag{8.1}$$

where $x = \log[A]$ and $m > 0$, m being the Hill coefficient.

The upper asymptote is a, the mid-point is $x_{50} = x_b$, the mid-point slope is $\left(\frac{dE}{dx}\right)_{x=x_{50}} = \frac{a \cdot m \cdot \ln 10}{4} = 0.576 \cdot a \cdot m$ (notice that for $m = 1$, rectangular hyperbola, the mid-point slope is equal to 0.576 after normalization by dividing by a) and the point of inflection is $x_I = x_b$. Because there is an identity between the inflection point and the mid-point, the Hill equation produces symmetric curves in all cases and, therefore, is not appropriate for fitting asymmetric E/x data.

8.2.2 But What to Do with Asymmetric Concentration-Effect Curves?

There are several empirical models capable of dealing with asymmetric E/x data: the Richards model, the Gompertz model and the modified Hill equation (Giraldo et al. 2002).

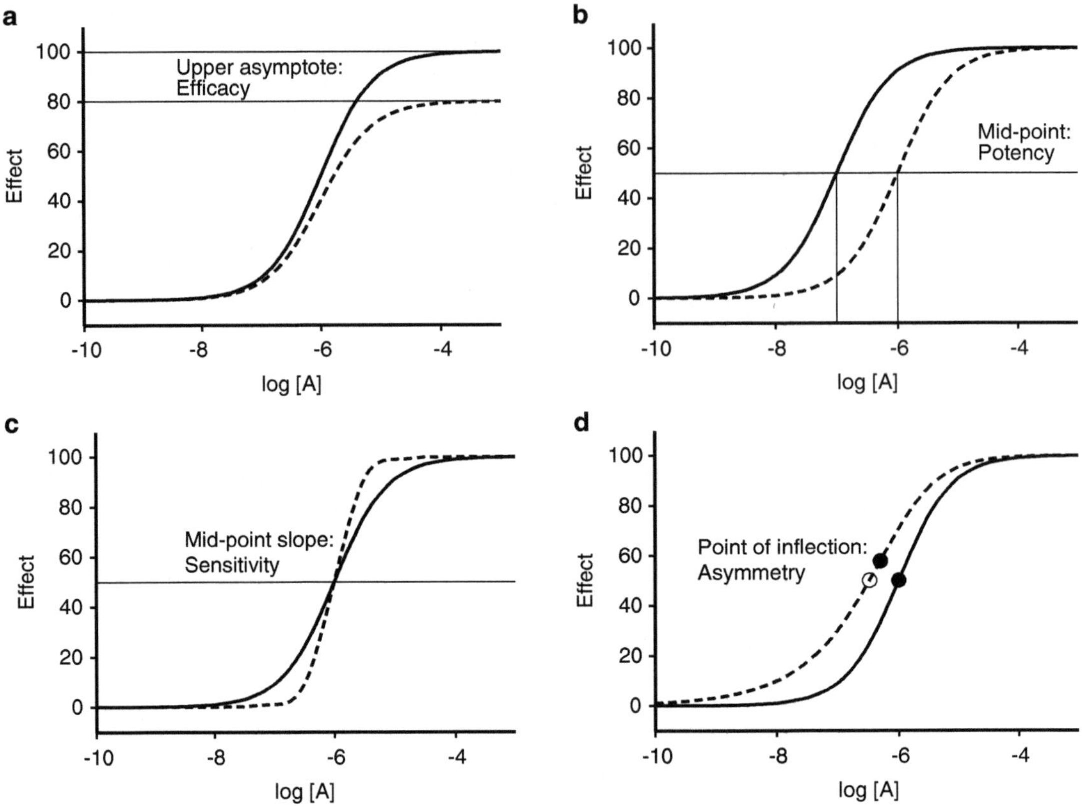

Fig. 8.2 Geometric parameters characterizing E/[A] curves. The data are represented using logarithm values for the X axis. For simplicity it is assumed that basal response is not present and then the effect for [A] = 0 is 0. (**a**) The upper asymptote (the value towards the effect tends as [A] increases; this geometric parameter reflects the efficacy of the system). (**b**). The mid-point (the value of log[A] for half maximum effect; this geometric parameter reflects the potency of the system). (**c**). The mid-point slope (this geometric parameter reflects the sensitivity of the system to small changes in [A] around the mid-point). (**d**). The point of inflection (point at which the curvature of the curve changes from convex to concave or vice versa; this geometric parameter determines the symmetry or asymmetry of the curve depending on whether the inflection point matches or not the mid-point). The *solid curve* is symmetric: the mid-point and the inflection point are coincident. The *dashed curve* is asymmetric: the mid-point and the inflection point are not coincident (*open* and *solid circles* are used for the mid-point and the inflection point, respectively)

8.2.2.1 The Richards Model

The Richards model (Richards 1959) is a generalization of the Hill equation by including an additional parameter (Eq. 8.2).

$$E = \frac{a}{\left(1 + 10^{m(x_b - x)}\right)^{s}} \tag{8.2}$$

with $s > 0$.

The upper asymptote is a, the mid-point is $x_{50} = x_b - \frac{1}{m}\log\left(2^{1/s} - 1\right)$, the mid-point slope is $\left(\frac{dE}{dx}\right)_{x=x_{50}} = \frac{a\cdot m\cdot \ln 10\cdot s\cdot\left(1-\frac{1}{2^{1/s}}\right)}{2}$, and the point of inflection is $x_I = x_b + \frac{1}{m}\log s$.

The new parameter, s, allows for asymmetry. If $s = 1$, Eq. 8.2 is equivalent to Eq. 8.1 and the theoretical curve is symmetric. Consistently with this feature, we see that if $s = 1$ then $x_I = x_{50} = x_b$. However, if $s \neq 1$ then $x_I \neq x_{50}$, and the theoretical curve is asymmetric. Interestingly, for $s \neq 1$, the degree of asymmetry of the curve, measured as the difference between x_I and x_{50}, relies on both s and m parameters, and $x_I - x_{50} = \frac{1}{m} \cdot \log\left(s \cdot \left(2^{1/s} - 1\right)\right)$. If $s > 1$ then $x_I < x_{50}$, the point of inflection is located before the mid-point, whereas if $s < 1$ then $x_I > x_{50}$, the point of inflection is placed after the mid-point. We see that for a given s value the degree of asymmetry decreases as the parameter m increases. In addition, it has been shown (Giraldo et al. 2002) that the degree of asymmetry is higher for Richards equations with $s < 1$ than for those equations with $s > 1$.

The Richards equation is able to model the degree of asymmetry present in E/x data by the inclusion of the s parameter into the modeling equation. However, by doing so, the parameter fitting becomes an overparameterized model arising from dependency among parameters, which introduces additional difficulties in data fitting (Van Der Graaf and Schoemaker 1999) (see Appendix for a discussion on data fitting). To account for asymmetry without increasing the number of parameters of the Hill equation led to the proposal of two new functions, the Gompertz model and the modified Hill equation.

8.2.2.2 The Gompertz Model

The Gompertz function (Gompertz 1825) may be written as

$$E = \frac{a}{e^{10^{m \cdot (x_I - x)}}} \tag{8.3}$$

The upper asymptote is a, the mid-point is $x_{50} = x_I - \frac{1}{m} \log(\ln 2)$, the mid-point slope is $\left(\frac{dE}{dx}\right)_{x = x_{50}} = \frac{a \cdot m \cdot \ln 10 \cdot \ln 2}{2} = 0.798 \cdot a \cdot m$, and the point of inflection is the parameter x_I. The Gompertz model is inherently asymmetric with $x_I < x_{50}$ for any value of m. The inflection point yields a fixed response value of $E = \frac{a}{e}$, lower than the response value for the mid-point, $E = \frac{a}{2}$, independently of the raw data. It can be shown (Giraldo et al. 2002) that, analogously to the Richards function, x_I tends to x_{50} as the parameter m increases. In addition, it can be proved that the Richards function tends to the Gompertz function as the asymmetric parameter s of the Richards function increases (Giraldo et al. 2002). Thus, the Gompertz function can be an alternative to model experimental asymmetric curves when the Richards function fails in curve fitting.

8.2.2.3 The Modified Hill Equation

The modified Hill equation (Sips 1948, 1950) (see (Boeynaems and Dumont 1980; Giraldo et al. 2002) for discussion) is defined as:

$$E = \frac{a}{\left(1 + 10^{x_b - x}\right)^p} \tag{8.4}$$

The modified Hill equation is equivalent to the Richards function if the m parameter of the latter model is fixed to 1. Consequently, the same happens for the geometric features of the E/x curves. The upper asymptote is a, the mid-point is $x_{50} = x_b - \log\left(2^{1/p} - 1\right)$, the mid-point slope is $\left(\frac{dE}{dx}\right)_{x = x_{50}} = \frac{a \cdot \ln 10 \cdot p \cdot \left(1 - \frac{1}{2^{1/p}}\right)}{2}$, and the point of inflection is $x_I = x_b + \log p$. We see that the parameter p contributes to both the steepness and the asymmetry of the curves. We have defined the degree of asymmetry of a curve as the difference between x_I and x_{50}. It can be seen that $x_I - x_{50} = \log\left(p \cdot \left(2^{1/p} - 1\right)\right)$. The modified Hill equation is symmetric, $x_I = x_{50} = x_b$, only for $p = 1$. If $p < 1$ then $x_I > x_{50}$ whereas if $p > 1$ then $x_I < x_{50}$.

Yet, in general and reducing the set of empirical models to the ones revised herein, how can we decide which empirical model follows the experimental data? The Hill equation, the Gompertz model and the modified Hill equation are nested within the Richards model and, therefore, the improvement in fitting that the four-parameter Richards model gains relative to the other three-parameter models can statistically be tested by an extra sum-of-squares F-test (Giraldo et al. 2002). In this context, the geometric features of the curve shape can be quantitatively assessed.

8.2.3 Generalizing the Hill Coefficient for Empirical Models

The Hill equation (Eq. 8.1) can be re-written as $\log \frac{E}{a - E} = m \cdot x - m \cdot x_b$, which is known as the Hill plot (Colquhoun 1971). The Hill plot is a straight line of slope m, commonly known as the Hill coefficient and denoted n_H. From the latter equation the Hill coefficient can be expressed as $n_H = \frac{d\log\frac{E}{a-E}}{dx}$, which in turn can be written (Giraldo 2003) as $n_H = \frac{a}{(a-E)\cdot E\cdot \ln 10}\cdot\frac{dE}{dx}$. This expression for n_H can be applied to any E/x model and thus it constitutes a practical way of generalizing the Hill coefficient. The parameter is independent of x when E/x corresponds to the Hill equation, however this may not happen in general. Accordingly, its value at the mid-point (Barlow and Blake 1989; Giraldo et al. 2002; Giraldo 2003) may solve the problem (Eq. 8.5).

$$n_{H50} = \frac{4\cdot\left(\frac{dE}{dx}\right)_{x50}}{a\cdot\ln 10} \qquad (8.5)$$

Equation 8.5 is the expression for the calculation of a generalized Hill coefficient under any empirical model. We see that is related with the slope (derivative) of the E/x curve at the mid-point and is normalized because it is divided by the asymptote a. Thus, as expressed, the Hill coefficient at the mid-point can be used as an index of the sensitivity of the pharmacological system to small changes in agonist concentration. Applying Eq. 8.5 to the above discussed empirical models yields, for the n_{H50} parameter, the values of m, $2\cdot m\cdot s\left(1-\frac{1}{2^{1/s}}\right)$, $2\cdot m\cdot\ln 2$ and $2\cdot p\cdot\left(1-\frac{1}{2^{1/p}}\right)$ under the Hill, Richards, Gompertz and modified Hill equation, respectively (Giraldo 2003).

8.2.4 Connecting Empirical and Mechanistic Models – The Asymmetry of the Curves and the Hill Coefficient

In an earlier work (Giraldo 2003) a connection between empirical and mechanistic models was shown by using a mechanistic model taken from the ion channel field, a ligand-gated ion channel with 4 binding sites. Nevertheless, the model can be applied to a general tetrameric receptor R that is activated after all the binding sites are occupied. The mechanistic model is

$$R \overset{K_1/4}{\rightleftharpoons} AR \overset{2K_2/3}{\rightleftharpoons} A_2R \overset{3K_3/2}{\rightleftharpoons} A_3R \overset{4K_4}{\rightleftharpoons} A_4R \overset{K_E}{\rightleftharpoons} A_4R^*$$

where K_1, K_2, K_3 and K_4 are the microscopic equilibrium dissociation constants and $K_E = \frac{[A_4R^*]}{[A_4R]}$ is the equilibrium constant for the opening/activation (ion channel/receptor) reaction. By defining the effect as the proportion of receptors in the open/active state, the following expression was obtained

$$E = \frac{[A]^4K_E}{\left(K_1K_2K_3K_4 + 4K_2K_3K_4[A] + 6K_3K_4[A]^2 + 4K_4[A]^3 + [A]^4(1+K_E)\right)}$$

By supposing that the efficacy is very low ($K_E \ll 1$) and no cooperativity between the binding sites ($K_i = K$), the previous mechanistic equation simplifies to

$$E = \frac{K_E}{\left(1+\frac{K}{[A]}\right)^4} = \frac{K_E}{(1+10^{\log K - x})^4}$$

where $x = \log[A]$. The latter equation corresponds to a Richards model (Eq. 8.2) with $a = K_E$, $m = 1$, $x_b = \log K$, and $s = 4$ or a Modified Hill equation with $p = 4$, with s and p defining the molecularity of the process. Thus, we see that empirical models (phenotype) may reflect some of the features that characterize mechanistic models (genotype). In addition, these values of s and $p \neq 1$ are indicative of asymmetry and may represent an example of the necessity of expanding the set of empirical models and consider that in some cases other models apart from the symmetric Hill model are needed. Applying the definition of the generalized Hill coefficient (Eq. 8.5), a value of $n_{H50} = 2\cdot 4\cdot\left(1-\frac{1}{2^{1/4}}\right) = 1.27$ is obtained, showing the influence of asymmetry in the slope at the mid-point.

We have shown an example in which a mechanistic model may be expressed as an empirical model under particular mechanistic features. It can be hypothesized that if a systematic analysis of mechanisms were done in a particular biological/pharmacological research area and the corresponding set of empirical equations were identified then we could proceed in the inverse order and try to propose some mechanistic conditions from the application of one or other empirical model. For instance, in the case of GPCRs asymmetric curves are found when total receptor and total G protein concentrations are not negligible one relative to the other ((Giraldo et al. 2002) and references therein). This indicates that the stoichiometry of the biological species in a proposed mechanistic GPCR system must be consistent with the symmetry of the experimental curves. Likewise many other molecular properties associated to GPCR function will be commented from the models presented on the next section.

8.3 What Can We Learn from Mechanistic Models?

We have seen that there is a relationship between the mechanism underlying experimental data and the shape of the curves they produce. Empirical models may be powerful enough to reveal that "something" at the biological level is happening if this affects any of the geometric characteristics of the curves. Yet, to properly analyze mechanistic hypotheses mechanistic models need to be used. On the following section a collection of mechanistic models of increasing complexity for the analysis of GPCR function will be shown and discussed.

8.3.1 GPCRs Present Constitutive Activity: The Two-State Model

Constitutive activity is a general property of GPCRs that leads to basal response in the absence of an agonist. This was first observed (Costa and Herz 1989) in experiments with δ opioid receptors showing that competitive antagonists can have negative efficacy and later confirmed from mutated GPCRs displaying functional response in the absence of agonist (Kjelsberg et al. 1992; Samama et al. 1993). These mutations were performed at the C-terminal portion of the third intracellular loop, a region involved in receptor-G protein coupling, of α_1- and β_2-adrenergic receptors. Mutated receptors showed a graded range of constitutive activity and a higher affinity for agonists, indicating that mutations released certain intramolecular constraints that restrained the receptor in its inactive conformation and induced receptor conformations that mimicked the active state of the wild type receptor (Kjelsberg et al. 1992; Samama et al. 1993).

Basal response can be added as an *ad hoc* parameter in any of the above described empirical models, for instance in the Hill equation

$$E = \text{Bottom} + \frac{\text{Top} - \text{Bottom}}{1 + 10^{m(x_b - x)}} \tag{8.6}$$

Fitting Eq. 8.6 to experimental data displaying basal response would allow the estimation of this property through the bottom parameter. However, because Eq. 8.6 is purely empirical the basal response estimate cannot be interpreted in terms of the receptor system. If this is our aim then a mechanistic model is needed. The simplest mechanistic model accounting for receptor constitutive activity is the two-state model of agonist action.

The two-state model ((del Castillo and Katz 1957; Monod et al. 1965) for ionic channels and (Karlin 1967; Thron 1973; Colquhoun 1973) for receptors; see (Leff 1995) for review) includes receptor constitutive activity by considering an equilibrium between two receptor states, an inactive R, and an active R*. The model (Box 8.1) contains four receptor species, the free receptor R and R* and the corresponding ligand-bound AR and AR* molecular entities. Because the four chemical equilibria are arranged in a thermodynamic cycle the four equilibrium constants regulating the relative populations between species are reduced to three independent constants L, K and α. The parameter L measures the constitutive activity of the receptor; the equilibrium association constant K relates to the mutual affinity between the ligand and the inactive receptor species; and α quantifies the intrinsic efficacy of the ligand. The parameter α appears in two of the equilibria, the one containing the binding of the agonist to the active receptor (αK) and the one measuring the induction of the activated receptor state in the ligand-receptor complex (αL). Either by the first equilibrium (conformational selection) or the second (conformational induction) the parameter α determines the agonist profile (full/partial agonist, neutral antagonist or inverse agonist) of the ligand.

Box 8.1: The Two-State Receptor Model

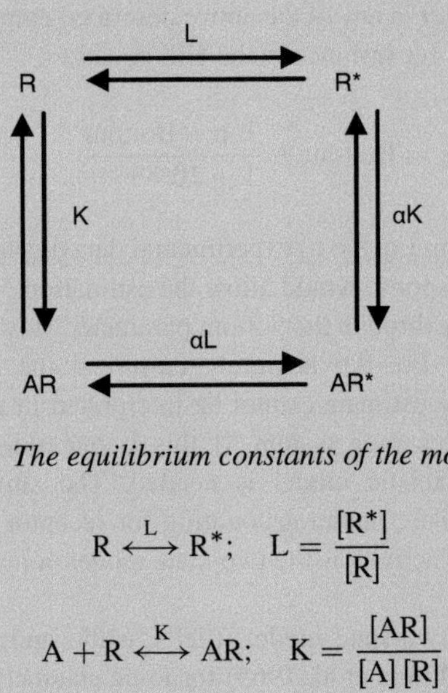

The equilibrium constants of the model

$$R \overset{L}{\longleftrightarrow} R^*; \quad L = \frac{[R^*]}{[R]}$$

$$A + R \overset{K}{\longleftrightarrow} AR; \quad K = \frac{[AR]}{[A]\,[R]}$$

$$AR \overset{\alpha L}{\longleftrightarrow} AR^*; \quad \alpha L = \frac{[AR^*]}{[AR]}$$

$$\alpha = \frac{[R]\,[AR^*]}{[R^*]\,[AR]}$$

$$A + R^* \overset{\alpha K}{\longleftrightarrow} AR^*; \quad \alpha K = \frac{\left[AR^*\right]}{[A]\left[R^*\right]}$$

$$\alpha = \frac{[R]\left[AR^*\right]}{\left[R^*\right][AR]}$$

The fraction of active receptors

$$f = \frac{[R]_{Active}}{[R]_T} = \frac{L\,(1 + \alpha K\,[A])}{1 + L + K\,[A]\,(1 + \alpha L)}$$

where $[R]_{Active} = [R^*] + [AR^*]$
and $[R]_T = [R] + [AR] + [R^*] + [AR^*]$

Geometric descriptors of the curves

- Left asymptote (Basal response: f for $[A] = 0$)

$$Basal = \frac{1}{1 + \frac{1}{L}}$$

- Right asymptote, the asymptotic f value as [A] increases (Top: $\lim_{[A]\to\infty} f$)

$$Top = \frac{1}{1 + \frac{1}{\alpha L}}$$

- The mid-point, the [A] value for half maximum f value

$$[A_{50}] = \frac{1 + L}{K\,(1 + \alpha L)}$$

$[A_{50}]$ is lower, equal and greater than 1/K for agonists ($\alpha > 1$), neutral antagonists ($\alpha = 1$) and inverse agonists ($\alpha < 1$), respectively.

In this modeling approach the pharmacologic response is analyzed by means of the fraction of active receptors (Eq. 8.7), which is obtained by applying the law of mass action. In the previous empirical models the geometric features of the E/[A] curves accounting for the efficacy (maximum response) and potency (location of the curve along the X-axis) of the drug-receptor pair were characterized by parameters lacking physical meaning. In a mechanistic model these geometric properties are expressed in terms of the equilibrium constants present in the model providing a biological context to the estimated parameters. In the two-state model of agonism (Box 8.1) the maximum response

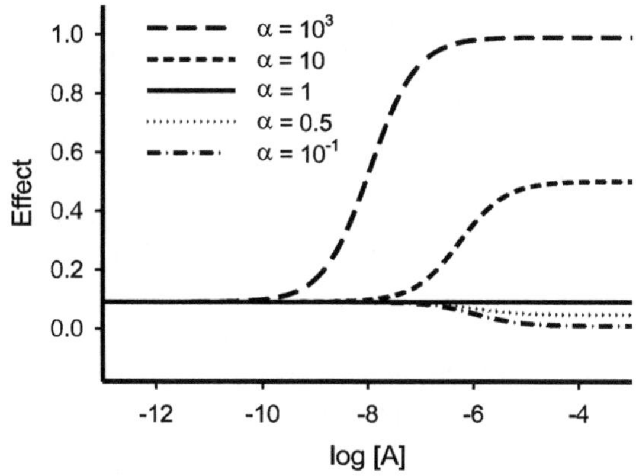

Fig. 8.3 The two-state model of agonism. Simulations are performed using Eq. 8.7 with $L = 10^{-1}$; $K = 10^6$; and $\alpha = 10^3$, 10, 1, 0.5 and 10^{-1}. The basal response is $E = L/(1 + L) = 0.0909$. Full agonism, neutral antagonism and inverse agonism is obtained with α greater, equal and lower than 1, respectively, with corresponding effect values greater, equal and lower than basal response, respectively

(Top) is determined by two constants, L, which determines in turn the basal response, and α, the intrinsic efficacy of the ligand in the receptor under study, whereas the potency ($[A_{50}]$) includes in addition an affinity term (K).

As mentioned above, the model can explain the behavior of full agonists, partial agonists, neutral antagonists and inverse agonists by the variation of the parameter α. Figure 8.3 illustrates this feature, five values of α (10^3, 10, 1, 0.5, 10^{-1}) were used, with constant values for L (10^{-1}) and K (10^6). For $\alpha = 1$, the effect is kept constant and coincident with the basal response ($L/(1 + L) = 0.0909$), meaning that the ligand (neutral antagonist) is not able to change the constitutive receptor effect. For $\alpha > 1$ (agonists) and $\alpha < 1$ (inverse agonists), asymptotic curves with effects greater and lower than the basal responses are obtained. Full and partial agonism (either with top close to one or significantly lower, respectively) is obtained with high and low $\alpha > 1$ values, respectively. As lower is $\alpha < 1$, higher is the capacity of the ligand to decrease the basal response (inverse agonism) with the limit being $E = 0$.

$$f = \frac{[R]_{Active}}{[R]_T} = \frac{L\,(1 + \alpha K\,[A])}{1 + L + K\,[A]\,(1 + \alpha L)} \tag{8.7}$$

As it has been recently discussed (Giraldo 2010), the concept of neutral antagonism is more a pharmacologic concept than a real property as, probably, many neutral antagonists are partial or inverse agonists whose positive or negative efficacies are too small to be detected with the current technology. As many classical equations of pharmacology were deduced assuming absence of basal response a revision of these equations is worth to be considered. In this context, an adaptation of the Schild and Cheng-Prusoff methods for binding affinity estimation has been used for modeling inverse agonists in the two-state model (Giraldo et al. 2007).

Remarkably, the two-state model of agonism has been used (Giraldo 2004) for the simulation of the different actions that a mutation may produce on receptor function thereby providing a theoretical framework for a quantitative analysis of receptor mutating effects.

8.3.1.1 Seeing the Two-State Model as an Empirical Model

A main concern is whether the two-state model matches any of the previously discussed empirical models and, if this were the case, what is the relation between the two fittings. It can be shown that Eq. 8.8 is equivalent to the Hill equation 8.6 with Bottom = c/b, Top = a, $x_b = \log(b)$ and $m = 1$.

$$E = \frac{c + a\,[A]}{b + [A]} \tag{8.8}$$

Equation 8.8 corresponds exactly to the mechanistic two-state model given by Eq. 8.7. It follows that if the Hill equation with $m = 1$ accurately fits experimental data then the empirical Hill model is compatible with a two-state mechanistic model. On the contrary, if a Hill equation with $m = 1$ does not correctly fit curve data other mechanistic models should be tested.

8.3.2 GPCRs Arrange into Dimers

The monomeric/dimeric nature of GPCRs has been the subject of passionate debate (Chabre and le Maire 2005; James et al. 2006; Bouvier et al. 2007). Thus, whereas on one hand there is experimental evidence indicating that a monomer receptor is able to couple to G proteins and transmit external signals inside the cell (Meyer et al. 2006; Whorton et al. 2007, 2008; Bayburt et al. 2007; White et al. 2007; Jastrzebska et al. 2004; Ernst et al. 2007; Kuszak et al. 2009; Rasmussen et al. 2011) on the other hand receptor dimerization/higher oligomerization has been shown to be a inherent property of GPCRs (Rios et al. 2001; Milligan 2004; Terrillon and Bouvier 2004; Jastrzebska et al. 2006; Carrillo et al. 2004; Ferre et al. 2010; Lohse 2010; Pin et al. 2007).

Yet, in the context of the present study, the point is what experimental findings can and cannot be described by the different mechanistic models currently at pharmacologists' disposition. Many models including receptor dimerization can be found in the literature, developed to explain E/[A] curve shapes other than that produced by the Hill equation with $m = 1$ (Karlin 1967; Colquhoun 1973; Wells 1992; Wreggett and Wells 1995; Chidiac et al. 1997; Armstrong and Strange 2001; Christopoulos and Kenakin 2002; Urizar et al. 2005; Albizu et al. 2006; Rovira et al. 2009). Thus, it seems important to characterize mechanistically what is found experimentally. The following selection of representative models will be discussed.

8.3.2.1 The Two-State Dimer Receptor Model

The two-state dimer receptor model (Franco et al. 2005, 2006) is presented in Box 8.2. The model is formally similar to the two-state model of agonism (Box 8.1); the receptor is constitutively active and in equilibrium between two receptor species, one inactive and the other active, but with the presence of two binding sites instead of a single one. Comparing the thermodynamic cycles of the monomeric and dimeric two-state models we see that the upper cycle of the latter is the same as the cycle of the former. The lower cycle of the two-state dimer model includes the binding of the second ligand to either the inactive or the active state of the singly-bound receptor and the conformational interconversion between the inactive and active states of the fully occupied dimer receptor. Two new constants appear in the model, μ and β. μ measures the binding cooperativity between the two bound molecules in the (R_2) inactive receptor state and β the differential affinity of the second ligand for the active receptor relative to the inactive one (conformational selection concept) or the differential capacity of activation of the doubly-bound receptor relative to the singly-bound one (conformational induction concept). Whereas α measures the intrinsic efficacy of the singly-bound ligand to activate the receptor relative to the free receptor itself β measures the intrinsic efficacy of the doubly-bound ligand to activate the receptor relative to the singly-bound one.

As in the monomeric two-state model the pharmacologic response is analyzed by the fraction of active receptors (Eq. 8.9), which in this (receptor dimer) case becomes a ratio between two second-degree polynomials. The empirical equation 8.9 turns into mechanistic when the a_i parameters are replaced by the equilibrium constants of the model (Box 8.2). The basal response depends only on the L parameter whereas α and β determine the maximum response. The potency of the agonist measured by $[A_{50}]$ includes in addition the binding constant K and the parameter μ.

$$f = \frac{[(RR)]_{Active}}{[(RR)]_T} = \frac{a_1 + a_2\,[A] + a_3[A]^2}{a_4 + a_5\,[A] + [A]^2} \qquad (8.9)$$

The interaction between the two binding sites causes the intricacy of the functional process and results in the E/[A] curve reshaping. Steeper and flatter curves than those corresponding to the two-state monomeric model (empirical Hill equation with a Hill coefficient of 1) and intermediate-plateau curves may be obtained as a result of the cooperativity between the two binding sites.

Box 8.2: The Two-State Dimer Receptor Model

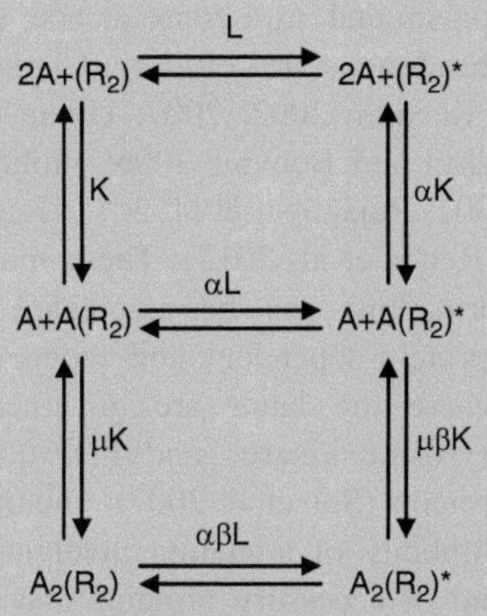

The equilibrium constants of the model

$$(RR) \overset{L}{\longleftrightarrow} (RR)^*; \quad L = \frac{[(RR)^*]}{[(RR)]}$$

$$A + (RR) \overset{K}{\longleftrightarrow} (RR)_A; \quad K = \frac{[(RR)_A]}{[A]\,[(RR)]}$$

$$(RR)_A \overset{\alpha L}{\longleftrightarrow} (RR)^*{}_A;$$

$$A + (RR)^* \overset{\alpha K}{\longleftrightarrow} (RR)^*{}_A;$$

$$\alpha = \frac{[(RR)^*{}_A]\,[(RR)]}{[(RR)^*]\,[(RR)_A]}$$

$$A + (RR)_A \overset{\mu K}{\longleftrightarrow} (RR)_{AA};$$

$$\mu = \frac{[(RR)_{AA}]\,[(RR)]}{[(RR)_A]\,[(RR)_A]}$$

$$(RR)_{AA} \overset{\alpha\beta L}{\longleftrightarrow} (RR)^*{}_{AA};$$

$$A + (RR)^*{}_A \overset{\mu\beta K}{\longleftrightarrow} (RR)^*{}_{AA};$$

$$\beta = \frac{[(RR)^*{}_{AA}]\,[(RR)_A]}{[(RR)_{AA}]\,[(RR)^*{}_A]}$$

The fraction of active receptors

$$f = \frac{[(RR)]_{Active}}{[(RR)]_T} = \frac{a_1 + a_2\,[A] + a_3[A]^2}{a_4 + a_5\,[A] + [A]^2}$$

where, $[(RR)]_{Active} = [(RR)^*] + [(RR)^*{}_A] + [(RR)^*{}_{AA}]$
and $[(RR)]_T = [(RR)] + [(RR)_A] + [(RR)_{AA}] + [(RR)^*] + [(RR)^*{}_A] + [(RR)^*{}_{AA}]$

$$a_1 = \frac{L}{K^2\mu\,(1+\alpha\beta L)}$$

$$a_2 = \frac{\alpha L}{K\mu\,(1+\alpha\beta L)}$$

$$a_3 = \frac{\alpha\beta L}{1+\alpha\beta L}$$

$$a_4 = \frac{1+L}{K^2\mu\,(1+\alpha\beta L)}$$

$$a_5 = \frac{1+\alpha L}{K\mu\,(1+\alpha\beta L)}$$

The geometric descriptors of the curves

- Left asymptote: Basal activity

$$\underset{\text{for } [A]=0}{f} = \frac{a_1}{a_4} = \frac{1}{1+\frac{1}{L}}$$

- Right asymptote: Efficacy

$$\lim_{[A]\to\infty} f = a_3 = \frac{1}{1+\frac{1}{\alpha\beta L}}$$

- The mid-point, the [A] value for half maximum f value

$$[A_{50}] = \frac{-b \pm \sqrt{b^2 - 4ac}}{2a}$$

With $a = a_1 - a_3a_4$; $b = a_1a_5 + a_3a_4a_5 - 2a_2a_4$; $c = -a_4(a_1 + a_3a_4)$ where the $\pm$ sign depends on A being an agonist or an inverse agonist.

Figure 8.4 shows two E/[A] curves resulting from the application of the two-state dimer receptor model. Two values of β (10^3, 10^{-3}) were used, with constant values of L (10^{-1}), K (10^6), α (10^3) and μ (1/4). The receptor is constitutive

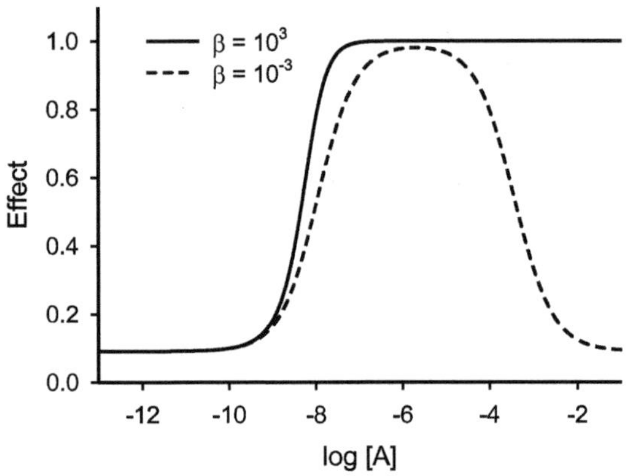

Fig. 8.4 The two-state dimer receptor model of agonism. Simulations were performed using Eq. 8.9 with $L = 10^{-1}$; $K = 10^{6}$; $\alpha = 10^{3}$; $\mu = 1/4$; and $\beta = 10^{3}$ and 10^{-3}. For $\beta = 10^{3}$ a sigmoid curve is obtained whereas for $\beta = 10^{-3}$ a *bell-shape curve* results whose right asymptote value is the same as the basal response because α and β cancel each other

active ($L > 1$). The ligand is an agonist ($\alpha > 1$) for the first binding site and shows no binding cooperativity for the inactive receptor ($\mu = 1/4$). The value of β reflects the capacity of activation of the fully occupied receptors relative to the singly occupied ones. For $\beta = 10^{3}$ a steep sigmoid curve is obtained whereas for $\beta = 10^{-3}$ a bell-shape curve results whose right asymptote $[1/(1 + 1/(\alpha\beta L))]$ reaches the value of the basal response $(1/1 + 1/L)$ because α and β cancel each other.

8.3.3 GPCRs May Signal Through Multiple Pathways: The Issue of Functional Selectivity

Both the two-state and the two-state dimer receptor models are on-off switch models, one single state for the inactive receptor (R and (RR), respectively) and one single state for the active receptor (R* and (RR)*, respectively). Clearly, this is an extreme simplification. The very flexible nature of GPCRs (as proteins) enables the presence of multiple receptor conformations, and nature has wisely chosen some of them for signaling specific pathways (Perez and Karnik 2005; Kobilka and Deupi 2007). The differential ligand-capacity for the selection of pathway-specific receptor active conformations has led to the concept of functional selectivity or biased agonism and, as a consequence, to the concept of pluridimensional efficacy (Clarke and Bond 1998; Kenakin 2002, 2007; Urban et al. 2006; Galandrin and Bouvier 2006; Violin and Lefkowitz 2007; Rajagopal et al. 2011; Kenakin et al. 2012; Reiter et al. 2012). These multiple active conformations can be associated with multiple G protein-dependent and independent pathways, where the latter are governed by β-arrestins, tyrosine kinases and PDZ-domain containing proteins (Sun et al. 2007). The impact that the multiplicity of signaling involving the wide spectrum of accessory proteins may have in drug development and therapeutics has been remarked (Mailman 2007; Rajagopal et al. 2010).

8.3.3.1 Distinguishing the Protomers Within a Dimer: The Symmetric/Asymmetric Three-State Dimer Receptor Model

Both the two-state and the two-state dimer model have been extended to include an additional active receptor state, R** and (RR)**, respectively, to account for an additional signaling pathway (Leff et al. 1997; Brea et al. 2009). Noticeably, a three-state dimer receptor model, including one inactive, (RR), and two active, (RR)* and (RR)**, receptor states, was able to explain the different behavior of typical and atypical $5\text{-}HT_{2A}$ receptor antagonists with respect to arachidonic acid and inositol phosphate pathways (Brea et al. 2009).

It is worth noting that, both the two-state (Franco et al. 2005, 2006) and the three-state (Brea et al. 2009) dimer receptor model consider the active receptor states as global entities and the two protomers within an active state identical. However, protomers within dimer active states do not need to be identical (e.g. for asymmetric active dimer state) as described by the new asymmetric/symmetric three-state dimer receptor model, AS3D (Rovira et al. 2010) (Box 8.3).

Box 8.3: The Asymmetric/Symmetric Three-State Dimer Receptor Model

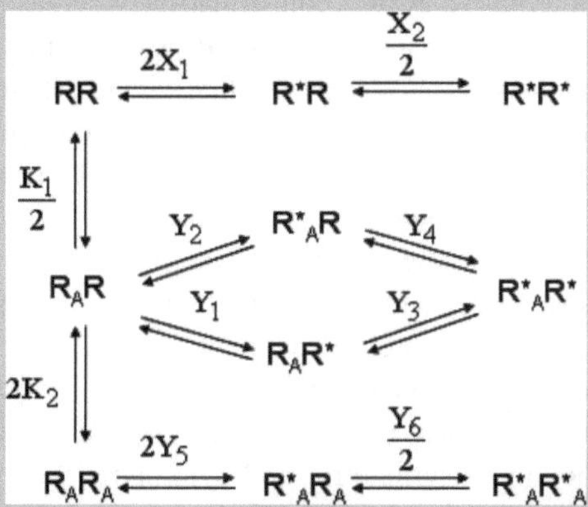

The equilibrium constants of the model

$$2X_1 = \frac{[R^*R]}{[RR]} \qquad \frac{X_2}{2} = \frac{[R^*R^*]}{[R^*R]}$$

$$\frac{K_1}{2} = \frac{[A]\,[RR]}{[R_AR]} \qquad 2K_2 = \frac{[A]\,[R_AR]}{[R_AR_A]}$$

$$Y_1 = \frac{[R_AR^*]}{[R_AR]} \qquad Y_2 = \frac{[R^*{}_AR]}{[R_AR]}$$

$$Y_3 = \frac{[R^*{}_AR^*]}{[R_AR^*]} \qquad Y_4 = \frac{[R^*{}_AR^*]}{[R^*{}_AR]}$$

$$2Y_5 = \frac{[R^*{}_AR_A]}{[R_AR_A]} \qquad \frac{Y_6}{2} = \frac{[R^*{}_AR^*{}_A]}{[R^*{}_AR_A]}$$

The fraction of active receptors

$$f_{R^*R} = \frac{[R^*R]_t}{[RR_t]} = 2\cdot\frac{a_1 + a_2\,[A] + a_3[A]^2}{a_4 + a_5\,[A] + [A]^2}$$

$$f_{R^*R^*} = \frac{[R^*R^*]_t}{[RR_t]} = 2\cdot\frac{b_1 + b_2\,[A] + b_3[A]^2}{b_4 + b_5\,[A] + [A]^2}$$

where,

$$[R^*R]_t = [R^*R] + [R^*{}_AR] + [R_AR^*] + [R^*{}_AR_A]$$

$$[R^*R^*]_t = [R^*R^*] + [R^*{}_AR^*] + [R^*{}_AR^*{}_A]$$

$$[RR_t] = [RR] + [R_AR] + [R_AR_A] + [R^*R] + [R^*{}_AR] + [R_AR^*] + [R^*{}_AR_A] + [R^*R^*] + [R^*{}_AR^*] + [R^*{}_AR^*{}_A]$$

$$a_1 = \frac{K_1K_2X_1}{1 + 2Y_5 + Y_5Y_6}$$

$$a_2 = \frac{K_2\,(Y_1 + Y_2)}{1 + 2Y_5 + Y_5Y_6}$$

$$a_3 = \frac{Y_5}{1 + 2Y_5 + Y_5Y_6}$$

$$b_1 = \frac{K_1K_2X_1X_2}{2\,(1 + 2Y_5 + Y_5Y_6)}$$

$$b_2 = \frac{K_2Y_1Y_3}{1 + 2Y_5 + Y_5Y_6}$$

$$b_3 = \frac{Y_5Y_6}{2\,(1 + 2Y_5 + Y_5Y_6)}$$

$$a_4 = b_4 = \frac{K_1K_2\,(1 + 2X_1 + X_1X_2)}{1 + 2Y_5 + Y_5Y_6}$$

$$a_5 = b_5 = \frac{2K_2\,(1 + Y_1 + Y_2 + Y_1Y_3)}{1 + 2Y_5 + Y_5Y_6}$$

The geometric descriptors of the curves

- Left asymptote: Basal activity

$$\underset{\text{for }[A]=0}{f_{R^*R}} = 2\frac{a_1}{a_4} = \frac{1}{1 + \frac{1}{2}\left(\frac{1}{X_1} + X_2\right)}$$

$$\underset{\text{for }[A]=0}{f_{R^*R^*}} = 2\frac{b_1}{b_4} = \frac{1}{1 + \frac{2}{X_2} + \frac{1}{X_1X_2}}$$

- Right asymptote: Efficacy

$$\lim_{[A]\to\infty} f_{R^*R} = 2a_3 = \frac{1}{1 + \frac{1}{2}\left(\frac{1}{Y_5} + Y_6\right)}$$

$$\lim_{[A]\to\infty} f_{R^*R^*} = 2b_3 = \frac{1}{1 + \frac{2}{Y_6} + \frac{1}{Y_5Y_6}}$$

(continued)

Box 8.3 (continued)

- The mid-point, the [A] value for half maximum f_{R*R} or f_{R*R*} values

$$[A_{50}] = \frac{-b \pm \sqrt{b^2 - 4ac}}{2a}$$

With $a = a_1 - a_3a_4$; $b = a_3a_4a_5 - 2a_2a_4 + a_1a_5$; and $c = -a_4(a_1 - a_3a_4)$ in the case of f_{R*R} and the same expressions but replacing a_i by b_i in the case of f_{R*R*}. The ± sign in $[A_{50}]$ equation results for the possibility of A being either a positive or an inverse agonist.

Following several experimental studies (Damian et al. 2006; Bayburt et al. 2007; Mancia et al. 2008; Pin et al. 2004; Xu et al. 2004; Hlavackova et al. 2005; Goudet et al. 2005), AS3D assumes that the asymmetric active receptor state (R*R) governs a G protein-dependent pathway whereas the symmetric active receptor state (R*R*) regulates a G protein-independent pathway. It is worth noting that both cis and trans activation, determined by Y_2 and Y_1 equilibrium constants (chemical equilibrium scheme in Box 8.3), are included in the AS3D model allowing the simulation of mechanistic processes under equilibrium conditions that may depend on the receptor system.

There are particular pharmacological subtleties that AS3D can represent. For example, an asymmetric R*R active receptor state allows to describe findings apparently paradoxical, such as an inverse agonist (with preferential affinity for the R molecular entity) binding to the protomer R in the R*R active dimer receptor species (if its concentration is significant) and contributing to an increase of the G protein-dependent functional response. Interestingly, this was recently found in a study involving dopamine class A dimers: maximal functional response was achieved by agonist binding to a single protomer; this is compatible with the proposal of an asymmetric R*R active species. However, addition of either an agonist or an inverse agonist to the second protomer had opposite effects; the agonist decreased the signaling whereas the inverse agonist enhanced it (Han et al. 2009). This is consistent with the agonist promoting an alternative R*R* conformation whereas the inverse agonist stabilizing the R*R conformation previously promoted by the first agonist (Rovira et al. 2010). Importantly, not only a conceptual or qualitative analysis is provided by the model but also a quantitative description.

Following the rationale of the previous mechanistic models, the pharmacologic responses of the two (G protein-dependent and G protein-independent) functional pathways are analyzed by the fractional receptor populations of the corresponding asymmetric and symmetric receptor species (Eqs. 8.10 and 8.11) (Rovira et al. 2010).

$$f_{R*R} = 2\frac{a_1 + a_2[A] + a_3[A]^2}{a_4 + a_5[A] + [A]^2} \tag{8.10}$$

$$f_{R*R*} = 2\frac{b_1 + b_2[A] + b_3[A]^2}{b_4 + b_5[A] + [A]^2} \tag{8.11}$$

As in the two-state dimer receptor model the pharmacologic response for each of the signaling pathways is a ratio between two second-degree polynomials. The empirical equations 8.10 and 8.11 turn into mechanistic when the a_i and b_i parameters are replaced by the equilibrium constants of the model (Box 8.3).

Figure 8.5 shows a simulation of the asymmetric R*R and symmetric R*R* pathways by using Eqs. 8.10 and 8.11 with $X_1 = 10^{-1}$; $X_2 = 1$; $K_1 = 10^{-8.5}$; $K_2 = 10^{-4}$; $Y_1 = 10^{-3}$; $Y_2 = 1$; $Y_3 = 10^2$; $Y_5 = 1$ and $Y_6 = 10^2$, where K_1 and K_2 are equilibrium dissociation constants for the binding of the first and second ligand to the inactive receptor, respectively, X_i are equilibrium constants for the induction of activated states for the free receptor and Y_i are equilibrium constants for the induction of activated states for the occupied receptor. The curve for the R*R-G protein-dependent signaling pathway is a bell-shape curve because, after reaching a maximum, the concentrations of the species associated with this pathway decrease as the concentrations of the R*R* associated-species of the G protein-independent pathway

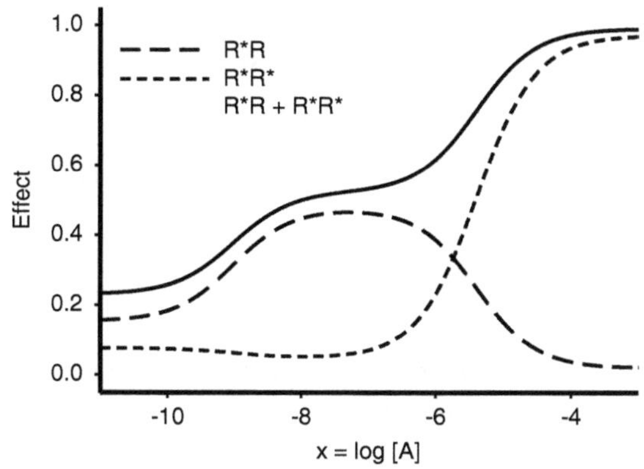

Fig. 8.5 The three-state asymmetric/symmetric dimer receptor model of agonism. Simulations were performed using Eqs. 8.10 and 8.11 with $X_1 = 10^{-1}$; $X_2 = 1$; $K_1 = 10^{-8.5}$; $K_2 = 10^{-4}$; $Y_1 = 10^{-3}$; $Y_2 = 1$; $Y_3 = 10^2$; $Y_5 = 1$ and $Y_6 = 10^2$. The curve for the R*R signaling pathway is a *bell-shape curve* because after reaching a maximum the concentrations of the species associated with this pathway decrease as the concentrations of the R*R* associated-species increase. The sum of the two effects, which may reflects the convergence of the two pathways at some downstream signaling point, displays a biphasic curve with an intermediate plateau

increase. A biphasic curve with an intermediate plateau results from the sum of the two effects, which may reflect the convergence of the two pathways at some downstream signaling point.

8.4 The Operational Model of Agonism: A Hybrid Between Empirical and Mechanistic Models

In the previous sections various examples of empirical and mechanistic models were described. The latter ones were characterized by the dependent variable being attached to the fraction of active receptors. However, the measured functional responses resulting from GPCR activation may reside at some downstream point in the signaling pathway. The operational model was proposed aiming to circumvent the complexity of the signal transduction process while including the essential information of the ligand-receptor recognition stage (Black and Leff 1983). As it can be seen in Box 8.4A the models consists of two steps, the binding of the agonist to the receptor, which is governed by an equilibrium constant, and the transduction of receptor occupancy into response, for which a logistic function was proposed. The final E/[A] equation (8.12) is the result of combining both expressions.

$$E = \frac{E_m \tau^n [A]^n}{(K_A + [A])^n + \tau^n [A]^n} \tag{8.12}$$

We have considered the operational model as hybrid because incorporates mechanism through the binding equilibrium (the equilibrium dissociation constant K_A) and empiricism by proposing a transduction function, whose form cannot necessarily be logistic in all cases and which includes two terms, E_m and K_E, which are not the product of a mechanistic development. E_m is defined as the maximum effect that the receptor system may yield (not confound with the maximum effect (Top) that the receptor may reach from the binding of a particular agonist) and K_E, the concentration of bound receptor for half E_m and then an index of the intrinsic efficacy of the agonist for a particular receptor. The parameter τ in Eq. 8.12 is the ratio between the total receptor concentration and K_E and reflects the operational efficacy of the agonist-receptor pair.

Box 8.4: The Operational Model of Agonism

A. The original proposal: Receptor constitutive activity is not included

$$A + R \overset{K_A}{\longleftrightarrow} AR \rightarrow E$$

The equilibrium constant of the model

$$K_A = \frac{[A][R]}{[AR]}$$

The function for the transduction of receptor occupation into response

$$E = \frac{E_m [AR]^n}{K_E{}^n + [AR]^n}$$

(continued)

Box 8.4 (continued)

The E/[A] functional response

$$E = \frac{E_m \tau^n [A]^n}{(K_A + [A])^n + \tau^n [A]^n}$$

where $\tau = \frac{[R_0]}{K_E}$ and $[R_0] = [R] + [AR]$.

Geometric descriptors of the curves

- Left asymptote (Basal response: f for $[A] = 0$)

$$\text{Basal} = 0$$

- Right asymptote, the asymptotic f value as [A] increases (Top: $\lim_{[A]\to\infty} f$)

$$\text{Top} = \frac{E_m}{1 + \frac{1}{\tau^n}}$$

- The mid-point, the [A] value for half maximum f value

$$[A_{50}] = \frac{K_A}{(2 + \tau^n)^{\frac{1}{n}} - 1}$$

In the particular case of $n = 1$

$$E = \frac{E_m \tau [A]}{K_A + (1 + \tau)[A]}$$

$$\text{Top} = \frac{E_m}{1 + \frac{1}{\tau}}$$

$$[A_{50}] = \frac{K_A}{1 + \tau}$$

B. An extension of the model: Receptor constitutive activity is included

$$R \rightarrow E$$

$$A + R \overset{K_A}{\longleftrightarrow} AR \rightarrow E$$

The equilibrium constant of the model

$$K_A = \frac{[A][R]}{[AR]}$$

The generation of stimulus by the receptor

$$S = [R] + \varepsilon [AR]$$

The function for the transduction of receptor stimulus into response

$$E = \frac{E_m S^n}{K_E{}^n + S^n}$$

The E/[A] functional response

$$E = \frac{E_m \chi^n (K_A + \varepsilon [A])^n}{(K_A + [A])^n + \chi^n (K_A + \varepsilon [A])^n}$$

With $\chi = \frac{[R_T]}{K_E}$ and $[R]_T = [R] + [AR]$

Geometric descriptors of the curves

- Left asymptote (Basal response: f for $[A] = 0$)

$$\text{Basal} = \frac{E_m}{1 + \frac{1}{\chi^n}}$$

- Right asymptote, the asymptotic f value as [A] increases (Top: $\lim_{[A]\to\infty} f$)

$$\text{Top} = \frac{E_m}{1 + \frac{1}{\varepsilon^n \chi^n}}$$

- The mid-point, the [A] value for half maximum effect

$$[A_{50}] = \frac{K_A \left(\sqrt[n]{1 + \varepsilon^n + 2\varepsilon^n \chi^n} - \sqrt[n]{2 + \chi^n + \varepsilon^n \chi^n} \right)}{\varepsilon \sqrt[n]{2 + \chi^n + \varepsilon^n \chi^n} - \sqrt[n]{1 + \varepsilon^n + 2\varepsilon^n \chi^n}}$$

In the particular case of $n = 1$

$$E = \frac{E_m \chi (K_A + \varepsilon [A])}{K_A (1 + \chi) + (1 + \varepsilon \chi)[A]}$$

(continued)

Box 8.4 (continued)

$$\text{Basal} = \frac{E_m}{1 + \frac{1}{\chi}}$$

$$\text{Top} = \frac{E_m}{1 + \frac{1}{\varepsilon\chi}}$$

$$[A_{50}] = \frac{(1 + \chi)\, K_A}{1 + \varepsilon\chi}$$

The operational model as it was originally proposed (Black and Leff 1983) did not contemplate receptor constitutive activity and thus basal response is equal to 0. The top value depends on E_m, the maximum effect of the receptor system, τ, the operational efficacy (an agonist-receptor dependent term), and n, the parameter enabling E/[A] curves to be steeper or flatter than rectangular hyperbola. The $[A_{50}]$ value, a parameter reflecting the potency of the agonist, includes in its definition K_A, the equilibrium dissociation constant of the ligand-receptor complex.

Figure 8.6 shows a simulation with $E_m = 100$; $n = 1$; $K_A = 10^{-6}$; and $\tau = 100$, 5 and 1, which may represent three ligands with the same affinity for the receptor but different intrinsic efficacy. The asymptotic maxima of the curves decrease as τ decreases. For the highest τ value the top of the curve ($E = 99$) is close to the E_m of the system indicating that the ligand is a full agonist; the other two curves with top significantly lower than E_m correspond to partial agonists.

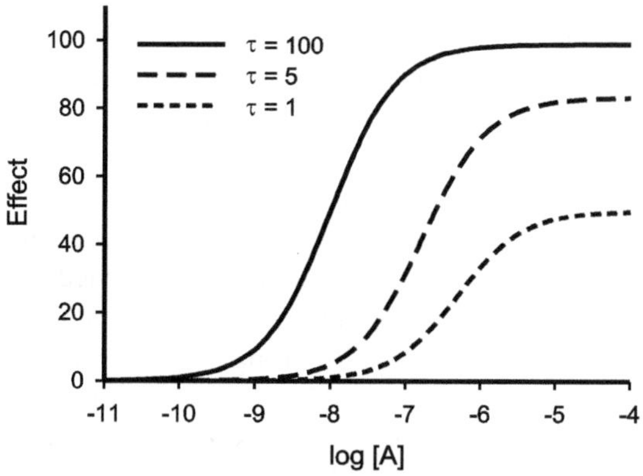

Fig. 8.6 The operational model of agonism. Simulations were performed using Eq. 8.12 with $E_m = 100$; $n = 1$; $K_A = 10^{-6}$; and $\tau = 100$, 5 and 1. Decreasing the τ values leads to curves with lower asymptotic maxima, 99, 83.3 and 50, respectively

8.4.1 The Operational Model of Agonism Including Receptor Constitutive Activity

The property of receptor constitutive activity has been recently added to the operational model of agonism (Slack and Hall 2012) (see also a previous analysis (Hall 2006) of the corresponding author of the former paper by using a limiting case of the ternary complex model (De Lean et al. 1980)). The model considers that the observed effect is a logistic function of the stimulus generated by the receptor with a maximum effect E_m and a transducer constant of stimulus into effect K_E; and that total receptor stimulus arises from the sum of that generated by the free receptors and that generated by the occupied ones. A proportionality constant $\varepsilon > 0$ is used for the stimulus generated by the occupied receptors. The value of ε measures the capability of the ligand-bound receptor to generate a stimulus relative to the free receptor: $\varepsilon > 1$, indicates that the ligand-bound generates greater stimulus than the free receptor and $\varepsilon < 1$, a smaller one. The E/[A] relationship is given by Eq. 8.13.

$$E = \frac{E_m \chi^n (K_A + \varepsilon\,[A])^n}{(K_A + [A])^n + \chi^n (K_A + \varepsilon\,[A])^n} \quad (8.13)$$

where $\chi = [R]_T/K_E$ measures the coupling efficiency of the signal transduction system. The parameters $[R]_T$, total receptor concentration, and K_A, agonist-receptor equilibrium dissociation constant, are defined as in the original operational model (Black and Leff 1983). The parameter χ resembles the earlier defined operational efficacy τ, but because the meaning is not exactly the same a different symbol was used (Slack and Hall 2012).

Figure 8.7 shows a simulation with $E_m = 100$; $n = 1$; $K_A = 10^{-6}$; $\chi = 0.5$ and $\varepsilon = 100$, 10, 1 and 10^{-1} assuming four ligands with different intrinsic efficacies. The receptor is constitutively active with a basal response $[Em/(1 + 1/\chi)]$ of

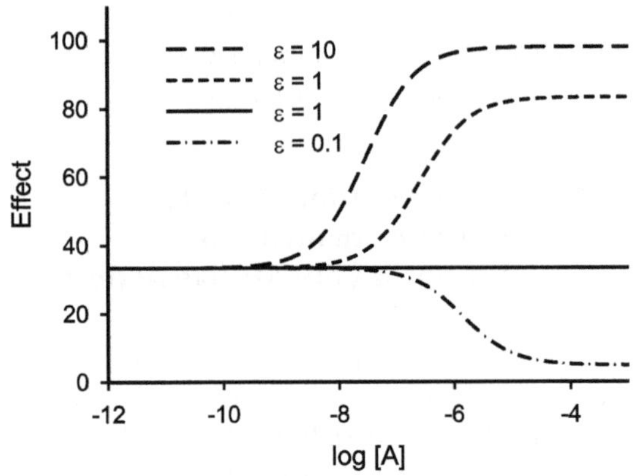

Fig. 8.7 The operational model of agonism with receptor constitutive activity. Simulations were performed with $E_m = 100$; $n = 1$; $K_A = 10^{-6}$; $\chi = 0.5$ and $\varepsilon = 100$, 10, 1 and 10^{-1} assuming four ligands with different intrinsic efficacy. The basal response is determined by χ. The intrinsic efficacy of the agonists featured by ε determines the height of the top relative to the basal response. The receptor is constitutively active with a basal response of 33.3 and top responses of 98.0 (full agonist), 83.3 (partial agonist), 33.3 (neutral antagonist) and 4.8 (inverse agonist) are obtained

33.3 and top responses $[Em/(1 + 1/(\varepsilon\chi))]$ of 98.0 (full agonist), 83.3 (partial agonist), 33.3 (neutral antagonist) and 4.8 (inverse agonist).

8.4.2 Quantification of Functional Selectivity by the Operational Model of Agonism

Functional selectivity can be accounted for by the operational model of agonism (for instance, in its original form (Black and Leff 1983)) by assuming that the operational efficacy parameter τ of an agonist-receptor pair varies with the signaling pathways (Kenakin et al. 2012). Thus, assuming in Eq. 8.12 that the parameters E_m and n are cell-specific (shared by all agonists for a common receptor through a given pathway), K_A is ligand-receptor specific and τ is both ligand- and pathway-specific, a transduction coefficient $\log(\tau/K_A)$ was defined to characterize agonism for a given pathway (Kenakin et al. 2012). A normalized transduction coefficient, $\Delta \log(\tau/K_A)_{j_1}$, was proposed to quantify the relative efficiency of an agonist compared to a reference agonist for a given pathway j_1. Finally, to determine the bias between two pathways (j_1 and j_2) for a given agonist relative to a reference agonist the quantity $\Delta\Delta \log(\tau/K_A)_{j_1-j_2} = \Delta \log(\tau/K_A)_{j_1} - \Delta \log(\tau/K_A)_{j_2}$ was suggested (Kenakin et al. 2012). From this bias parameter, the authors expected that a scale for the quantification of ligand selectivity between signaling pathways could be constructed. It is worth emphasizing that for the scale to be useful, the differences in receptor densities between cell types should be cancelled. In this case, the bias factor would reflect only the differences in efficacy and affinity of the ligands associated with their selectivity for the pathways. It is important to note that quantification of bias, as proposed in this study (Kenakin et al. 2012), can be extremely important in the communication between pharmacologists and medicinal chemists and, consequently, in guiding drug discovery in a more productive and safer direction.

8.5 Concluding Remarks

In the present study, a selection of empirical and mechanistic models has been shown and their most salient features analyzed in detail. In addition, the operational model (a hybrid between empirical and mechanistic models) has been included. The pros and cons of these approaches have been commented and the connection between them through the shape of the curves that they produce discussed. In this regard, it is worth noting that a mechanistic model can be very useful for simulation or for fitting experimental data; however, some caution is needed because different mechanistic models may produce similar curves and be apparently well suited for reality description, being possible that the chosen mechanistic proposal be wrong (see, for example comments in (Giraldo 2008)). A careful comparison of all possible models appropriate for a particular experimental situation should always be made in order to find the true model with the true explanation.

Acknowledgements This work was supported by the Spanish projects SAF2010-19257, TIN2009-13618, TIN2012-33116, Fundació La Marató de TV3 (110230) and RecerCaixa 2010ACUP 00378. The 2nd author has been supported by the Ramón y Cajal Program of the Spanish Ministry of Economy and Competitiveness.

A.1 Appendix

A.1.1 Fitting Mathematical Models to Experimental Data

In a pharmacological assay, the input agonist concentration, [A], and its associated output effect, E, are recorded under particular experimental conditions. Experimental data are given by a list of varying effects, E_i, $i = 1,\ldots,N$, obtained using different agonist concentrations, $[A]_i$, $i = 1,\ldots,N$. A mathematical model of pharmacological effect is a mathematical equation $E = f([A])$, in which E represents the pharmacological effect, [A] is the concentration of the drug and f is a mathematical function containing a number of parameters.

The problem of fitting a mathematical model to a given experimental data list consists in finding the model parameters that produce the function $E = f([A])$ best approximating the experimental pairs $([A]_i, E_i)$, $i = 1,\ldots N$ in the sense that the model values are as close as possible to the experimental measurements. In other words, the differences $|f([A]_i)-E_i|$ are as small as possible. In mathematical terms this is expressed in the following sum of squares which exclusively depends on the values of the model parameters:

$$\text{Energy(Parameters)} = \sum_{i=1}^{N} |f([A]_i) - E_i|^2 \tag{8.14}$$

which should be minimum for the optimal model parametric values (least squares fitting). We would like to note that such set of optimal parameters produce a curve that visually fits the profile of the scatter plot given by the experimental pairs $([A]_i, E_i)$, $i = 1,\ldots N$.

In the case of pharmacological models, the dependency between the concentration and the effect given by $E = f([A])$ is non-linear with respect the model parameters and, thus, Eq. 8.14 is known as non-linear regression.

A.1.2 Traditional Gradient-Based Nonlinear Regression Algorithms Versus Stochastic Approaches

Given that the best set of parameter estimates must be a minimum of the energy given by Eq. 8.14, they can be found by means of optimization techniques. There are two main families of optimization strategies: local and global.

Local approaches search for the closest local minimum given an initial seed point. They mainly rely on the gradient of the function to be minimized and update the initial guess using the direction of this gradient because is the direction of maximum decrease of the energy (gradient descent). Gradient descent methods are efficient minimizers as far as the energy function can be differentiated with respect the parameters (which is always the case in a least squares fitting of pharmacological models) and they provide an optimal solution as far as the energy function is convex (i.e. has a unique local minimum). In the case of multiple minima (which is a typical case for mechanistic pharmacological models), the gradient descent strongly depends on the initial seed. This limitation can be partially overcome by running the gradient descent algorithm using different initial seed points. This solution has two main shortcomings. First it increases the computational time of the minimization process and, second, given that there is no a-priory knowledge on the number and distribution of the local minima, there is not a clear strategy for choosing an initial set of seeds guaranteeing that the optimal solution (global minimum) will be reached.

Global approaches are a way of searching for global optimal solutions (global minimum) without the need of a special definition of the initial

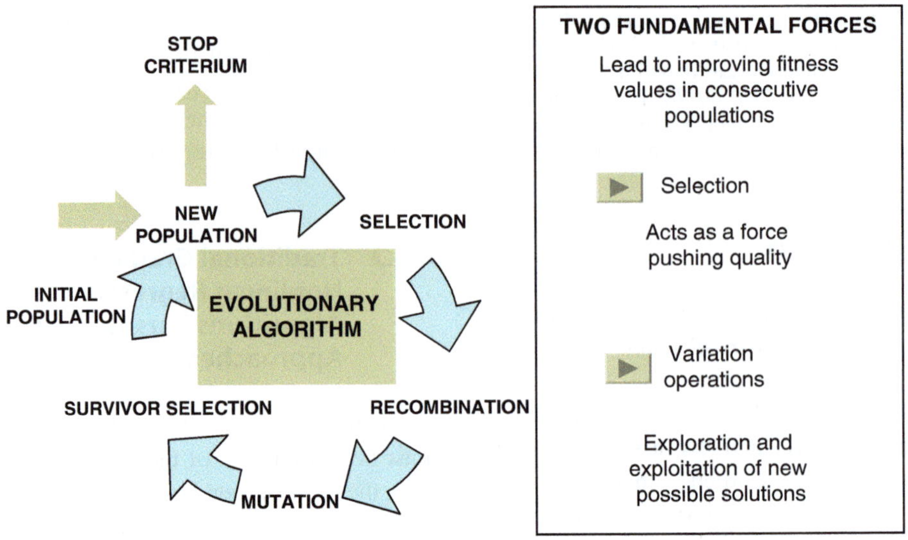

Fig. A.1 Scheme of the mechanisms ruling an Evolutionary Algorithm (EA). EAs maintain a population of possible solutions and their fitness on the search space. In each iteration, selection determines which individuals are chosen to produce new solutions (exploitation). Recombination and mutation combine and perturb the individuals (exploration). Finally, the survival step decides who survives in the new population. This process iterates until the stop criterion occurs

set of seeds. For special search spaces and cost functions, there is a solid mathematical theory ensuring convergence of some global algorithms (simulated annealing (Ashyraliyev et al. 2009)). Although, there is no proof of convergence for the general case. Evolutionary Algorithms (EAs) have proven their ability for optimizing non-analytic multi-modal functions in a wide range of real life problems, such as parameter estimation (Ravikumar and Panigrahi 2010), pattern and text recognition (Rizki et al. 2002) and image processing (Cagnoni et al. 2008). EAs are a class of stochastic optimization methods that simulate the process of natural evolution. Unlike gradient descent methods that evolve a single initial value each time, EAs maintain a population of possible solutions that evolve according to rules of selection and other operators, such as recombination and mutation (Holland 1975).

Each individual in the population receives a measure of its fitness in the environment. Selection focuses attention on high fitness individuals, thus, exploiting the available fitness information. Selection determines which individuals are chosen to produce offsprings. Recombination and mutation perturb those individuals, providing general heuristics for exploration. Recombination produces new individuals in combining the information contained in the parents and offsprings are mutated by small perturbations with low probability. Finally, survival step decides who survives (among parents and offsprings) to form the new population. This process iterates until stop criterion occurs. Figure A.1 outlines the scheme of a standard EA.

References

Albizu L, Balestre MN, Breton C et al (2006) Probing the existence of G protein-coupled receptor dimers by positive and negative ligand-dependent cooperative binding. Mol Pharmacol 70:1783–1791

Armstrong D, Strange PG (2001) Dopamine D2 receptor dimer formation: evidence from ligand binding. J Biol Chem 276:22621–22629

Ashyraliyev M, Fomekong-Nanfack Y, Kaandorp JA, Blom JG (2009) Systems biology: parameter estimation for biochemical models. FEBS J 276:886–902

Barlow R, Blake JF (1989) Hill coefficients and the logistic equation. Trends Pharmacol Sci 10:440–441

Bayburt TH, Leitz AJ, Xie G, Oprian DD, Sligar SG (2007) Transducin activation by nanoscale lipid bilayers containing one and two rhodopsins. J Biol Chem 282:14875–14881

Black JW, Leff P (1983) Operational models of pharmacological agonism. Proc R Soc Lond B Biol Sci 220: 141–162

Boeynaems JM, Dumont JE (1980) Outlines of receptor theory. Elsevier/North-Holland, Amsterdam, pp 73–75

Bouvier M, Heveker N, Jockers R, Marullo S, Milligan G (2007) BRET analysis of GPCR oligomerization: newer does not mean better. Nat Methods 4:3–4

Brea J, Castro M, Giraldo J et al (2009) Evidence for distinct antagonist-revealed functional states of 5-hydroxytryptamine(2A) receptor homodimers. Mol Pharmacol 75:1380–1391

Cagnoni S, Lutton E, Olague G (2008) Genetic and evolutionary computation for image processing and analysis. Hindawi Publishing Corporation, New York

Carrillo JJ, Lopez-Gimenez JF, Milligan G (2004) Multiple interactions between transmembrane helices generate the oligomeric alpha1b-adrenoceptor. Mol Pharmacol 66:1123–1137

Chabre M, le Maire M (2005) Monomeric G-protein-coupled receptor as a functional unit. Biochemistry 44:9395–9403

Chidiac P, Green MA, Pawagi AB, Wells JW (1997) Cardiac muscarinic receptors. Cooperativity as the basis for multiple states of affinity. Biochemistry 36: 7361–7379

Christopoulos A, Kenakin T (2002) G protein-coupled receptor allosterism and complexing. Pharmacol Rev 54:323–374

Clarke WP, Bond RA (1998) The elusive nature of intrinsic efficacy. Trends Pharmacol Sci 19:270–276

Colquhoun D (1971) Lectures on biostatistics. Oxford University Press, London, pp 361–364

Colquhoun D (1973) The relationship between classical and cooperative models for drug action. In: Rang HP (ed) A symposium on drug receptors. University Park Press, Baltimore, pp 149–182

Costa T, Herz A (1989) Antagonists with negative intrinsic activity at delta opioid receptors coupled to GTP-binding proteins. Proc Natl Acad Sci U S A 86: 7321–7325

Damian M, Martin A, Mesnier D, Pin JP, Baneres JL (2006) Asymmetric conformational changes in a GPCR dimer controlled by G-proteins. EMBO J 25:5693–5702

De Lean A, Stadel JM, Lefkowitz RJ (1980) A ternary complex model explains the agonist-specific binding properties of the adenylate cyclase-coupled β-adrenergic receptor. J Biol Chem 255: 7108–7117

del Castillo J, Katz B (1957) Interaction at end-plate receptors between different choline derivatives. Proc R Soc Lond B 146:369–381

Ernst OP, Gramse V, Kolbe M, Hofmann KP, Heck M (2007) Monomeric G protein-coupled receptor rhodopsin in solution activates its G protein transducin at the diffusion limit. Proc Natl Acad Sci U S A 104:10859–10864

Ferre S, Woods AS, Navarro G, Aymerich M, Lluis C, Franco R (2010) Calcium-mediated modulation of the quaternary structure and function of adenosine A2A-dopamine D2 receptor heteromers. Curr Opin Pharmacol 10:67–72

Franco R, Casadó V, Mallol J et al (2005) Dimer-based model for heptaspanning membrane receptors. Trends Biochem Sci 30:360–366

Franco R, Casadó V, Mallol J et al (2006) The two-state dimer receptor model: a general model for receptor dimers. Mol Pharmacol 69:1905–1912

Galandrin S, Bouvier M (2006) Distinct signaling profiles of beta1 and beta2 adrenergic receptor ligands toward adenylyl cyclase and mitogen-activated protein kinase reveals the pluridimensionality of efficacy. Mol Pharmacol 70:1575–1584

Giraldo J (2003) Empirical models and Hill coefficients. Trends Pharmacol Sci 24:63–65

Giraldo J (2004) Agonist induction, conformational selection, and mutant receptors. FEBS Lett 556:13–18

Giraldo J (2008) On the fitting of binding data when receptor dimerization is suspected. Br J Pharmacol 155:17–23

Giraldo J (2010) How inverse can a neutral antagonist be? Strategic questions after the rimonabant issue. Drug Discov Today 15:411–415

Giraldo J, Vivas NM, Vila E, Badia A (2002) Assessing the (a)symmetry of concentration-effect curves: empirical versus mechanistic models. Pharmacol Ther 95:21–45

Giraldo J, Serra J, Roche D, Rovira X (2007) Assessing receptor affinity for inverse agonists: Schild and Cheng-Prusoff methods revisited. Curr Drug Targets 8: 197–202

Gompertz B (1825) On the nature of the function expressive of the law of human mortality. Philos Trans R Soc Lond 36:513–585

Goudet C, Kniazeff J, Hlavackova V et al (2005) Asymmetric functioning of dimeric metabotropic glutamate receptors disclosed by positive allosteric modulators. J Biol Chem 280:24380–24385

Hall DA (2006) Predicting dose-response curve behavior: mathematical models of allosteric receptor-ligand interactions. In: Bowery NG (ed) Allosteric receptor modulation in drug targeting. Taylor & Francis, New York, pp 39–77

Han Y, Moreira IS, Urizar E, Weinstein H, Javitch JA (2009) Allosteric communication between protomers of dopamine class A GPCR dimers modulates activation. Nat Chem Biol 5:688–695

Hill AV (1913) The combinations of haemoglobin with oxygen and with carbon monoxide. Biochem J 7: 471–480

Hlavackova V, Goudet C, Kniazeff J et al (2005) Evidence for a single heptahelical domain being turned on upon activation of a dimeric GPCR. EMBO J 24: 499–509

Holland JH (1975) Adaptation in natural and artificial systems. University Michigan Press, Ann Arbor

James JR, Oliveira MI, Carmo AM, Iaboni A, Davis SJ (2006) A rigorous experimental framework for detecting protein oligomerization using bioluminescence resonance energy transfer. Nat Methods 3:1001–1006

Jastrzebska B, Maeda T, Zhu L et al (2004) Functional characterization of rhodopsin monomers and dimers in detergents. J Biol Chem 279:54663–54675

Jastrzebska B, Fotiadis D, Jang GF, Stenkamp RE, Engel A, Palczewski K (2006) Functional and structural characterization of rhodopsin oligomers. J Biol Chem 281:11917–11922

Karlin A (1967) On the application of "a plausible model" of allosteric proteins to the receptor for acetylcholine. J Theor Biol 16:306–320

Kenakin T (2002) Drug efficacy at G protein-coupled receptors. Annu Rev Pharmacol Toxicol 42:349–379

Kenakin T (2007) Collateral efficacy in drug discovery: taking advantage of the good (allosteric) nature of 7TM receptors. Trends Pharmacol Sci 28:407–415

Kenakin T, Watson C, Muniz-Medina V, Christopoulos A, Novick S (2012) A simple method for quantifying functional selectivity and agonist bias. ACS Chem Neurosci 3:193–203

Kjelsberg MA, Cotecchia S, Ostrowski J, Caron MG, Lefkowitz RJ (1992) Constitutive activation of the alpha 1B-adrenergic receptor by all amino acid substitutions at a single site. Evidence for a region which constrains receptor activation. J Biol Chem 267: 1430–1433

Kobilka BK, Deupi X (2007) Conformational complexity of G-protein-coupled receptors. Trends Pharmacol Sci 28:397–406

Kuszak AJ, Pitchiaya S, Anand JP, Mosberg HI, Walter NG, Sunahara RK (2009) Purification and functional reconstitution of monomeric mu-opioid receptors: allosteric modulation of agonist binding by Gi2. J Biol Chem. doi:10.1074/jbc.M109.026922

Leff P (1995) The two-state model of receptor activation. Trends Pharmacol Sci 16:89–97

Leff P, Scaramellini C, Law C, McKechnie K (1997) A three-state receptor model of agonist action. Trends Pharmacol Sci 18:355–362

Lohse MJ (2010) Dimerization in GPCR mobility and signaling. Curr Opin Pharmacol 10:53–58

Mailman RB (2007) GPCR functional selectivity has therapeutic impact. Trends Pharmacol Sci 28: 390–396

Mancia F, Assur Z, Herman AG, Siegel R, Hendrickson WA (2008) Ligand sensitivity in dimeric associations of the serotonin 5HT2c receptor. EMBO Rep 9: 363–369

Meyer BH, Segura JM, Martinez KL et al (2006) FRET imaging reveals that functional neurokinin-1 receptors are monomeric and reside in membrane microdomains of live cells. Proc Natl Acad Sci U S A 103:2138–2143

Milligan G (2004) G protein-coupled receptor dimerization: function and ligand pharmacology. Mol Pharmacol 66:1–7

Monod J, Wyman J, Changeux JP (1965) On the nature of allosteric transitions: a plausible model. J Mol Biol 12:88–118

Perez DM, Karnik SS (2005) Multiple signaling states of G-protein-coupled receptors. Pharmacol Rev 57: 147–161

Pin JP, Kniazeff J, Binet V et al (2004) Activation mechanism of the heterodimeric GABA(B) receptor. Biochem Pharmacol 68:1565–1572

Pin JP, Neubig R, Bouvier M et al (2007) International Union of Basic and Clinical Pharmacology. LXVII. Recommendations for the recognition and nomenclature of G protein-coupled receptor heteromultimers. Pharmacol Rev 59:5–13

Rajagopal S, Rajagopal K, Lefkowitz RJ (2010) Teaching old receptors new tricks: biasing seven-transmembrane receptors. Nat Rev Drug Discov 9:373–386

Rajagopal S, Ahn S, Rominger DH et al (2011) Quantifying ligand bias at seven-transmembrane receptors. Mol Pharmacol 80:367–377

Rasmussen SG, DeVree BT, Zou Y et al (2011) Crystal structure of the beta(2) adrenergic receptor-Gs protein complex. Nature 477:549–557

Ravikumar V, Panigrahi BK (2010) Comparative study of evolutionary computing methods for parameter estimation of quality signals. Int J Appl Evol Comput 1:28–59

Reiter E, Ahn S, Shukla AK, Lefkowitz RJ (2012) Molecular mechanism of β-arrestin-biased agonism at seven-transmembrane receptors. Annu Rev Pharmacol Toxicol 52:179–197

Richards FJ (1959) A flexible growth function for empirical use. J Exp Bot 10:290–300

Rios CD, Jordan BA, Gomes I, Devi LA (2001) G-protein-coupled receptor dimerization: modulation of receptor function. Pharmacol Ther 92:71–87

Rizki MM, Zmuda MA, Tamburino LA (2002) Evolving pattern recognition systems. IEEE Trans Evol Comput 6:594–609

Rovira X, Vivo M, Serra J, Roche D, Strange PG, Giraldo J (2009) Modelling the interdependence between the stoichiometry of receptor oligomerization and ligand binding for a coexisting dimer/tetramer receptor system. Br J Pharmacol 156:28–35

Rovira X, Pin JP, Giraldo J (2010) The asymmetric/symmetric activation of GPCR dimers as a possible mechanistic rationale for multiple signalling pathways. Trends Pharmacol Sci 31:15–21

Samama P, Cotecchia S, Costa T, Lefkowitz RJ (1993) A mutation-induced activated state of the β_2-adrenergic receptor. Extending the ternary complex model. J Biol Chem 268:4625–4636

Sips R (1948) On the structure of a catalyst surface. J Chem Phys 16:490–495

Sips R (1950) On the structure of a catalyst surface. II. J Chem Phys 18:1024–1026

Slack R, Hall D (2012) Development of operational models of receptor activation including constitutive receptor activity and their use to determine the efficacy of the chemokine CCL17 at the CC chemokine receptor CCR4. Br J Pharmacol 166:1774–1792

Sun Y, McGarrigle D, Huang XY (2007) When a G protein-coupled receptor does not couple to a G protein. Mol Biosyst 3:849–854

Terrillon S, Bouvier M (2004) Roles of G-protein-coupled receptor dimerization. EMBO J 5:30–34

Thron CD (1973) On the analysis of pharmacological experiments in terms of an allosteric receptor model. Mol Pharmacol 9:1–9

Urban JD, Clarke WP, von Zastrow M et al (2006) Functional selectivity and classical concepts of quantitative pharmacology. J Pharmacol Exp Ther 320:1–13

Urizar E, Montanelli L, Loy T et al (2005) Glycoprotein hormone receptors: link between receptor homodimerization and negative cooperativity. EMBO J 24: 1954–1964

Van Der Graaf PH, Schoemaker RC (1999) Analysis of asymmetry of agonist concentration-effect curves. J Pharmacol Toxicol Methods 41:107–115

Violin JD, Lefkowitz RJ (2007) Beta-arrestin-biased ligands at seven-transmembrane receptors. Trends Pharmacol Sci 28:416–422

Wells JW (1992) Analysis and interpretation of binding at equilibrium. In: Hulme EC (ed) Receptor-ligand interactions. A practical approach. Oxford University Press, Oxford, pp 289–395

White JF, Grodnitzky J, Louis JM et al (2007) Dimerization of the class A G protein-coupled neurotensin receptor NTS1 alters G protein interaction. Proc Natl Acad Sci U S A 104:12199–12204

Whorton MR, Bokoch MP, Rasmussen SG et al (2007) A monomeric G protein-coupled receptor isolated in a high-density lipoprotein particle efficiently activates its G protein. Proc Natl Acad Sci U S A 104:7682–7687

Whorton MR, Jastrzebska B, Park PS et al (2008) Efficient coupling of transducin to monomeric rhodopsin in a phospholipid bilayer. J Biol Chem 283:4387–4394

Wreggett KA, Wells JW (1995) Cooperativity manifest in the binding properties of purified cardiac muscarinic receptors. J Biol Chem 270:22488–22499

Xu H, Staszewski L, Tang H, Adler E, Zoller M, Li X (2004) Different functional roles of T1R subunits in the heteromeric taste receptors. Proc Natl Acad Sci U S A 101:14258–14263

Part IV

Bioinformatics Tools and Resources for GPCRs

GPCR & Company: Databases and Servers for GPCRs and Interacting Partners

9

Noga Kowalsman and Masha Y. Niv

Abstract

G-protein-coupled receptors (GPCRs) are a large superfamily of membrane receptors that are involved in a wide range of signaling pathways. To fulfill their tasks, GPCRs interact with a variety of partners, including small molecules, lipids and proteins. They are accompanied by different proteins during all phases of their life cycle. Therefore, GPCR interactions with their partners are of great interest in basic cell-signaling research and in drug discovery.

Due to the rapid development of computers and internet communication, knowledge and data can be easily shared within the worldwide research community via freely available databases and servers. These provide an abundance of biological, chemical and pharmacological information.

This chapter describes the available web resources for investigating GPCR interactions. We review about 40 freely available databases and servers, and provide a few sentences about the essence and the data they supply. For simplification, the databases and servers were grouped under the following topics: general GPCR–ligand interactions; particular families of GPCRs and their ligands; GPCR oligomerization; GPCR interactions with intracellular partners; and structural information on GPCRs. In conclusion, a multitude of useful tools are currently available. Summary tables are provided to ease navigation between the numerous and partially overlapping resources. Suggestions for future enhancements of the online tools include the addition of links from general to specialized databases and enabling usage of user-supplied template for GPCR structural modeling.

N. Kowalsman • M.Y. Niv (✉)
Institute of Biochemistry, Food Science and Nutrition, The Robert H. Smith Faculty of Agriculture, Food and Environment and The Fritz Haber Center for Molecular Dynamics, The Hebrew University of Jerusalem, Jerusalem, Israel
e-mail: niv@agri.huji.ac.il

M. Filizola (ed.), *G Protein-Coupled Receptors - Modeling and Simulation*, Advances in Experimental Medicine and Biology 796, DOI 10.1007/978-94-007-7423-0_9,

Keywords
G protein-coupled receptors • Servers • Databases • Models • Bitter taste receptors • Protein-protein interactions • Ligand sets • Membrane mismatch • Oligomerization • Repository • Interface • Curation • Xray crystallography • Homology • Partners • Database cross referencing

Abbreviations

2D	2-dimensional
3D	3-dimensional
5-HT	5-hydroxytryptamine
BLAST	Basic Local Alignment Search Tool
CPI	chemical–protein interaction
GIPs	GPCR-interacting proteins
GpHR	glycoprotein-hormone receptor
GPCR	G-protein-coupled receptor
GDP	guanosine diphosphate
GTP	guanosine-5′-triphosphate
KEGG	Kyoto Encyclopedia of Genes and Genomes
OR	olfactory receptor
PDZ	PSD-95/Discs-large/ZO-1
QSAR	quantitative structure–activity relationship
TM	transmembrane
UniProtKB	Universal Protein Resource Knowledgebase

9.1 Introduction

G-protein-coupled receptors (GPCRs) constitute a large superfamily of membrane receptors that are involved in a wide range of signaling pathways, celebrated in the 2012 Nobel Prize in Chemistry, awarded jointly to Robert J. Lefkowitz and Brian K. Kobilka for "*for studies of G-protein-coupled receptors*". GPCRs are activated by diverse ligands, including amines, odorants, fatty acids, peptides and neurotransmitters (Congreve et al. 2011a; Jacoby et al. 2006). These receptors undergo several stages in their life cycle, starting from biosynthesis in the endoplasmic reticulum (ER), maturation in the Golgi apparatus, and transport to the cell surface, followed by signal transduction and receptor internalization. At each of these stages, they are accompanied by specialized interacting protein partners and small molecules (Maurice et al. 2011; Bockaert et al. 2010). The repertoire of GPCR-interacting proteins (GIPs) is large and varied. It includes heat-shock proteins, kinases (for example GRKs), GPCR-associated sorting proteins (GASPs), PSD-95/Discs-large/ZO-1 (PDZ) domain-containing proteins, G-protein α subunit and many more (Maurice et al. 2011; Magalhaes et al. 2012), see Fig. 9.1.

The involvement of GPCRs in numerous pathways and pathological conditions makes them one of the most important classes of pharmacological targets (Overington et al. 2006; Schlyer and Horuk 2006). Many pathological conditions have been shown to be connected to interactions between GPCRs and other proteins. For example, the interaction between serotonin receptor 5-hydroxytryptamine 1B (5-HT_{1B}) and an altered form of p11 protein has been shown to be related to depression (Bockaert et al. 2010). The interaction between the PDZ protein MUPP1 and gamma-aminobutyric acid receptor B ($GABA_B$) fine-tunes receptor signaling and is potentially involved in diseases such as epilepsy (Magalhaes et al. 2012). Serotonin 5-HT_{2A} receptors (2AR) – metabotropic glutamate receptor 2 (mGluR2) complex has been suggested to be involved in the altered cortical processes of schizophrenia (Gonzalez-Maeso et al. 2008). It is therefore not surprising that the interactions between GPCRs and other molecules, such as natural ligands, synthetic agonists and antagonists, other GPCRs, G-proteins and

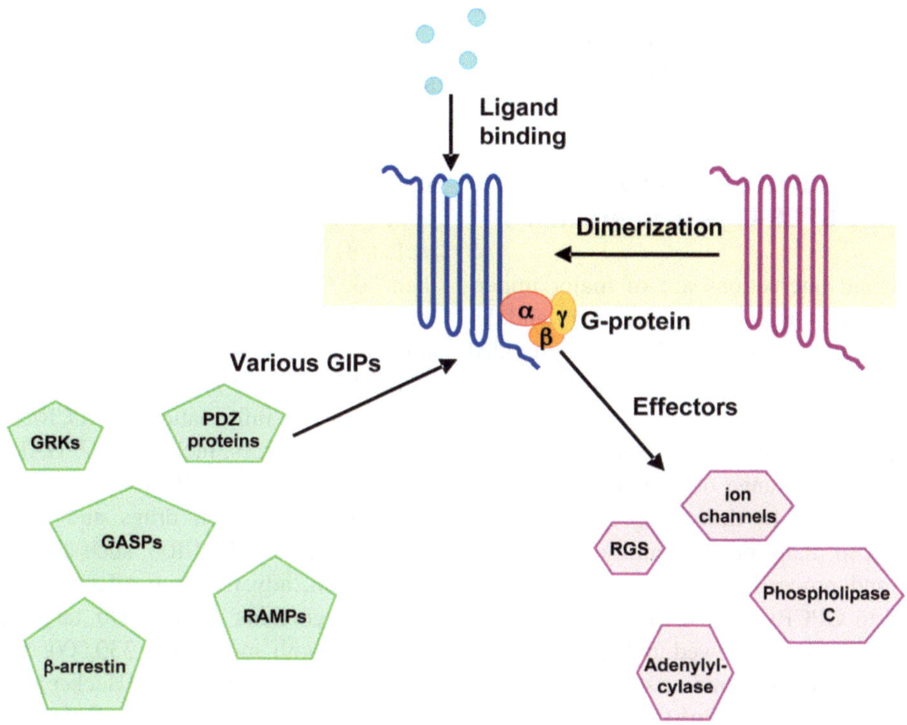

Fig. 9.1 Schematic view of a prototypical GPCR and some possible interacting partners. This cartoon shows the main interactions of a GPCR: ligand binding, oligomerization with other GPCRs, coupling to G-proteins (that interact with additional effectors) and binding with additional GIPs from various protein families. The GPCRs are shown in *blue* and *purple lines*, the ligands are depicted as *cyan circles*, the G-protein units are shown as *red*, *yellow* and *orange ellipses*, the effectors are marked as *pink hexagons* and the additional GIPs are shown as *green pentagons*. The membrane is illustrated as a *large yellow rectangle*

GIPs, are key aspects of GPCR-targeted drug development. Moreover, taking into account GPCR interactions may help in designing more specific, and perhaps more potent drugs. For instance, synthetic bivalent and multivalent drugs for GPCR oligomers represent attempts in this direction (Shonberg et al. 2011).

The rapid development of the internet and increase in information-sharing resources (such as electronic journals, databases and servers) make it easier for research groups and organizations to disseminate data to the research community. This leads also to an increase in the freely available data on GPCRs. A vast repertoire of web-based databases and servers is being developed in this field, raising the need to highlight and organize these data to make it easily accessible.

In this chapter, we focus on freely available databases and servers that supply information on GPCR interactions. We organize these resources in the following order: (i) general receptor–ligand and GPCR–ligand interaction databases, (ii) GPCR–ligand interaction databases dedicated to particular GPCR families, (iii) databases focused on GPCR oligomerization, (iv) databases dedicated to GPCR interactions with GIPs and signaling pathways, and (v) databases and servers addressing structural information on GPCRs. We conclude by outlining the challenges and opportunities in establishing, maintaining, and using GPCR–partner databases and servers, and highlight topics that, in our opinion, could further enhance GPCR interactions-related online tools.

9.2 Databases and Servers Related to GPCR Ligands and Interactions

9.2.1 General GPCR Ligands and GPCR-Ligand Interactions

GPCR–ligand interactions are of major importance for GPCR-targeted drug discovery and development. Indeed, recent studies have provided a vast abundance of data on GPCRs and their ligands.

Several freely available databases offer information on a wide range of GPCR–ligand interactions. We begin by describing databases that are general to many protein families, including GPCRs, and proceed to databases and servers dedicated to GPCRs and GPCR–ligand interactions. The databases reviewed in this chapter, and their url addresses are summarized in Table 9.1.

Databases such as **PubChem** (Bolton et al. 2008), **PubChem BioAssay** (Wang et al. 2012), **Psychoactive Drug Screening Program Ki (PDSP Ki)** (Roth et al. 2000), **BindingDB** (Liu et al. 2007) and **SuperTarget II** (Hecker et al. 2012), supply information on many ligands of numerous receptors, including GPCRs. These databases include links to information on bioassays, K_i and IC_{50} values and more.

PubChem is a very large database based on highly automated processing that contains more than 30 million compounds, some of which are GPCR ligands. For each compound, PubChem provides information on basic physical and chemical properties (such as molecular weight, hydrophobicity (AlogP), molecular formula), 2-dimensional (2D) and 3-dimensional (3D) structures that can be downloaded, synonyms, related literature, classification, patents, a list of vendors, and links to other (structurally related) compounds (Bolton et al. 2008). Many of the other databases mentioned in this chapter supply links to PubChem. The PubChem BioAssay database holds records on bioactivity screens of chemical substances. It provides searchable data from each bioassay, including information on the conditions and notes specific to a particular screening protocol. PubChem and PubChem BioAssay are linked (Wang et al. 2012). PDSP Ki contains more than 55,000 receptor K_i values. In addition to the K_i values it also supplies references to the relevant experiment (Roth et al. 2000). BindingDB focuses on interactions of proteins considered to be drug targets (GPCRs included) with small molecules. It contains more than 900,000 entries of binding data, for more than 6,000 protein targets and almost 400,000 small molecules and includes all data from PDSP Ki database (Liu et al. 2007). BindingDB allows broadening the information on GPCRs by linking to different sources, including IUPHAR-DB (see below). SuperTarget II holds information on the relations between drugs and their protein targets (including GPCRs), such as drug-related information, adverse drug effects and more. It currently contains over 6,000 target proteins, annotated with more than 330,000 relations to almost 200,000 compounds (Hecker et al. 2012).

The **Therapeutic Target Database (TTD)** aims to facilitate target-oriented drug discovery. The TTD database connects proteins (GPCRs included) with pathological conditions and known drugs. TTD contains about 2,000 targets and close to 18,000 drugs (Chen et al. 2002; Zhu et al. 2012). It links to the Kyoto Encyclopedia of Genes and Genomes (KEGG) pathway database which provides information on the pathways in which the proteins are involved (see Sect. 9.2.4). For example, searching the serotonin receptor 5-HT_{1A} (Universal Protein Resource Knowledgebase (UniProtKB) entry P08908) retrieves a table with an entry on 5-HT_{1A} listing several diseases and drugs entries. Each entry leads to a new information page. The 5-HT_{1A} information page contains synonyms, list of eight related diseases, lists of known related drugs, quantitative structure–activity relationship (QSAR) models, agonists and antagonists and more.

IUPHAR-DB and **GPCRDB** provide general information on GPCRs. Here we focus on data provided on GPCR interactions with various ligands. IUPHAR-DB contains curated information on protein superfamilies that are major biological targets of licensed medicinal drugs. This includes almost 360 GPCRs from human and

Table 9.1 Information on GPCR ligands and general GPCR–ligand interactions can be obtained via these databases

Database	URL	Species	Information on binding constants	Additional information	References
PubChem and PubChem BioAssay	http://pubchem.ncbi.nlm.nih.gov	Various species	Ki, Kd, IC_{50}	Contains information on millions of small compounds, including their biological activities.	Bolton et al. (2008), Wang et al. (2012)
PDSP Ki	http://pdsp.med.unc.edu/pdsp.php	Various species	Ki, Kd, IC_{50}	Contains published and internally derived K_i values for a large number of drugs and drug candidates (for GPCRs, ion channels, transporters and enzymes).	Roth et al. (2000)
BindingDB	http://www.bindingdb.org/bind/index.jsp	Various species	Ki, Kd, IC_{50}	Holds measured binding affinities for proteins (including GPCRs) and small-molecules.	Liu et al. (2007)
TTD	http://xin.cz3.nus.edu.sg/group/ttd/ttd.asp	Human	None	Connects proteins (including GPCRs) with pathological conditions and known drugs.	Chen et al. (2002), Zhu et al. (2012)
SuperTarget II	http://bioinf-apache.charite.de/supertarget_v2/	Various species	Ki, Kd, IC_{50} if available	Contains information on drug protein targets (GPCRs included), for example, drug-related information and adverse drug effects, gene ontologies, pathways and more.	Hecker et al. (2012)
IUPHAR-DB	http://www.iuphar-db.org	Human and rodents	Ki, Kd and IC_{50} if possible	Supplies pharmacological, pathophysiological, chemical, genetic, functional and anatomical information for GPCRs and ion channels.	Harmar et al. (2009)
GPCRDB	http://www.gpcr.org/7tm	Various species	Ki, Kd and IC_{50} if possible	Contains sequences, ligand binding constants, oligomerization data and mutations; computationally derived data, i.e. multiple sequence alignments and homology models.	Horn et al. (1998, 2003), Vroling et al. (2011)
GPCR SARfari	https://www.ebi.ac.uk/chembl/sarfari/gpcrsarfari	Various species	Ki, Kd, IC_{50}	Contains information on the ligands and binding constants for family A GPCRs and basic statistics on the properties of the ligands.	Bellis et al. (2011)
GLIDA	http://pharminfo.pharm.kyoto-u.ac.jp/services/glida/	Human, mouse and rat	Not directly	Provides information on the GPCRs and ligands. Enables ligand-based searches and searches for binding patterns of similar GPCRS to sets of ligands.	Okuno et al. (2006, 2008)
CPI-Predictor	http://www.lmmd.org/online_services/cpi_predictor/	Human	Not relevant	A server that predicts given a ligand to which GPCR it will bind, if at all.	Cheng et al. (2012)

The data from all these databases can be downloaded. Some parts of the data in GPCR IUPHAR-DB cannot be downloaded but can be given upon request and they are free for academic users. GLIDA is deposited in PubChem and can be downloaded from there rodent sources and thousands of their ligands. It integrates pharmacological (including the K_i and IC_{50} values of ligands, when available), pathophysiological, chemical, genetic, functional and anatomical information (Harmar et al. 2009). GPCRDB specializes in GPCRs from different species. It contains data on sequences, ligand-binding constants, oligomerization and mutations from more than 12,000 experiments. In addition, it stores computationally derived data, such as multiple sequence alignments and homology models (Horn et al. 1998, 2003; Vroling et al. 2011).

GPCR SARfari (part of ChEMBL database) and **GPCR-Ligand Database (GLIDA)** are dedicated to interactions between GPCRs and their ligands. GPCR SARfari integrates target, 3D structure, compounds of experimental relevance and screening data (including binding constants) in a single chemogenomics database for class A GPCRs. It provides information on the molecular weight, AlogP, polar surface area and affinities of the compounds that bind a specific GPCR. In total, GPCR SARfari contains information on over 142,000 compounds (Bellis et al. 2011). GLIDA holds thousands of human, rat and mouse GPCR entries and tens of thousands of ligand entries. It supplies general information on GPCRs and their ligands and links to additional information databases such as GPCRDB, IUPHAR-DB and PubChem. GLIDA allows a user-friendly ligand-based search of the database that is based on clustering of ligand structures. In addition, it can create correlation maps between similar GPCRs and their ligands (Okuno et al. 2006, 2008). Each database contains information that is not found in the other: GPCR SARfari supplies binding constants and allows filtering of the search results based on preferred values of activity types (such as $LogIC_{50}$, Binding, activity percent and many more). GLIDA provides interaction-correlation maps between similar GPCRs and their ligands. For example, for $5\text{-}HT_{1A}$, GPCR SARfari returns more than 3,600 binding compounds and more than 5,000 entries on bioactivity data for ligands, whereas GLIDA supplies data on 1,110 agonists and 604 antagonists.

A multitarget quantitative structure–activity relationship (mt-QSAR) method was developed recently using a dataset of about 82,000 chemical-protein interaction pairs between almost 51,000 compounds and 136 GPCRs and is available via the mt-QSAR-based web server named **CPI-Predictor**. The user inserts the compound of interest (either by structure or in SMILES format) and the server returns target GPCRs predicted to bind this compound. This predictor has the potential to facilitate applications in network pharmacology and drug repositioning (Cheng et al. 2012).

9.2.2 Ligand Interactions of Particular GPCR Families

In addition to the general GPCR–ligand databases, databases and servers specialized in specific GPCR families have been established (summarized in Table 9.2). The dedicated efforts and manual curation of some of these sites provide data that are unavailable in the more general resources, justifying the existence of these specialized databases alongside the more general ones.

Glycoprotein-Hormone Receptor (GpHR) Information System (GRIS) (Van Durme et al. 2006) and **Sequence-Structure-Function-Analysis of Glycoprotein Hormone Receptors (SSFA-GPHR)** (Kleinau et al. 2007) are dedicated to GpHRs, including thyrotropin receptor (TSHR), follitropin receptor (FSHR) and lutropin/choriogonadotropin receptor (LHR/CGR). GRIS contains GpHR sequences, models, and information about mutations and their phenotypes. It allows the users to perform residue and rotamer analyses using the 3D models and to predict the structural effects of mutations *in silico* (Van Durme et al. 2006). SSFA-GPHR integrates large amounts of information on mutations in GpHRs and uses 2D snake-plot diagrams and 3D models to supply information on the mutation's location. In addition, it includes information on the effects of mutations on receptor expression, binding and activity (Kleinau et al. 2007).

Table 9.2 Databases for ligand interactions of particular GPCR families

Database	URL	Species	Additional information	References
GRIS	http://gris.ulb.ac.be/ (the site is not always accessible)	Various species	Contains heterogeneous data on GpHRs such as sequences, models, mutations and their phenotypes.	Van Durme et al. (2006)
SSFA-GPHR	http://www.ssfa-gphr.de/	Human and rat	Integrates large amounts of information on mutations in GpHRs, including their effects on GpHR receptor expression, binding and activity.	Kleinau et al. (2007)
OdorDB, ORDB	http://senselab.med.yale.edu/odordbhttp://senselab.med.yale.edu/ordb	Various species	Supplies basic information on the odorants and receptors, and ORDB link between odorants and olfactory receptors. ORDB was developed over the years and now includes a range of chemosensory receptors.	Crasto et al. (2002, 2003)
DoOR	http://neuro.uni-konstanz.de/DoOR/default.html	*Drosophila melanogaster*	Provides response profile of odor receptors and odorants.	Galizia et al. (2010)
ODORactor	http://mdl.shsmu.edu.cn/ODORactor/	Human and mouse	Predicts olfactory receptors for small molecules.	Liu et al. (2011)
HORDE	http://genome.weizmann.ac.il/horde/	Human. Supply orthologues in chimpanzee, dog, opossum and platypus	Provides information on the genomic localization and cluster organization of 855 human olfactory receptor genes.	Olender et al. (2004), Safran et al. (2003)
SuperScent	http://bioinf-applied.charite.de/superscent/index.php?site=home	Various species	A database for volatile scents. The compounds can be searched by structure, similarity or properties. Each scent entry includes basic details and supplier information.	Dunkel et al. (2009)
SuperSweet	http://bioinf-applied.charite.de/sweet	Human	Dedicated to sweet and predicted-to-be sweet molecules. Provides homology models of the sweet receptor and its predicted complexes with the molecules.	Ahmed et al. (2011)
BitterDB	http://bitterdb.agri.huji.ac.il/bitterdb/	Human models	Links bitter compounds and bitter taste receptors. Compounds can be searched by structure, properties, etc. Supplies 2D snake-plot representations of the receptors.	Wiener et al. (2012)

The data from all these databases can be downloaded except in the cases: OdorDB and ORDB. Structures and sequences can be downloaded from GRIS. In SuperScent, SuperSweet and BitterDB some of the data on each molecule can be downloaded separately. In SSFA-GPHR the search results as 2D Snake plot or 3D homology model can be downloaded

Several specialized databases are devoted to sensory receptors, including **OdorDB**, **Olfactory Receptor Database (ORDB)** (Crasto et al. 2002, 2003), **Database of Odor Responses (DoOR)** (Galizia et al. 2010), **ODORactor** (Liu et al. 2011), **The Human Olfactory Data Explorer (HORDE) database** (Olender et al. 2004; Safran et al. 2003) **SuperScent** (Dunkel et al. 2009), **SuperSweet** (Ahmed et al. 2011) and **BitterDB** (Wiener et al. 2012).

The database for odor molecules, OdorDB, and the database for olfactory receptors (ORs), ORDB, are related and link odor ligands to the ORs they activate. ORDB was expanded and contains additional chemosensory receptors, for example taste papilla receptors. The chemosensory genes and proteins can be searched using a variety of classification entries such as organism, tissue, strain and the labs that do functional analysis on the receptor. The odors in OdorDB can be searched by different traits (for example cyclic structures, functional groups). It also lists the receptors found to be involved with each odor (Crasto et al. 2002, 2003). DoOR database supplies the response profile of ORs in *Drosophila*. It allows searching by receptor or by odorant and retrieves the normalized response graphs of each. DoOR provides a downloadable map of the *Drosophila* antennal lobe's physiological response to an odorant (Galizia et al. 2010). The server ODORactor predicts whether a molecule (inserted in SMILES format or via drawing its 2D structure) is an odorant, and its possible associated ORs. The server provides links to other data sources for information about the receptors (Liu et al. 2011). HORDE is a database of human OR genes, supplying information on genomic and cluster organizations of 855 human OR genes and their orthologues in other mammals (Olender et al. 2004; Safran et al. 2003). The SuperScent database contains more than 2,000 volatile scents. The scents are classified and can be searched by structure, or by properties such as name, molecular weight, chemical group and more. Each scent entry includes basic details and supplier information and is linked to the relevant PubChem entry for additional information (Dunkel et al. 2009). SuperSweet is a database for natural and artificial sweet molecules, extracted from the literature and publicly available databases such as PubChem and PDB. This dataset was expanded by using similarity search methods, hence it also contains molecules that are predicted, but not proven, to be sweet. In addition, SuperSweet contains a homology model of the $T1R_3$-$T1R_2$ sweet receptor dimer. For some of the sweet compounds, the server provides docking poses in the model structure (Ahmed et al. 2011).

BitterDB was developed and is maintained by our group (Wiener et al. 2012) (http://bitterdb.agri.huji.ac.il/bitterdb/). This database includes a collection of known bitter compounds, gathered from the literature and from Merck index, and their 25 associated human bitter taste receptors. It encourages users to add their own information on bitter or tasteless molecules. BitterDB provides several ways to investigate the bitter world: search for bitter compounds by different physicochemical properties, search for bitter molecules with structures similar to a query compound, cross-link between a bitter compound and the receptor that binds it, option to use Basic Local Alignment Search Tool (BLAST) against human bitter receptor sequences and more. It supplies basic information on the bitter compounds and receptors and a 2D snake-plot representation of the receptors.

9.2.3 GPCR Oligomerization

In the last decade, it has become clear that many GPCRs may function as homo or heterodimers or higher order oligomers (Gonzalez-Maeso et al. 2008; Gorinski et al. 2012; Lohse 2010). Oligomerization has implications for trafficking, signaling and pharmacology of many members of the GPCR family (Lohse 2010; Rediger et al. 2011) thus increasing the importance of GPCR oligomerization for general understanding and drug discovery.

Two main web resources deal with GPCR oligomerization: the **GPCR Oligomerization Knowledge Base Project (GPCR-OKB)** and the **GPCRs Interaction Partners (GRIP)** server and database (Table 9.3).

Table 9.3 Databases and servers dealing with GPCR oligomerization

Database	URL	Species	Additional information	References
GPCR-OKB	http://data.gpcr-okb.org/gpcr-okb/	Various species	Contains information about GPCR oligomers, including phenotypic changes, evidence for physiological/clinical relevance, effects of oligomer-specific ligands and proposed mechanisms of activation.	Khelashvili et al. (2010)
GRIP server	http://grip.cbrc.jp/GRIP/	Not relevant	Predicts interfaces for oligomerization sites based on sequence alignment of the query GPCR and its homologs, and a template structure (either rhodopsin or β2 adrenergic receptor).	Nemoto et al. (2009)
GRIP database	http://grip.cbrc.jp/GDB/index.html	Uses human sequences as representative	Holds information on GPCR oligomerization and suggested interfaces based on published experiments and GRIP server predictions.	Nemoto et al. (2011)
CTMDapp Downloadable executable	http://memprotein.org/new-continuum-molecular-dynamics-ctmd-software-made-available-by-computational-modeling-core	Not relevnt	A novel approach that can predict oligomerization interfaces on membrane proteins monomers, GPCRs included. The predictions are based on energetic cost calculations of the hydrophobic mismatch-driven deformation in membrane bilayers generated by multihelical membrane proteins.	Mondal et al. (2011)
BioGRID	http://thebiogrid.org/	Various species	Contains genetic and protein interaction data (GPCRs included) from various species, including humans. For a specific protein, the database supplies a list of proteins that have been found to interact with it and the appropriate reference.	Stark et al. (2006)
HPRD	http://www.hprd.org/	Human	Holds information on domain architecture, post-translational modifications, interaction networks and disease association for each protein in the human proteome (including GPCRs). The database supplies vast information on a query protein, including a list of interacting partners (some of them additional GPCRs).	Keshava Prasad et al. (2009)

The data from these databases except GRIP database can be downloaded. HPRD is free for academic users

GPCR-OKB compiles data on oligomerization for all types of GPCRs. It allows browsing, visualization, and structured querying of computational and experimental information on GPCR oligomerization. The database provides information on the oligomer, including the proposed interface (at TM and residue levels), phenotypic changes, *in-vivo* evidence and biological relevance. In addition, GPCR-OKB presents the effects of ligands that are oligomer-specific and their proposed mechanisms of activation. For additional data on the protomers, it links to GPCRDB (Khelashvili et al. 2010).

The GRIP database (Nemoto et al. 2011) includes two sections: an experimental information section which contains experimentally identified GPCR oligomers, and their experimentally suggested interfaces for the oligomerization, if known; the second section contains predictions on GPCR oligomerization made by the GRIP server. The prediction relies on multiple sequence alignment-based conservation scores and their mapping on the surface of the template structure of either rhodopsin or β2 adrenergic receptor. The output includes a model structure and a list of interacting residues (Nemoto et al. 2009). The user can choose to view results based on experiments, predictions or both. It is known that $5HT_{1A}$ can form homo and hetero oligomers (Lukasiewicz et al. 2007; Renner et al. 2012; Woehler et al. 2009) and we have recently identified, using computational and experimental methods, residues involved in $5HT_{1A}$ homodimerization (Gorinski et al. 2012). GPCR-OKB finds seven entries for $5\text{-}HT_{1A}$ oligomerization either with another $5HT_{1A}$ or additional GPCRs. No entries for $5HT_{1A}$ oligomerization were found in the GRIP database.

Recently, a new Continuum-Molecular Dynamics (CTMD) software tool was developed and made available for download (Mondal et al. 2011). This novel tool energetically quantifies the hydrophobic mismatch-driven deformation in membrane bilayers generated by multihelical membrane proteins. The method accounts for both the membrane remodeling energy and the energy contribution from any incomplete alleviation of the hydrophobic mismatch by membrane remodeling and can thus predict optimal monomeric orientations within oligomers. Results may also explain how distinct ligand-induced conformations of GPCRs result in differential effects on the membrane environment. The CTMD software executable can be downloaded from the Membrane Protein Structural Dynamics Consortium (MPSDC) Gateway http://memprotein.org/.

The **Biological General Repository for Interaction Datasets (BioGRID)** (Stark et al. 2006) and the **Human Protein Reference Database (HPRD)** (Keshava Prasad et al. 2009) hold information on protein interactions, including GPCR–GPCR interactions. BioGRID contains genetic and protein interaction data from various species, including humans. It stores over 500,000 curated interactions taken from datasets and individual focused studies, derived from over 30,000 publications. The user can search for a specific GPCR and obtain, for each species, a list of proteins that have been found to interact with it, along with the relevant reference (Stark et al. 2006).

HPRD contains information regarding domain architecture, post-translational modifications, interaction networks and disease association for each protein in the human proteome. Querying for a specific GPCR returns a list of interacting partners (some of them other GPCRs) and the relevant references. HPRD is manually curated (Keshava Prasad et al. 2009).

BioGRID supplies five entries of interacting partners for $5\text{-}HT_{1A}$, but only when using the synonym HTR1A, Three of these are other GPCRs. Searching HPRD for $5\text{-}HT_{1A}$ (using the name 5-hydroxytryptamine receptor, 1A) returns eight entries of interacting partners, including $5HT_{1A}$ itself.

9.2.4 Intracellular Partners and Signaling Pathways

During their life cycle, GPCRs are accompanied by a range of specialized GIPs that assist nascent receptors in proper folding, target them to the appropriate subcellular compartments and help them fulfill their signaling tasks (Maurice et al. 2011), see Fig. 9.1. GIPs have an important role in regulating receptor–ligand interaction. The interaction between GPCRs and GIPs creates larger GPCR-associated protein complexes, placing the GPCRs at the center of protein networks that participate in the dynamic regulation of the receptor (Daulat et al. 2009; Ritter and Hall 2009).

GIPs databases and servers deal with G-proteins (Elefsinioti et al. 2004; Satagopam et al. 2010; Theodoropoulou et al. 2008) and PDZ domains (Beuming et al. 2005), and these are summarized in Table 9.4 and described below.

The heterotrimeric G-protein complex is vital for GPCR signal transduction. It is composed of α, β, and γ subunits, each containing several types (in humans there are 21 Gα, 6 Gβ and 12 Gγ subunits) (Oldham and Hamm 2008). The interaction pattern of GPCRs with G-proteins is complex: some GPCRs can bind several types of G-proteins, yielding various signal-transduction

Table 9.4 Databases of GPCRs interactions with other proteins and signal-transduction

Database	URL	Species	Additional information	References
gpDB	http://biophysics.biol.uoa.gr/gpDB/	Various species	Provides information on the coupling between GPCR and G-protein α subunit and the interactions between the G-protein units and different effectors. Last updated in 2008.	Elefsinioti et al. (2004)
Human-gpDB	http://biophysics.biol.uoa.gr/human_gpdb/	Human	Provides information on GPCR–G-protein interactions and coupling between GPCR and G-protein α subunit and the interactions between the G-protein units and different effectors. Allows 2D and 3D visualization of the connections between GPCRs, related drugs, G-proteins and effectors.	Satagopam et al. (2010)
PRED-COUPLE 2.00	http://athina.biol.uoa.gr/bioinformatics/PRED-COUPLE2/	Not relevant	Predicts the G-protein families that couple to a specific GPCR.	Sgourakis et al. (2005)
PDZBase	http://icb.med.cornell.edu/services/pdz/start	Various species	Provides information on more than 300 interactions involving PDZ domains, some of them with GPCRs. Information is obtained from *in-vivo* or *in-vitro* experiments.	Beuming et al. (2005)
REACTOME	http://www.reactome.org/ReactomeGWT/entrypoint.html	Various species	REACTOME is a database of pathways and reactions in human biology. Allows detailed visualization and navigation of biological pathways.	Croft et al. (2011)
KEGG pathway database	http://www.genome.jp/kegg/pathway.html	Various species	Links between different KEGG databases. Contains genomic information, pathways of pathological conditions and synthesis of drugs.	Kanehisa and Goto (2000), Kanehisa et al. (2012)
The UCSD Signaling Gateway	http://www.signaling-gateway.org/molecule/	Various species	Dedicated to proteins involved in signaling (GPCRs included). Each protein entry contains interacting partners and other information.	Li et al. (2002), Saunders et al. (2008)

The data can be downloaded from all databases except gpDB and human-gpDB. For downloading the entire KEGG dataset academic users can subscribe to the KEGG ftp site (managed by NPO Bioinformatics, Japan) and non-academic users requires a commercial license agreement

responses; and the same type of G-protein can bind to different GPCRs. In addition, the G-protein's subunits activate or inhibit a variety of effector proteins that modulate the cellular signal-transduction functions (Bosier and Hermans 2007; Wettschureck and Offermanns 2005).

Two databases are available for G-protein complexes: the general **gpDB** (Elefsinioti et al. 2004; Theodoropoulou et al. 2008) and the more recent **Human-gpDB** (Satagopam et al. 2010). Both provide information on the coupling between GPCR and G-protein α subunit and on the interactions between the G-protein subunits and different effectors, and supply additional information on the interacting partners. The human-gpDB allows 2D and 3D visualization of the network of connections between GPCRs, related drugs, G-proteins and effectors. It also offers links to many additional GPCR web resources (Satagopam et al. 2010).

The server **PRED-COUPLE 2.00** predicts coupling specificity of GPCRs to G-protein families. The PRED-COUPLE algorithm combines a data-mining exploratory approach with the discriminative power of profile hidden Markov models. The method was trained using two datasets of GPCRs with known non-promiscuous binding to $G\alpha_i/G\alpha_o$, $G\alpha_q/G\alpha_{11}$

or $G\alpha_s$ and validated on two other datasets of GPCRs. The user inserts a GPCR sequence and gets a normalized probability score (from 0 to 1) for coupling to each one of the four G-protein families. A higher score represents a higher coupling probability. Probability scores below 0.3 are considered negative predictions (Sgourakis et al. 2005). For example, the PRED-COUPLE retrieves only $G\alpha_i/G\alpha_o$ as having significant probability for coupling to the human 5-HT_{1A} receptor. $G\alpha_i/G\alpha_o$ is indeed the main coupling G-protein α subunit of 5-HT_{1A} (Pucadyil et al. 2005).

PDZ domains consist of 80–90 amino acids that usually bind to a specific peptide sequence at the C terminus of their interacting proteins. Many GPCRs contain a PDZ-binding motif at their C terminus that allows them to bind PDZ proteins. Through these interactions, the PDZ proteins participate in GPCR regulation of signaling, trafficking, and function (Magalhaes et al. 2012; Romero et al. 2011). **PDZBase** is a specific database for interactions involving PDZ domains. The database can be searched by PDZ name, accession code, ligand accession code, specific sequence motif in either the PDZ or the ligand, and specific residue in the PDZ domain. PDZBase contains more than 300 interactions obtained from either *in-vivo* (coimmunoprecipitation) or *in-vitro* (GST-fusion or related pull-down) experiments (Beuming et al. 2005).

BioGRID and HPRD (mentioned in Sect. 9.2.3) provide references for interactions between GPCRs and other proteins, such as G-proteins, PDZ domain-containing proteins and additional GIPs (Stark et al. 2006; Keshava Prasad et al. 2009).

General information on GPCR-signaling pathways and the molecules which make up these pathways can be found in the **REACTOME** server (Croft et al. 2011), the **Kyoto Encyclopedia of Genes and Genomes (KEGG) pathway database** (Kanehisa and Goto 2000; Kanehisa et al. 2012) or the **UCSD Signaling Gateway** (Li et al. 2002; Saunders et al. 2008). REACTOME is a curated database of pathways and reactions (pathway steps) in human biology (where 'reaction' includes many biological events that entail changes in state, such as binding, activation, translocation and degradation, in addition to classical biochemical reactions). It allows detailed visualization and navigation of known pathways. It also returns inferred reactions for orthologous proteins of over 20 non-human species (Croft et al. 2011). The KEGG pathway database is part of the KEGG integrated database resource, which consists of 17 categorized databases (Kanehisa and Goto 2000; Kanehisa et al. 2012). KEGG pathway links the different KEGG databases, thus providing information on related drugs and additional, mostly genomic, information on GPCRs. It allows examination of the pathways of pathological conditions and diseases, showing the roles of different proteins, including GPCRs, in those conditions (Kanehisa and Goto 2000; Kanehisa et al. 2012). The UCSD Signaling Gateway provides information on more than 8,000 proteins (GPCRs included) involved in cellular signaling. The data on each protein is represented by a "Molecule Page". The molecule page is divided into two main parts; one contains information contributed by authors, such as a short review on the main functional and biological properties of the molecule, including key citations, an interactive overview of the different states of the protein and the transitions between them, a list of known functions and more. The second part holds data acquired automatically from public databases, for example sequence information, links with information on interacting partners, graphical representations of protein domains and motifs, links to pathway information in KEGG and REACTOME (Li et al. 2002; Saunders et al. 2008).

As an example, the REACTOME server results in 12 entries of reaction, pathway, complex, or protein connected to 5-HT_{1A} for the 5-HT_{1A} receptor query. Each entry consists of a few lines of information and links to an information page that includes a pathway scheme of that reaction, pathway, etc. Querying the KEGG pathway database on 5-HT_{1A}, results in an entry with a thumbnail map and information on serotonergic synapses. The map is interactive and allows obtaining more information specific for 5-

$HT_{1A\text{-}F}$ and other serotonin receptors. Searching the UCSD Signaling Gateway for 5-HT_{1A}, results in two entries: one for the mouse receptor and one for the human receptor. Each entry links to a "molecule page" with information on the receptor and to annotated data, including pathways. SuperTarget II (mentioned in Sect. 9.2.2) also provides pathway information and is linked to the KEGG pathway database (Hecker et al. 2012). Searching SuperTarget II for 5-HT_{1A} results in target details page that contains a link to KEGG pathways related information, list of drugs related to 5-HT_{1A}, synonyms, similar targets and list of proteins known to bind 5-HT_{1A}. Each of these can open into a separate page.

9.2.5 Structural Information on GPCRs

Structural information on GPCRs is invaluable for GPCR signaling research and drug design (Shoichet and Kobilka 2012; Granier and Kobilka 2012). GPCRs share a seven-transmembrane (TM) helical fold but the structure and length of their N and C termini and loops vary greatly. For many years, challenges in expressing and crystallizing GPCRs prevented deciphering their structure. However, the last few years have witnessed great breakthroughs in GPCR X-ray crystallography, paving the way for solving the structures of 17 GPCRs, some in their active state (bound to agonists), some in their inactive state (bound to antagonists) and some captured in both (Shoichet and Kobilka 2012; Katritch et al. 2012; Jacobson and Costanzi 2012). Most recently, an active structure of peptide-bound neurotensin receptor (White et al. 2012) and an NMR structure of chemokine receptor were solved (Park et al. 2012).

Despite the outstanding progress in GPCR X-ray crystallography, structures were solved so far for a small portion of all GPCR types. Thus, modeling the structures of additional GPCRs, to be used in research and drug development, remains indispensable. With X-ray structures on the rise, the models can now be better assessed and evaluated (Mobarec et al. 2009; Yarnitzky et al. 2010; Costanzi 2012; Levit et al. 2012).

In some cases, there have been successes in predicting interactions with ligands using GPCR models (Kufareva et al. 2011; Levit et al. 2011; Provasi et al. 2009) and in numerous cases, homology-based virtual screening campaigns have yielded high hit rates and contributed to drug discovery (Evers and Klebe 2004; Congreve et al. 2011b; Vilar et al. 2011; Carlsson et al. 2011). Here we review freely available web resources for experimentally solved and computationally modeled GPCR structures (summarized in Table 9.5).

The main universal depository of X-ray- and NMR-solved structures of macromolecules is the **Protein Data Bank (PDB)** (Berman et al. 2000). A user can create a personalized PDB (MyPDB) account to save previous searches, get alerts on new structure releases, save notes on structures and more. Solved GPCR structures can be retrieved from PDB or alternatively from the **Membrane Proteins of Known 3D Structure** database, http://blanco.biomol.uci.edu/mpstruc. The latter database provides useful information on integral membrane proteins to which crystallographic, or sometimes NMR structures, have been determined to a resolution sufficient to identify TM helices of helix-bundle membrane proteins (typically up to 4–4.5 Å). The **Orientations of Proteins in Membranes (OPM)** database extracts the structural data from the PDB and predicts the orientation of the protein (GPCRs included) in the membrane at the residue level and the tilt of the protein in degrees. The proteins are classified by species and localization in membrane types (Lomize et al. 2006).

An important resource is UniProt (http://www.uniprot.org/) (The UniProt Consortium 2012). UniProt is composed from several databases that provide information on protein sequence and annotation. This information is very valuable for protein modeling, including GPCRs.

Several databases that were mentioned in the previous sections provide 3D models for visualizations of query proteins (GPCRDB, GPCR

Table 9.5 Databases and servers holding GPCR structures

Database	URL	Additional information	References
PDB	http://www.rcsb.org/pdb/home/home.do	The main universal depository of X-ray- and NMR-solved macromolecule structures.	Berman et al. (2000)
Membrane Proteins of Known 3D Structure database	http://blanco.biomol.uci.edu/mpstruc	Provides information about integral membrane proteins whose structures have been determined.	None available
OPM	http://opm.phar.umich.edu/	Predicts the orientation of the proteins in the membrane for membrane protein structures found in the PDB.	Lomize et al. (2006)
Human GPCR models repository	http://cssb.biology.gatech.edu/skolnick/files/gpcr/gpcr.html	A repository of homology models for all 907 human GPCRs.	Zhang et al. (2006)
GPCRRD	http://zhanglab.ccmb.med.umich.edu/GPCRRD/	A resource for modeling GPCRs. Composed of three main parts: database of experimental restraints that can be used in the modeling, GPCR-ITASSER modeling server and 1,028 pre-computed models for human GPCRs.	Zhang and Zhang (2010)
GPCR-SSFE	http://www.ssfa-7tmr.de/ssfe/	Provides homology models of the TM helices for more than 5,000 family A GPCRs and the templates used for modeling the TMs.	Worth et al. (2009)
GPCR-ModSim	http://gpcr.usc.es/	A server for GPCR modeling that aims to integrate: multiple sequence alignment, modeling, loop refinement and global refinement stages.	Rodriguez et al. (2012)
GPCRautomodel	http://genome.jouy.inra.fr/GPCRautomdl/cgi-bin/welcome.pl	A modeling server devoted to olfactory receptor modeling and docking odorants to these models. The server can model other types of GPCRs as well.	Launay et al. (2012)
GLL and GDD	http://cavasotto-lab.net/Databases/GDD/	Sets of ligands for GPCR targets and the decoys produced for these ligands.	Gatica and Cavasotto (2012)
EVfold_membrane	http://evfold.org/transmembrane/	A program for membrane protein folding. The models created by this program can be downloaded and the software code for creating evolutionary constraints can be obtained from the authors.	Hopf et al. (2012)

SARfari, SSFA-GPHR, GRIS and SuperSweet). Resources that are specifically dedicated to providing models or servers for modeling GPCRs are described next.

A repository of homology models for human GPCRs (Zhang et al. 2006) was recently updated and now contains almost 1,000 human GPCR model structures combined with virtual screening of the ZINC (Irwin et al. 2012) database (Zhou and Skolnick 2012). The **I-TASSER** server enables modeling of protein structures (including GPCRs) using multiple templates (Zhang 2008). The recent **GPCR Research Database (GPCRRD)** is designed to facilitate the construction of 3D structural models of GPCRs. The GPCRRD contains three main parts: (i) the GPCR experimental restraints database. This contains experimental results that have been systematically collected from the literature and other GPCR-related web resources and can be used as restraints in modeling; (ii) the GPCR-ITASSER, which is a GPCR-specialized version of the I-TASSER server, and (iii) pre-computed models of 1,028 putative human GPCRs (from 2010) using GPCR-ITASSER. The user can either search or browse the databases for restraints and models (Zhang and Zhang 2010).

The **GPCR-Sequence-Structure-Feature-Extractor (GPCR-SSFE)** database provides initial homology models of the TM helices for more than 5,000 family A GPCRs from various species. The templates are selected for each TM separately. The models and sequence alignments can be viewed and downloaded, as can the templates that were used to create the TM helices. The site also provides a link for loop modeling (Worth et al. 2009). The new **GPCR Modeling and Simulation toolkit (GPCR-ModSim)** server aims to integrate all stages of modeling: multiple sequence alignment, modeling, loop refinement and global refinement (using molecular dynamics). GPCR-ModSim allows the user to decide on every step involved in the modeling pipeline: selection of the template from lists of either "active" or "inactive" and "inactive-like" structures, manual editing of the query-template alignment, selection of the most appropriate 3D-model, the definition of the loop regions to be refined and, if the user wishes, global refinement. The user can choose several templates and based on each of them alone the server will create homology models (Rodriguez et al. 2012).

GPCRautomodel is another modeling server for GPCRs. It was created mainly for modeling ORs but it can model other GPCRs as well. The user can decide between seven GPCR templates for the modeling. The server also enables docking of small molecules to the resulting GPCR model. For each set of models generated, one template structure is used. The website supplies a library of odorant compounds that can be docked but the user can also insert his or her own molecule (Launay et al. 2012). Note that to use GPCR-ModSim and GPCRautomodel (ligand-binding prediction), the user must first get a MODELLER (Sali and Blundell 1993) license key, which is freely available. Both servers have a link to the MODELLER registration site.

Of significance for docking studies, some chemical libraries may have a strong bias toward GPCR-like ligands because GPCRs are important drug targets (Kolb et al. 2009). Recently, a **GPCR ligand library (GLL)** for 147 targets, selecting 39 decoy molecules for each ligand, was collected in the **GPCR Decoy Database (GDD)**. Decoys were chosen to ensure a ligand–decoy similarity of six physical properties, while enforcing ligand–decoy chemical dissimilarities. The docking performance of the GDD was evaluated on 19 GPCRs, showing a marked decrease in enrichment compared to bias-uncorrected decoy sets. Both the GLL and GDD are freely available to the scientific community (Gatica and Cavasotto 2012).

A new algorithm for membrane protein folding has been recently published, **EVfold_membrane**, with very impressive success in *de-novo* 3D-structure predictions (Hopf et al. 2012). The authors blind-tested the EVfold_membrane protocol on α-helical transmembrane proteins representing 23 diverse families with known structures. In 21 of the cases the Cα-rmsd was 2.6–4.8 Å for more than 70 % of the length (Hopf et al. 2012). There is no server for **EVfold_membrane**, but the receptors models, the 3D coordinates and the evolutionary constraints used in the paper can be downloaded. The software code for the evolutionary constraint calculation can be obtained upon request. This program may help in modeling class B and class C GPCRs, which is challenging duc to thc lack of X-ray structures of the TM part for these families.

9.3 Discussion and Outlook

A large variety of freely available databases and servers supply information on GPCR interactions and binding partners. These databases provide an important tool in understanding biological pathways, chemoinformatics and development of new drugs, and in enabling "systems chemistry" alongside the more established "systems biology". The value of these resources is reflected in the significant number of publications citing these databases and servers during 2011–2012 (Fig. 9.2).

Interestingly, the specialized GPCR interaction databases supply links to the more general and typically bigger ones, but the opposite does not hold true. In many cases, the specialized databases were created by dedicated groups for their own research and for disseminating data to other groups. These specialized resources can contribute valuable insights and information and

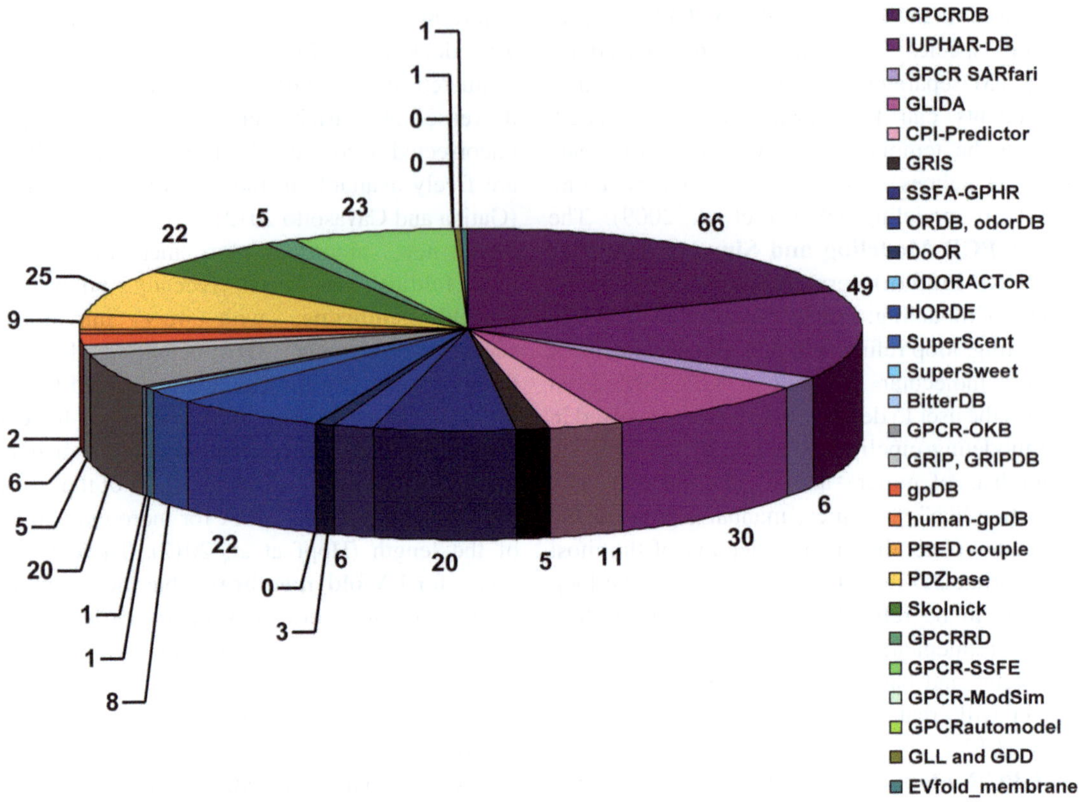

Fig. 9.2 Number of publications citing the papers that describe the databases and servers reviewed in this chapter, during 2011–2012. The figure shows the total number of citations for papers connected to the databases and servers discussed in Sects. 9.2.2, 9.2.3, 9.2.4, and 9.2.5 in the years 2011–2012 (up to October). The numbers of citations were taken from the Web Of Knowledge[SM] (Thomson Reuters) using the "Create citations report" option

are often manually curated. In our view, introducing links from the general databases to the specialized ones would be useful for the GPCR community. Very recently, links from the large and widely used ZINC database of compounds (Irwin et al. 2012) to the specialized BitterDB database of bitter compounds were introduced.

Among the specialized databases, only few are dedicated to GIPs. It is known that GPCRs interact with many different GIPs throughout their life cycle and these interactions are vital for the maturation and proper signaling of the GPCRs. We believe that the research community and drug discovery can greatly benefit from additional web resources on specialized GIPs and we hope that in the future, more websites on GIPs and GPCR effectors will be developed.

Computational screening and modeling are often used by the research community and drug discovery in general, and in the GPCRs field in particular. Computational resources are particularly powerful in cases where experimental results are limited. For example, the number of solved GPCR structures is less than 2 % of all GPCR types. Modeling of GPCRs helps researchers in cases where there is no available structure. However, the user must bear in mind that the models are only predictions, and are not necessarily fully accurate. Therefore, the structural models must be validated prior to their use, for example by checking structural characteristics such as Ramachandran plots (Ramachandran et al. 1963), i.e. using PROCHECK (Laskowski et al. 1993) and by comparing residues that are known to be important in GPCRs against solved GPCR structures (Levit et al. 2012), The most powerful confirmation of a model's accuracy stems from comparisons to experimental data. This may include indirect inference from mutagenesis and subsequent binding or functional assays (Levit et al. 2012), from screening enrichment data or, ultimately, comparing against newly solved structure (Kufareva et al. 2011).

As repositories of GPCR models rapidly become outdated as new templates are being published every few months, it is usually advisable to create new models, rather than relying on pre-computed ones. However, for comparative studies that call for many (not necessarily fine-tuned) structures underlie the need for pre-computed models, and the high number of models repositories' citations, 108 for (Zhang et al. 2006) and 26 for (Worth et al. 2009), illustrate their extensive usage by the scientific community.

With the increasing number of specialized modeling servers, community-wide experiments for the modeling of GPCRs, such as GPCR Dock (Kufareva et al. 2011), are highly desirable and will help evaluate and further improve the different modeling approaches.

Moreover, to increase the accuracy of the model predictions using the different modeling algorithms, it is advantageous to allow users to introduce GPCR templates of choice in the modeling server, as more and more GPCR structures become available.

The abundance of databases and servers results in repetitions and partial redundancy. For example, a recently published analysis of commercial and public bioactivity databases (including PDSP Ki and PubChem) showed that a large proportion of the information is shared by different databases; however, the data overlap between these databases is not complete (Tiikkainen and Franke 2012). We saw a similar picture for GPCR SARfari and GLIDA databases.

To simplify database usage, initiatives such as BioDBCore are calling to create uniform and centralized database descriptions (Gaudet et al. 2011). The BioDBCore initiative calls for standardization across the biological databases using a unified system for describing these databases in a single, centralized location (Gaudet et al. 2011). Having a united platform for all databases and servers will make them even more accessible to users from all fields. However, this kind of platform should be practical in the sense that it should allow small labs that do not have a lot of resources to participate and contribute their knowledge and databases to it. Furthermore, it is not yet clear how partial overlaps and partial redundancies will be treated. It seems that the best approach a user can currently take is to check the relevance and usefulness of each individual resource for his or her particular research needs. We hope that this review and the summary in Tables 9.1, 9.2, 9.3, 9.4, and 9.5 will make this task an easier one.

Acknowledgments We thank Dr. Tali Yarnitzky for critical reading of the manuscript and Ayelet Kowalsman for helping with the figures. Grants from the Deutsche Forschungsgemeinschaft (DFG) ME 1024/8and from the Nutrigenomics and Functional Foods Research Center at The Robert H. Smith Faculty of Agriculture, Food and Environment, The Hebrew University of Jerusalem are gratefully acknowledged.

References

Ahmed J, Preissner S, Dunkel M, Worth CL, Eckert A, Preissner R (2011) SuperSweet – a resource on natural and artificial sweetening agents. Nucleic Acids Res 39:D377–D382

Bellis LJ, Akhtar R, Al-Lazikani B, Atkinson F, Bento AP et al (2011) Collation and data-mining of literature bioactivity data for drug discovery. Biochem Soc Trans 39:1365–1370

Berman HM, Westbrook J, Feng Z, Gilliland G, Bhat TN et al (2000) The Protein Data Bank. Nucleic Acids Res 28:235–242

Beuming T, Skrabanek L, Niv MY, Mukherjee P, Weinstein H (2005) PDZBase: a protein-protein interaction database for PDZ-domains. Bioinformatics 21: 827–828

Bockaert J, Perroy J, Becamel C, Marin P, Fagni L (2010) GPCR interacting proteins (GIPs) in the nervous system: Roles in physiology and pathologies. Annu Rev Pharmacol Toxicol 50:89–109

Bolton E, Wang Y, Thiessen PA, Bryant SH (2008) PubChem: Integrated Platform of Small Molecules and Biological Activities. In: Wheeler RA, Spellmeyer DC (eds) Annual reports in computational chemistry. Elsevier Science, Amsterdam, pp 217–241

Bosier B, Hermans E (2007) Versatility of GPCR recognition by drugs: from biological implications to therapeutic relevance. Trends Pharmacol Sci 28: 438–446
Carlsson J, Coleman RG, Setola V, Irwin JJ, Fan H et al (2011) Ligand discovery from a dopamine D3 receptor homology model and crystal structure. Nat Chem Biol 7:769–778
Chen X, Ji ZL, Chen YZ (2002) TTD: Therapeutic Target Database. Nucleic Acids Res 30:412–415
Cheng F, Zhou Y, Li J, Li W, Liu G, Tang Y (2012) Prediction of chemical-protein interactions: multitarget-QSAR versus computational chemogenomic methods. Mol Biosyst 8:2373–2384
Congreve M, Langmead CJ, Mason JS, Marshall FH (2011a) Progress in structure based drug design for G protein-coupled receptors. J Med Chem 54: 4283–4311
Congreve M, Langmead C, Marshall FH (2011b) The use of GPCR structures in drug design. In: Neubig RR (ed) Pharmacology of G protein coupled receptors, Advances in Pharmacology. Elsevier Inc., Amsterdam. pp 1–36
Costanzi S (2012) Homology modeling of class a G protein-coupled receptors. Methods Mol Biol 857:259–279
Crasto C, Marenco L, Miller P, Shepherd G (2002) Olfactory Receptor Database: a metadata-driven automated population from sources of gene and protein sequences. Nucleic Acids Res 30:354–360
Crasto CJ, Liu N, Shepherd GM (2003) Databases for the Functional Analyses of Olfactory Receptors. In: Kotter R (ed) Neuroscience databases: a practical guide. Kluwer Academic Publishers, Düsseldorf, pp 37–50
Croft D, O'Kelly G, Wu G, Haw R, Gillespie M et al (2011) Reactome: a database of reactions, pathways and biological processes. Nucleic Acids Res 39: D691–D697
Daulat AM, Maurice P, Jockers R (2009) Recent methodological advances in the discovery of GPCR-associated protein complexes. Trends Pharmacol Sci 30:72–78
Dunkel M, Schmidt U, Struck S, Berger L, Gruening B et al (2009) SuperScent – a database of flavors and scents. Nucleic Acids Res 37:D291–D294
Elefsinioti AL, Bagos PG, Spyropoulos IC, Hamodrakas SJ (2004) A database for G proteins and their interaction with GPCRs. BMC Bioinformatics 5:208
Evers A, Klebe G (2004) Successful virtual screening for a submicromolar antagonist of the neurokinin-1 receptor based on a ligand-supported homology model. J Med Chem 47:5381–5392
Galizia CG, Munch D, Strauch M, Nissler A, Ma S (2010) Integrating heterogeneous odor response data into a common response model: A Door to the complete olfactome. Chem Senses 35:551–563
Gatica EA, Cavasotto CN (2012) Ligand and decoy sets for docking to G protein-coupled receptors. J Chem Inf Model 52:1–6
Gaudet P, Bairoch A, Field D, Sansone SA, Taylor C et al (2011) Towards BioDBcore: a community-defined information specification for biological databases. Nucleic Acids Res 39:D7–D10
Gonzalez-Maeso J, Ang RL, Yuen T, Chan P, Weisstaub NV et al (2008) Identification of a serotonin/glutamate receptor complex implicated in psychosis. Nature 452:93–97
Gorinski N, Kowalsman N, Renner U, Wirth A, Reinartz MT et al (2012) Computational and Experimental Analysis of the TM4/TM5 Dimerization Interface of the Serotonin 5-HT1A Receptor. Mol Pharm 82: 448–463
Granier S, Kobilka B (2012) A new era of GPCR structural and chemical biology. Nat Chem Biol 8:670–673
Harmar AJ, Hills RA, Rosser EM, Jones M, Buneman OP et al (2009) IUPHAR-DB: the IUPHAR database of G protein-coupled receptors and ion channels. Nucleic Acids Res 37:D680–D685
Hecker N, Ahmed J, von Eichborn J, Dunkel M, Macha K et al (2012) SuperTarget goes quantitative: update on drug-target interactions. Nucleic Acids Res 40:D1113–D1117
Hopf TA, Colwell LJ, Sheridan R, Rost B, Sander C, Marks DS (2012) Three-Dimensional Structures of Membrane Proteins from Genomic Sequencing. Cell 149:1607–1621
Horn F, Weare J, Beukers MW, Horsch S, Bairoch A et al (1998) GPCRDB: an information system for G protein-coupled receptors. Nucleic Acids Res 26: 275–279
Horn F, Bettler E, Oliveira L, Campagne F, Cohen FE, Vriend G (2003) GPCRDB information system for G protein-coupled receptors. Nucleic Acids Res 31: 294–297
Irwin JJ, Sterling T, Mysinger MM, Bolstad ES, Coleman RG (2012) ZINC: A Free Tool to Discover Chemistry for Biology. J Chem Inf Model 52:1757–1768
Jacobson KA, Costanzi S (2012) New insights for drug design from the X-ray crystallographic structures of GPCRs. Mol Pharm 82:361–371
Jacoby E, Bouhelal R, Gerspacher M, Seuwen K (2006) The 7 TM G-Protein-Coupled Receptor Target Family. ChemMedChem 1:760–782
Kanehisa M, Goto S (2000) KEGG: kyoto encyclopedia of genes and genomes. Nucleic Acids Res 28: 27–30
Kanehisa M, Goto S, Sato Y, Furumichi M, Tanabe M (2012) KEGG for integration and interpretation of large-scale molecular data sets. Nucleic Acids Res 40:D109–D114
Katritch V, Cherezov V, Stevens RC (2012) Diversity and modularity of G protein-coupled receptor structures. Trends Pharmacol Sci 33:17–27
Keshava Prasad TS, Goel R, Kandasamy K, Keerthikumar S, Kumar S et al (2009) Human Protein Reference Database–2009 update. Nucleic Acids Res 37: D767–D772

Khelashvili G, Dorff K, Shan J, Camacho-Artacho M, Skrabanek L et al (2010) GPCR-OKB: the G Protein Coupled Receptor Oligomer Knowledge Base. Bioinformatics 26:1804–1805

Kleinau G, Brehm M, Wiedemann U, Labudde D, Leser U, Krause G (2007) Implications for molecular mechanisms of glycoprotein hormone receptors using a new sequence-structure-function analysis resource. Mol Endocrinol 21:574–580

Kolb P, Rosenbaum DM, Irwin JJ, Fung JJ, Kobilka BK, Shoichet BK (2009) Structure-based discovery of beta2-adrenergic receptor ligands. Proc Natl Acad Sci USA 106:6843–6848

Kufareva I, Rueda M, Katritch V, Stevens RC, Abagyan R, GPCR Dock Participants (2011) Status of GPCR Modeling and Docking as Reflected by Community-wide GPCR Dock 2010 Assessment. Structure 19: 1108–1126

Laskowski RA, Macarthur MW, Moss DS, Thornton JM (1993) Procheck – a Program to Check the Stereochemical Quality of Protein Structures. J Appl Crystallogr 26:283–291

Launay G, Teletchea S, Wade F, Pajot-Augy E, Gibrat JF, Sanz G (2012) Automatic modeling of mammalian olfactory receptors and docking of odorants. Protein Eng Des Sel 25:377–386

Levit A, Yarnitzky T, Wiener A, Meidan R, Niv MY (2011) Modeling of human prokineticin receptors: interactions with novel small-molecule binders and potential off-target drugs. PLoS One 6:e27990

Levit A, Barak D, Behrens M, Meyerhof W, Niv MY (2012) In: Vaidehi N, Klein-Seetharaman J (eds) Homology model-assisted elucidation of binding sites in GPCRs: methods and protocols, Methods in molecular biology. Springer Science + Business Media, pp 179–205

Li J, Ning Y, Hedley W, Saunders B, Chen Y et al (2002) The Molecule Pages database. Nature 420:716–717

Liu T, Lin Y, Wen X, Jorissen RN, Gilson MK (2007) BindingDB: a web-accessible database of experimentally determined protein-ligand binding affinities. Nucleic Acids Res 35:D198–D201

Liu X, Su X, Wang F, Huang Z, Wang Q et al (2011) ODORactor: a web server for deciphering olfactory coding. Bioinformatics 27:2302–2303

Lohse MJ (2010) Dimerization in GPCR mobility and signaling. Curr Opin Pharmacol 10:53–58

Lomize MA, Lomize AL, Pogozheva ID, Mosberg HI (2006) OPM: orientations of proteins in membranes database. Bioinformatics 22:623–625

Lukasiewicz S, Blasiak E, Faron-Gorecka A, Polit A, Tworzydlo M et al (2007) Fluorescence studies of homooligomerization of adenosine A2A and serotonin 5-HT1A receptors reveal the specificity of receptor interactions in the plasma membrane. Pharmacol Rep 59:379–392

Magalhaes AC, Dunn H, Ferguson SS (2012) Regulation of GPCR activity, trafficking and localization by GPCR-interacting proteins. Br J Pharmacol 165: 1717–1736

Maurice P, Guillaume JL, Benleulmi-Chaachoua A, Daulat AM, Kamal M, Jockers R (2011) GPCR-interacting proteins, major players of GPCR function. Adv Pharmacol 62:349–380

Mobarec JC, Sanchez R, Filizola M (2009) Modern homology modeling of G-protein coupled receptors: which structural template to use?. J Med Chem 52:5207–5216

Mondal S, Khelashvili G, Shan J, Andersen OS, Weinstein H (2011) Quantitative modeling of membrane deformations by multihelical membrane proteins: application to G-protein coupled receptors. Biophys J 101:2092–2101

Nemoto W, Fukui K, Toh H (2009) GRIP: a server for predicting interfaces for GPCR oligomerization. J Recept Signal Transduct Res 29:312–317

Nemoto W, Fukui K, Toh H (2011) GRIPDB – G protein coupled Receptor Interaction Partners DataBase. J Recept Signal Transduct Res 31:199–205

Okuno Y, Yang J, Taneishi K, Yabuuchi H, Tsujimoto G (2006) GLIDA: GPCR-ligand database for chemical genomic drug discovery. Nucleic Acids Res 34: D673–D677

Okuno Y, Tamon A, Yabuuchi H, Niijima S, Minowa Y et al (2008) GLIDA: GPCR–ligand database for chemical genomics drug discovery–database and tools update. Nucleic Acids Res 36:D907–D912

Oldham WM, Hamm HE (2008) Heterotrimeric G protein activation by G-protein-coupled receptors. Nat Rev Mol Cell Biol 9:60–71

Olender T, Feldmesser E, Atarot T, Eisenstein M, Lancet D (2004) The olfactory receptor universe–from whole genome analysis to structure and evolution. Genet Mol Res 3:545–553

Overington JP, Al-Lazikani B, Hopkins AL (2006) How many drug targets are there?. Nat Rev Drug Discov 5:993–996

Park SH, Das BB, Casagrande F, Tian Y, Nothnagel HJ et al (2012) Structure of the chemokine receptor CXCR1 in phospholipid bilayers. Nature 491:779–783

Provasi D, Bortolato A, Filizola M (2009) Exploring molecular mechanisms of ligand recognition by opioid receptors with metadynamics. Biochemistry 48:10020–10029

Pucadyil TJ, Kalipatnapu S, Chattopadhyay A (2005) The serotonin1A receptor: a representative member of the serotonin receptor family. Cell Mol Neurobiol 25: 553–580

Ramachandran GN, Ramakrishnan C, Sasisekharan V (1963) Stereochemistry of polypeptide chain configurations. J Mol Biol 7:95–99

Rediger A, Piechowski CL, Yi CX, Tarnow P, Strotmann R et al (2011) Mutually opposite signal modulation by hypothalamic heterodimerization of ghrelin and melanocortin-3 receptors. J Biol Chem 286: 39623–39631

Renner U, Zeug A, Woehler A, Niebert M, Dityatev A et al (2012) Heterodimerization of serotonin receptors 5-HT1A and 5-HT7 differentially regulates receptor signalling and trafficking. J Cell Sci 125:2486–2499

Ritter SL, Hall RA (2009) Fine-tuning of GPCR activity by receptor-interacting proteins. Nat Rev Mol Cell Biol 10:819–830

Rodriguez D, Bello X, Gutierrez-de-Teran H (2012) Molecular Modelling of G Protein-Coupled Receptors Through the Web. Mol Infor 31:334–341

Romero G, von Zastrow M, Friedman PA (2011) Role of PDZ proteins in regulating trafficking, signaling, and function of GPCRs: means, motif, and opportunity. Adv Pharmacol 62:279–314

Roth BL, Lopez E, Patel S, Kroeze WK (2000) The multiplicity of serotonin receptors: Uselessly diverse molecules or an embarrassment of riches?. Neuroscientist 6:252–262

Safran M, Chalifa-Caspi V, Shmueli O, Olender T, Lapidot M et al (2003) Human Gene-Centric Databases at the Weizmann Institute of Science: GeneCards, UDB, CroW 21 and HORDE. Nucleic Acids Res 31:142–146

Sali A, Blundell TL (1993) Comparative protein modelling by satisfaction of spatial restraints. J Mol Biol 234:779–815

Satagopam VP, Theodoropoulou MC, Stampolakis CK, Pavlopoulos GA, Papandreou NC et al (2010) GPCRs, G-proteins, effectors and their interactions: human-gpDB, a database employing visualization tools and data integration techniques. Database (Oxford) 2010: baq019

Saunders B, Lyon S, Day M, Riley B, Chenette E et al (2008) The Molecule Pages database. Nucleic Acids Res 36:D700–D706

Schlyer S, Horuk R (2006) I want a new drug: G-protein-coupled receptors in drug development. Drug Discov Today 11:481–493

Sgourakis NG, Bagos PG, Papasaikas PK, Hamodrakas SJ (2005) A method for the prediction of GPCRs coupling specificity to G-proteins using refined profile Hidden Markov Models. BMC Bioinformatics 6:104

Shoichet BK, Kobilka BK (2012) Structure-based drug screening for G-protein-coupled receptors. Trends Pharmacol Sci 33:268–272

Shonberg J, Scammells PJ, Capuano B (2011) Design strategies for bivalent ligands targeting GPCRs. ChemMedChem 6:963–974

Stark C, Breitkreutz BJ, Reguly T, Boucher L, Breitkreutz A, Tyers M (2006) BioGRID: a general repository for interaction datasets. Nucleic Acids Res 34: D535–D539

The UniProt Consortium (2012) Reorganizing the protein space at the Universal Protein Resource. Nucleic Acids Res 40:D71–D75

Theodoropoulou MC, Bagos PG, Spyropoulos IC, Hamodrakas SJ (2008) gpDB: a database of GPCRs, G-proteins, effectors and their interactions. Bioinformatics 24:1471–1472

Tiikkainen P, Franke L (2012) Analysis of commercial and public bioactivity databases. J Chem Inf Model 52:319–326

Van Durme J, Horn F, Costagliola S, Vriend G, Vassart G (2006) GRIS: glycoprotein-hormone receptor information system. Mol Endocrinol 20:2247–2255

Vilar S, Ferino G, Phatak SS, Berk B, Cavasotto CN, Costanzi S (2011) Docking-based virtual screening for ligands of G protein-coupled receptors: not only crystal structures but also in silico models. J Mol Graph Model 29:614–623

Vroling B, Sanders M, Baakman C, Borrmann A, Verhoeven S et al (2011) GPCRDB: information system for G protein-coupled receptors. Nucleic Acids Res 39:D309–D319

Wang Y, Xiao J, Suzek TO, Zhang J, Wang J et al (2012) PubChem's BioAssay Database. Nucleic Acids Res 40:D400–D412

Wettschureck N, Offermanns S (2005) Mammalian G proteins and their cell type specific functions. Physiol Rev 85:1159–1204

White JF, Noinaj N, Shibata Y, Love J, Kloss B et al (2012) Structure of the agonist-bound neurotensin receptor. Nature 490:508–513

Wiener A, Shudler M, Levit A, Niv MY (2012) BitterDB: a database of bitter compounds. Nucleic Acids Res 40:D413–D419

Woehler A, Wlodarczyk J, Ponimaskin EG (2009) Specific oligomerization of the 5-HT1A receptor in the plasma membrane. Glycoconj J 26:749–756

Worth CL, Kleinau G, Krause G (2009) Comparative sequence and structural analyses of G-protein-coupled receptor crystal structures and implications for molecular models. PLoS One 4:e7011

Yarnitzky T, Levit A, Niv MY (2010) Homology modeling of G-protein-coupled receptors with X-ray structures on the rise. Curr Opin Drug Discov Devel 13:317–325

Zhang Y (2008) I-TASSER server for protein 3D structure prediction. BMC Bioinformatics 9:40

Zhang J, Zhang Y (2010) GPCRRD: G protein-coupled receptor spatial restraint database for 3D structure modeling and function annotation. Bioinformatics 26:3004–3005

Zhang Y, Devries ME, Skolnick J (2006) Structure Modeling of All Identified G Protein-Coupled Receptors in the Human Genome. PLoS Comput Biol 2:e13

Zhou H, Skolnick J (2012) FINDSITE(X): A Structure-Based, Small Molecule Virtual Screening Approach with Application to All Identified Human GPCRs. Mol Pharm 9:1775–1784

Zhu F, Shi Z, Qin C, Tao L, Liu X et al (2012) Therapeutic target database update 2012: a resource for facilitating target-oriented drug discovery. Nucleic Acids Res 40:D1128–D1136

Bioinformatics Tools for Predicting GPCR Gene Functions

10

Makiko Suwa

Abstract

The automatic classification of GPCRs by bioinformatics methodology can provide functional information for new GPCRs in the whole '*GPCR proteome*' and this information is important for the development of novel drugs. Since GPCR proteome is classified hierarchically, general ways for GPCR function prediction are based on hierarchical classification. Various computational tools have been developed to predict GPCR functions; those tools use not simple sequence searches but more powerful methods, such as alignment-free methods, statistical model methods, and machine learning methods used in protein sequence analysis, based on learning datasets. The first stage of hierarchical function prediction involves the discrimination of GPCRs from non-GPCRs and the second stage involves the classification of the predicted GPCR candidates into family, subfamily, andsub-subfamily levels. Then, further classification is performed according to their protein-protein interaction type: binding G-protein type, oligomerized partner type, etc. Those methods have achieved predictive accuracies of around 90 %. Finally, I described the future subject of research of the bioinformatics technique about functional prediction of GPCR.

Keywords

Bioinformatics • Function • G-protein coupled receptor • G-protein • Dimerization • Class • family • subfamily • GPCR discrimination • Hierarchical GPCR classification • PPT Network classification • Transmembrane helix • Alignment free method • Machine learning method • Statistical model method • Motif based search • Sequence similarity search

M. Suwa (✉)
Department of Chemistry and Bioscience, College of Science and Technology, Aoyama Gakuin University, 5-10-1 Fuchinobe, Chuo-ku, Sagamihara-shi, Kanagawa 252-5258, Japan

Computational Biology Research Center (CBRC), National Institute of Advanced Industrial Science and Technology (AIST), 4-2-7 Aomi, Kotou-ku, Tokyo, Japan
e-mail: suwa@chem.aoyama.ac.jp; m-suwa@aist.go.jp

M. Filizola (ed.), *G Protein-Coupled Receptors - Modeling and Simulation*, Advances in Experimental Medicine and Biology 796, DOI 10.1007/978-94-007-7423-0_10,

Abbreviations

GPCR	G-protein coupled receptor
TM	Transmembrane
TMH	Transmembrane helix
HMM	Hidden Markov Model
SVM	Support Vector Machine
NN	Neural Network
CA	Cell Automaton
kNN	K Nearest Neighbor
SOM	Self-organization Map
PPI	Protein-Protein Interaction

10.1 Introduction

G-protein-coupled receptors (GPCRs) are membrane proteins that are characterized by seven transmembrane (TM) helices. The binding of ligands, including biological amines, peptides, hormones, and odorant substances etc., from the extracellular side to GPCRs induces signal transduction to the cell interior by activating G-proteins ($G_{i/o}$, $G_{q/11}$, G_s, and $G_{12/13}$) that are conjugated at the cytoplasmic side of the cell. This process is followed by different types of signal transduction (Vauquelin and Mentzer 2007; Kobilka 2007).

GPCRs exist in almost all biological cells and abnormal events occurring in GPCR signal transduction have been related to various serious conditions, such as allergy, heart trouble, cancer, high blood pressure, and inflammation. As approximately 40 % (Wise et al. 2002) of drugs distributed around the world are designed to control the function of GPCRs, many researchers are looking into ways to reveal GPCR function. Of the approximately 800 GPCR genes (Vassilatis et al. 2003) encoded in human, most are predicted to be olfactory receptors and the remaining 200–300 genes are said to have the potential of being drug design targets. Of these, 100–150 are called 'orphan' GPCRs because their endogenous ligands are still unknown. As the discovery of a ligand that activates orphan GPCRs may be directly linked to new drug development, there is intense competition among pharmaceutical companies to search for ligands. Newly de-orphanized GPCRs include the orexin receptor (Sakurai et al. 1998), the apelin receptor (Tatemoto et al. 1998), the melanin-concentrating hormone receptor (Saito et al. 1999), and the ghrelin receptor (Kojima et al. 1999) etc. Their discovery has made a huge impact on both academe and industry (Civelli et al. 2006).

For a long time, experimental difficulties in structure determination and gene expression had prevented researchers from fully understanding the functional mechanism and classification of orphan GPCRs. Nevertheless, a treasure trove of genome information of many species has been unearthed and several structures of GPCRs have been identified (Rosenbaum et al. 2009; Kobilka and Schertler 2008; Palczewski et al. 2000; Rasmussen et al. 2011; Dorota et al. 2012). In this context, it is now possible to overview the sequences to understand the general rule of GPCR function by using bioinformatics tools. The automatic classification of GPCRs by bioinformatics tools can provide functional information of newly determined GPCRs in the whole *'GPCR proteome'* and this information is important for the development of novel drugs by pharmaceutical companies. This review introduces several useful bioinformatics tools for GPCR analysis.

10.2 Overview of GPCRs: Known Classes and Repertoires of GPCRs

GPCRs are classified into six classes (Class A, Class B, Class C, Class D, Class E, and Class F GPCRs) as defined in GPCRDB (Horn et al. 1998, 2000, 2003), based on their amino acid sequences and functional similarities.

Class A: This class, often called the "rhodopsin-like family," is the largest, accounting for around 80 % of all GPCR genes (Fridmanis et al. 2007). More than half of the members of this class are predicted to be olfactory receptors and the remaining are receptors with known endogenous ligands and orphan receptors. The members are classified into family level (e.g., adrenergic receptor), subfamily level (e.g., β adrenergic receptor), and sub-subfamily level (e.g., β2 adrenergic receptor). Although they show quite low sequence similarity to the receptors of the other families, the solved structures (Rosenbaum et al. 2009; Kobilka and Schertler 2008; Palczewski et al. 2000; Rasmussen et al. 2011; Dorota et al. 2012) indicate that all Class A GPCRs share a common structure (Fig. 10.1a) having seven TM helices together with the eighth helix at the C terminus, an S-S bond between TM helix 3(TMH3) and TMH4 at the extracellular side of membrane, a palmitoylated cysteine at the C terminal tail, and Class A specific conserved residues. The conserved residues in the extracellular region affect ligand binding selectivity, whereas those in the cytoplasmic region affect coupling selectivity with G-proteins. Many articles have reported the positions of conserved residues (Nygaard et al. 2009; Mizadegan et al. 2003; Karmik et al. 2003; Madabushi et al. 2004; Muramatsu and Suwa 2006; Wess 1998; Suwa et al. 2011) and the mutation experiments of key residues that have significant influence on the selectivity of ligand binding and G-protein coupling. Here, strongly conserved residues in all Class A GPCR sequences are observed in each TM helix region: the DRY motif in TMH3, the NPxxY pattern in TMH7, etc. The characterization of subfamilies and their ligand binding sites is important as Class A GPCRs include a very large number of drug-targeted receptors.

Class B: This class has around 70 members that are further divided into several subclasses (Parthier et al. 2009; Chapter et al. 2010; Dong et al. 2008, 2009). The first subclass B1 has approximately 20 members (mainly secretin/glucagon/VIP receptors) regulated by peptide hormones. Those receptors activate adenylyl cyclase and the phosphatidyl inositol calcium pathway by coupling to G_s and G_q proteins, respectively. Class B receptors have seven TM helices and a long N-terminal domain of approximately 120 residues stabilized by several S-S bonds (Fig. 10.1b). Despite the quite different amino acid sequences between subclass B1 GPCRs and Class A GPCRs, they have a common feature in the extracellular/cytoplasmic loop and the TM helices. The second subclass B2 includes more than 30 members, an example of which is CD97. The members of this subclass have very long N-terminal domains with repeats of well-known protein modules, such as the EGF domain and the cadherin repeat, which may be involved in cell adhesion. Only a few are known to associate with G-proteins and most of them are orphan receptors.

Class C: This includes the metabotropic glutamate receptor family, GABA receptors, calcium-sensing receptors, and taste receptors (Pin et al. 2003; Brauner-Osborne et al. 2007). Several ligands of Class C GPCRs have been identified. These receptors are characterized by a large extracellular N-terminal domain that is often shaped like a clam with approximately 600 residues to which ligands bind (Fig. 10.1c). This domain is connected to a TM helix by a loop that contains cysteine-rich domains (CRDs). It is suggested that most of the Class C GPCRs form homodimers or heterodimers (Pin et al. 2003; Brauner-Osborne et al. 2007).

Class D: This includes the pheromone receptor family (Eilers et al. 2005), such as STE2 and STE3. The amino acid sequences of these receptors have seven hydrophobic domains, similar to Classes A, B, and C GPCRs. Therefore, it is believed to be a GPCR family.

Class E: This includes the cAMP receptor family (Eichinger and Noegel 2005; Troemel et al. 1995; Prabhu and Eichinger 2006) from slime mold. The amino acid sequences of these receptors contain seven hydrophobic domains that are considered TM helices. These receptors are also believed to interact with some G-proteins. They do not have significant similarity to Class A GPCR sequences and are characterized by their unique signatures.

Class F: This class includes the frizzled/smoothened family (Malbon 2004; Ruiz-Gómez et al. 2007). The frizzled family has a

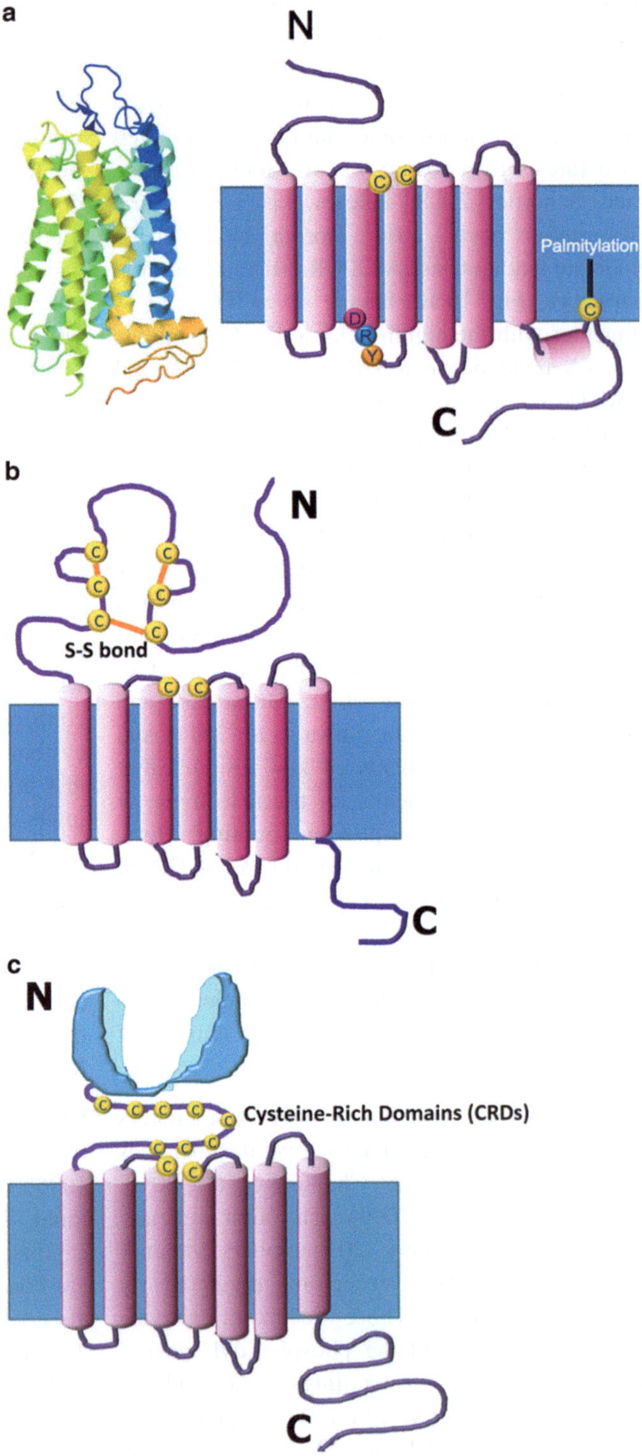

Fig. 10.1 (**a**) Schematic diagram of Class A GPCR structure (*right*) together with representative PDB structure (*left*): bovine rhodopsin (PDB:1f88). (**b**) Schematic diagram of Class B GPCR structure. This figure shows a subclass B1 GPCR (such as secretin receptor). It has a long N-terminal loop that is stabilized by S–S bonds (*yellow circles* represent cysteine). In the case of subclass B2 members, the N-terminal region consists of the repeat of well-known protein modules. (**c**) Schematic diagram of Class C GPCR structure. It has a large extracellular domain that is shaped like a clam. The loop region connecting this *clam-shaped* domain to the TM helix is called the cysteine-rich domains (CRDs) (*yellow circles* represent cysteine)

cysteine-rich domain that forms disulfide bridges in the extracellular region, which is composed of mainly alpha helices. This family is related to the Wnt signaling pathway. On the other hand, the smoothened family encoded by the SMO gene is related to pathways participating in many developmental processes.

Classes A, B, C, and F GPCRs are found in mammals, whereas Class D GPCRs are found only in fungi and Class E GPCRs are exclusive to *Dictyostelium*. These classes are further divided into family, subfamily, and sub-subfamily (sometimes called subtypes) based on the functions of the GPCRs and their specific ligands.

Class A GPCRs bind to several types of G-proteins ($G_{i/o}$, $G_{q/11}$, G_s, $G_{12/13}$). Class B GPCRs are classified into two types: those that bind to G_s and those that bind to $G_{q/11}$. Class C GPCRs bind to $G_{i/o}$ and $G_{q/11}$. There are also "promiscuous" receptors that couple with multiple types of G-proteins. The G-proteins of Class F GPCRs (frizzled/smoothened family) are still not known.

Fredriksson et al. (2003) created a phylogenetic tree of approximately 800 known human GPCR sequences and proposed a new taxonomy called "*GRAFS*," which consisted of five main receptor groups (metabotropic glutamate receptors, rhodopsin-like receptors, adhesion type receptors, frizzled/smoothened family, and secretin receptors). This taxonomy was applied to 13 kinds of eukaryotes (Fredriksson and Schiöth 2005). Concerning the comprehensive GPCR gene analysis of individual species, many publications that deal with insect (Hill et al. 2002), plant (Josefsson 1999), human and mouse (Vassilatis et al. 2003; Thora et al. 2006), and human and dog (Haitina et al. 2009) are available.

10.3 GPCR Function Prediction by Using Bioinformatics Tools

10.3.1 Strategy for Function Prediction

GPCR function is characterized by two aspects: the type of binding ligand and the type of binding G-protein. In general, an established way to predict protein function is to classify proteins into groups whose members are linked by sequence similarity using a conventional sequence search method. However, in the case of GPCRs, the function-similarity relationship is unclear. For example, it is suggested in ref. (Gaulton and Attwood 2003) that, (1) some homologous GPCR pairs that bind to the same ligands bind to different types of G-protein; (2) those pairs that bind to the same type of G-protein bind to a different ligand; and (3) some GPCR pairs bind to both the same ligand and the same G-protein even though they show less than 25 % sequence similarity.

Given this situation, various computational tools have been developed to understand GPCR function. Those tools use not simple sequence searches but more powerful methods, such as alignment-free methods, statistical model methods (hidden Markov models (HMMs), etc. See Sect. 10.3.2.), and machine learning methods (support vector machines (SVMs), etc. See Sect. 10.3.2.). Karchin et al. compared several methods and suggested that SVM achieved the highest performance (Karchin et al. 2002).

The aforementioned tools are applied to the hierarchical classification of GPCRs (Fig. 10.2). The first stage involves the discrimination of GPCRs from non-GPCRs. At the second stage, the predicted GPCR candidates are classified into family, subfamily, and sub-subfamily levels. At the final stage, further classification is performed according to protein-protein interaction (PPI) type: binding G-protein type and oligomerization partner type. Classification accuracy is evaluated based on the known hierarchical families of GPCRs, which are available in several databases.

10.3.2 Overview of Bioinformatics Algorithms Used in Protein Sequence Analysis

We present an overview of the general bioinformatics algorithms used in GPCR sequence analysis in this section.

Table 10.1 summarizes the bioinformatics tools according to the three layers of function

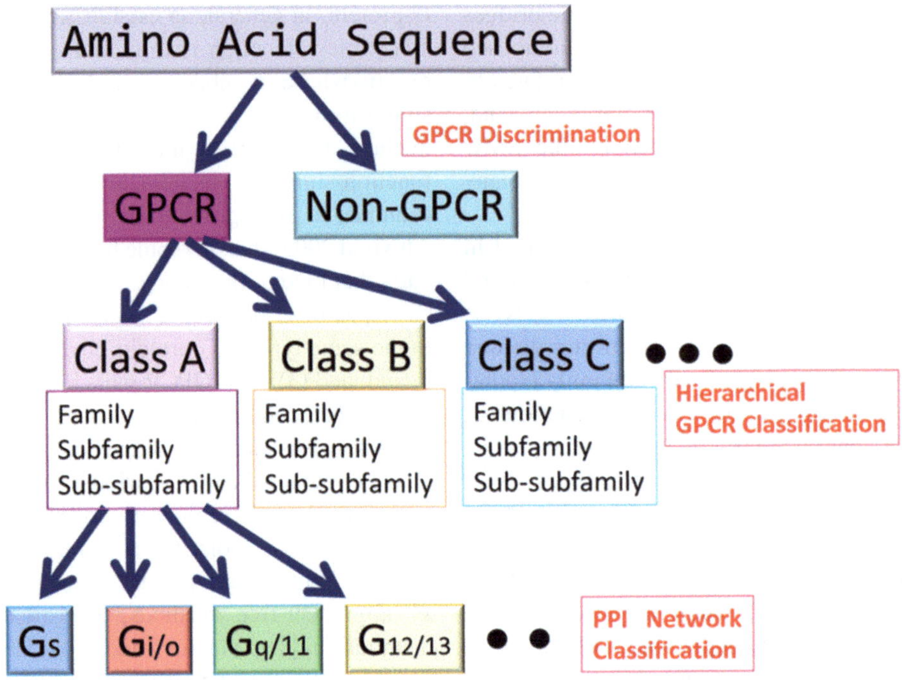

Fig. 10.2 Schematic diagram of hierarchical GPCR function prediction. The flow of the prediction is as follows: (1) GPCR discrimination from other proteins at the first stage; (2) hierarchical classification of GPCRs at the second stage; and (3) PPI (protein-protein interaction) network classification at the third stage

prediction. The tool has been classified according to the classification hierarchy in Fig. 10.2. About the tool that has been developed specifically for GPCR was listed along with the URL of the WEB. (Please note that there is a possibility that the URL will change in the future.)

Further information about the features of each tool are shown in the following text. (Sects. 10.3.2, 10.3.3, 10.3.4, 10.3.5, and 10.3.6)

Sequence similarity search method involves searching sequence databases by means of alignment to a query sequence. Statistical assessment scoring (e.g., occurrence score E-value, P-value) is conducted to determine how well sequences in the database align to the query sequence. The following are popular protein or nucleotide sequence search tools. **BLAST** (Altschul et al. 1990) is the most commonly used sequence similarity search tool. It uses heuristics to perform rapid local alignment searches. **PSI-BLAST** (Altschul et al. 1997) can help find remote relationships by constructing a position-specific scoring matrix (PSSM) through multiple alignments of sequences gathered by the repeated use of **BLAST** search. **FASTA** (Pearson and Lipman 1988) is another commonly used sequence similarity search tool that employs heuristics for rapid local alignment search. **SSEARCH** is a search tool that uses the Smith-Waterman algorithm (Smith and Waterman 1981), an optimal local alignment algorithm. These methods are used to classify GPCRs into groups whose members are linked by sequence similarity.

Motif-based approach: A motif is short nucleotide/amino acid sequence pattern that has a significant biological meaning. It is extracted from a multiple sequence alignment as a conserved region. The main role of the motif is to discriminate a functional region from a query sequence. Furthermore, it is often used to find a remote homologue, as proteins in the same family conserve functional regions although the whole sequences show low similarity.

Table 10.1 Bioinformatics tools categorized based on functional prediction methodology

	Motif based method	Alignment–free method	Statistical model method and machine learning method
GPCR discrimination		TMFinder (Charles et al. 2001) SOSUI (Hirokawa et al. 1998) QFC method (Kim et al. 2000) TopPred (Elofson and von Heijne 2007) SCANPI (Bernsel et al. 2008)	TMHMM (Krogh et al. 2001) HMMTOP (Tusnády et al. 2001) 7TMHMM (Möller et al. 2001a) http://tp12.pzr.uni-rostock.de/~moeller/7tmhmm/ GPCRHMM (Markus et al. 2006) http://gpcrhmm.sbc.su.se/
Hierarchical GPCR classification	PROSITE (Sigrist et al. 2010)	GPCR Tree (Davies et al. 2008a) http://igrid-ext.cryst.bbk.ac.uk/gpcrtree/	Method of Huang et al. 2004 PRED-GPCR (Papasaikas et al. 2004) http://athina.biol.uoa.gr/bioinformatics/PRED-GPCR/
	PRINTS (Attwood et al. 2012)	Method of Lapnish et al. (2005)	GPCRsclass (Bhasin and Raghava 2004, 2005) http://www.imtech.res.in/raghava/gpcrsclass/ GPCRpred (Bhasin and Raghava 2004, 2005) http://www.imtech.res.in/raghava/gpcrpred/
	Pfam (Finn et al. 2010) gpcrmotif (Gangal and Kumar 2007)	Method of 7TMRMine (Lu et al. 2009)	GPCR-CA (Xiao and Qiu 2010) GPCR-SVMFS (Li et al. 2010) GPCR-Mpredictor http://111.68.99.218/gpcr-mpredictor/
PPI Network classification		Method of Filizola and Weinstein (Filizola and Weinstein 2005)	GRIFFIN (Yabuki et al. 2005) http://griffin.cbrc.jp PRED-COUPLE (Sgurakis et al. 2005) http://athina.biol.uoa.gr/bioinformatics/PRED-COUPLE2/ GRIP (Nemoto et al. 2009) http://grip.cbrc.jp/GRIP/

PROSITE (Sigrist et al. 2010) database uses regular expression or profile expression to describe motifs. For example, the regular expression of a certain motif is M – [GS] – x – {RQ}, where M represents Met, [GS] represents Gly or Ser, {RQ} represents neither Arg nor Gln, and x represents any amino acid residue. This expression can generate a large number of variations. In contrast, a profile expression is described by the position-specific matrix of the appearance probability of 20 types of amino acids (or 4 types of nucleotides). **PRINTS** (Attwood et al. 2012) is a collection of so-called "fingerprints," a group of conserved motifs taken from a multiple sequence alignment. Together, the motifs form the characteristic signature of an aligned protein family. **Pfam** (Finn et al. 2010) is a database that includes functional annotations and is described by HMM (see description below). Using the search tool **HMMER** (Eddy 1998), the calibrated profile HMM of each family is used to score the sequences of all other families. **InterPro** (Hunter et al. 2009) is a motif database that integrates PROSITE, PRINTS, Pfam, and many other motif databases.

Hidden Markov Model (HMM) (Gollery 2008): Let us imagine a sequence of signals that are output one after another. When the stochastic process of the signal output value is determined

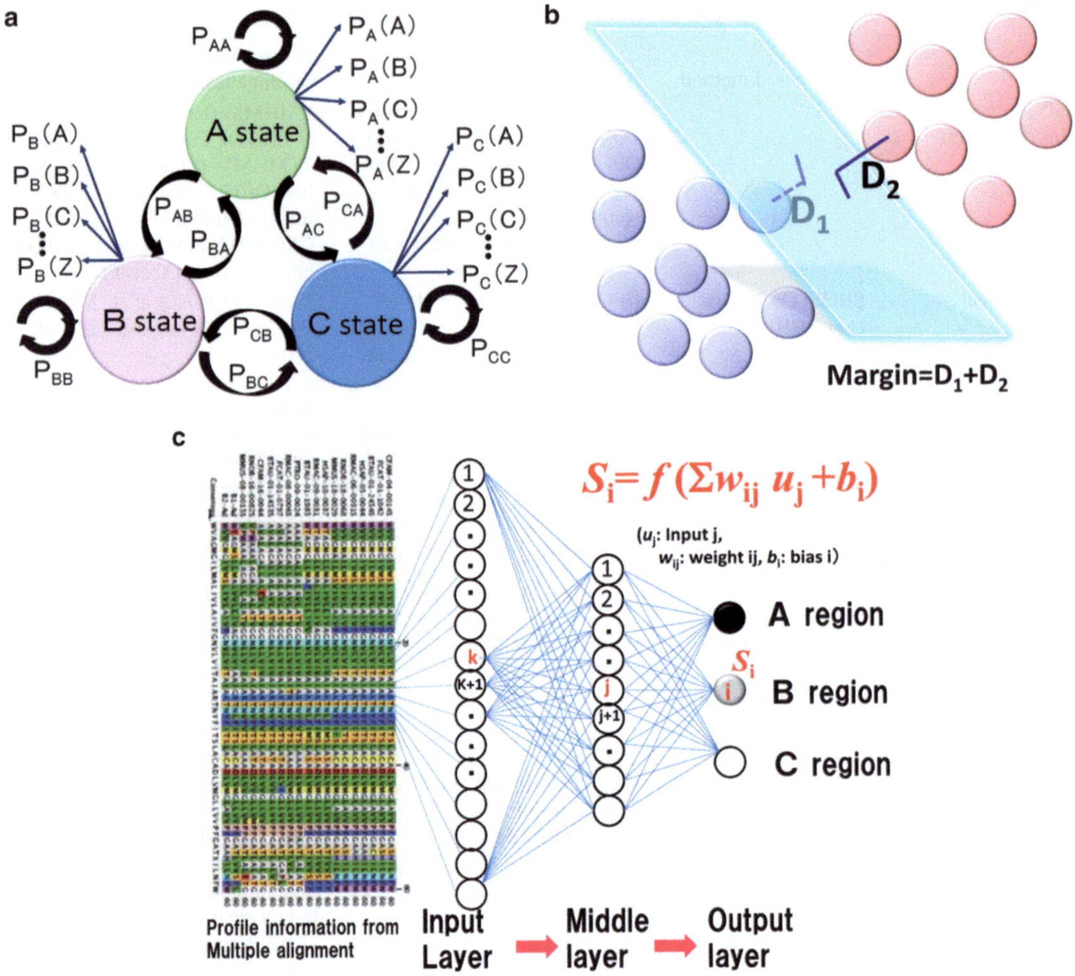

Fig. 10.3 (**a**) Schematic diagram of HMM. This is an example of a statistical model that expresses the internal state transition, in which a sign (*alphabets* in this case) is output in a sequence, according to the internal states (A, B, C states). For example, P_{AB} indicates the probability of transition from A state to B state. $P_A(C)$ indicates the probability that letter C is output from A state. (**b**) Schematic diagram of SVM. This is a machine learning method that performs linear shape distinction of a point in N-dimensional space in two groups. We can choose a distinction plane that maximizes the sum of the minimum altitude lengths (D1 + D2) among all the points of the two groups (*blue* and *red circles*), in the case of a three-dimensional space. As regards the points in the N-dimensional space, although it is difficult to imagine, it is possible to choose such a hyperplane mathematically. (**c**) Schematic diagram of neural network. A model that changes the bond weight of each synapse by learning the artificial nerve system formed in the brain. In this figure, the input layer is the amino acid appearance frequency of each site that is obtained from a multiple sequence alignment and such information is learned through the middle layer. The final network information is integrated as the strength of the last output (A, B or C region)

only by the present value regardless of the past output value, this process is called the **Markov process**. This model has several internal states that make transitions mutually within this process and output a signal value according to each state. Figure 10.3a is an example of three internal transition states (A, B, C) which determine the occurrence probability of each signal (alphabets in this case) and make transitions according to the probability *eg*. from A state and B state (P_{AB}). Because only a series of signals can be observed from the outside, without knowing the internal transition state, it is called as "*hiding*". For the analysis of GPCR sequences using HMM, the character of each amino acid is used as the signal value and transition probability of internal states and occurrence probability of characters are calculated from multiple sequence alignment of known GPCRs.

Support Vector Machine (SVM) is a machine learning method (Cristianini and Shawe-Taylor 2000) that distinguishes coordinates (X_1 ... X_N) in two groups in N-dimensional space. For example, when this discrimination method is performed in the three-dimensional space (Fig. 10.3b), several distinction planes can be drawn between the two groups. Here, altitude lines relative to the distinction plane are drawn from all the points in the two groups, and the minimum altitude length is chosen for each group. Thus, a straight distinction plane is selected so that the sum of the minimum altitude lengths (**margin** = D1 + D2 in Fig. 10.3b) becomes the maximum. Although it is difficult to imagine an N-dimensional space, the solution of hyper plane is obtained mathematically in this space. In the case of SVM application to GPCR classification, a GPCR is described by an N-dimensional coordinate vector composed of N feature elements, such as physicochemical parameters. As SVM is used to divide data into two classes, several discrimination planes should be combined in order to classify GPCRs into several families.

Neural Network (NN) is a model (Bishop 1994) in which artificial neurons form a network by combining with artificial synapses that change the weight of binding to neighboring synapses (Fig. 10.3c). Thus, it is able to solve problems by learning data and simulating the characteristics of the brain. Generally, NN has layers corresponding to the input, middle, and output layers as shown in Fig. 10.3c. Learning the relationship between the input dataset and the correct answer is conducted by changing the binding weight of each synapse pair. It is well used in the discrimination problem in which input data have many dimensions and linear separation is impossible. In case of NN application to GPCR classification, multiple alignment of a specific family is used as input data (Fig. 10.3c) and the learned binding weight network of synapses discriminates the specific family.

Self-Organization Map (SOM) is a teacherless NN that consists of an input layer and an output layer (Kohonen 1990). First, an N-dimensional vector (at the input layer) and several nodes arranged on a two-dimensional plane (output layer) are given. Randomly generated N-dimensional vectors (dignity vectors) are linked to each node. Learning is performed by calculating the Euclidean distance between the input vector and the dignity vectors and then each input vector is linked to the node that has the nearest dignity vector. After repeating the renewal of dignity vectors as the averaged vector composed of the input data and the predefined dignity vector, the input data are categorized into several nodes. SOM can change the nonlinear relationship that exists between data of a high dimension into a two-dimensional image that can simply represent a geometric relationship. When SOM is applied to GPCR classification, a GPCR is described as the N-dimensional vector of several parameters.

***K* Nearest Neighbor (kNN)** (Nearest-Neighbor Methods in Learning and Vision 2005) is a machine learning method for classifying N-dimensional vectors into several labeled classes. Using user-defined number *k*, the Euclidean distances from a query vector (unlabeled vector) assign k nearest neighborhood vectors (labeled vectors) sequentially. In the classification step, a query vector is classified to a class whose label is most frequent among the *k* nearest vectors. A result may change with what kind of number is chosen about *k*. When *k*NN is applied to GPCR classification, a GPCR is described as the N-dimensional coordinate vector composed of N feature elements, such as physicochemical parameters.

Cellular Automaton (CA) (Wolfram 1984) is a collection of "*colored*" cells on a grid of specified shape that evolves through a number of discrete time steps according to a set of rules based on such states as the *"on"* or *"off"* of neighboring cells. The grid can be in any finite number of dimensions. For each cell, a set of cells called its ***neighborhood*** is defined relative to the specified cell. An initial state (time $t = 0$) is selected by assigning a state for each cell. A new ***generation*** is created according to some fixed *rule* (generally, a mathematical function) that determines the new state of each cell in terms of the current state of the cell and the states of the cells in its neighborhood. Typically, the rule for updating the state of cells is the same for each cell and does not change over time, and is applied to the whole grid simultaneously. In the case of GPCR classification, a GPCR is described as the N-dimensional coordinate vector composed of N feature elements, and each kind of class/family corresponds to a two-dimensional CA image.

Decision tree (Menzies and Hu 2003) is a decision support tool that uses a tree-like graph to compare competing alternatives and assign occurrence values to those alternatives. In the tree-like graph, the leaves represent alternatives to be selected and the branches represent conjunctions of features that lead to those classifications. The machine learning technique for inducing a decision tree from data is called decision tree learning. More descriptive names for such tree models include the classification tree or the regression tree. Decision trees are commonly used in operations research, specifically in decision analysis, to help identify the most likely strategy to reach a goal. When used for GPCR family classification, the leaves represent each family and the branches indicate some conserved character for each GPCR family.

Alignment-free method (sometimes called the physicochemical method) does not require the consensus sequence obtained from a multiple sequence alignment. Here, each amino acid of a single sequence is translated into some physicochemical parameters and this information is used for function prediction. This method is often useful to annotate a function to an unknown sequence that cannot be assigned to any known family.

10.3.3 GPCR Classification Database

Many useful GPCR databases are available for the classification of GPCR sequences, including **GPCRDB** (Horn et al. 1998, 2000, 2003), **IUPHAR** (GPCR database) (Harmar et al. 2009), **GPCR-PD**^{TM} (Hodges et al. 2002), **ORDB** (Crasto et al. 2002), **GPCR safari** (https://www.ebi.ac.uk/chembl/sarfari/gpcrsafari), **SEVENS** (Suwa and Ono 2009, 2010; Ono et al. 2005), **gpDB** (Theodoropoulou et al. 2008), **Human-gpDB** (Satagopam et al. 2010), **GLIDA** (Okuno et al. 2008), and **GPCR-OKB** (Kerashvili et al. 2010). GPCRDB is the most popular and includes known GPCR sequences from UniProt and GENBANK. IUPHAR and GPCR-PD^{TM} accumulate literature information as well as sequence information. ORDB focuses on the olfactory receptor, a subfamily of GPCRs. SEVENS integrates GPCR genes that have been comprehensively predicted from the genome sequences of 58 eukaryotes. Furthermore, unique data resources regarding the interaction of a GPCR with G-proteins and effectors are summarized in gpDB and Human-gpDB, whereas the relationships between GPCRs and their ligands are included in GLIDA, and GPCR oligomer information is found in GPCR-OKB. These databases summarize well-organized GPCR genes useful as training data for tuning bioinformatics tools.

10.3.4 Tools for GPCR Discrimination from Other Proteins

Predicting the TM helix region is the most fundamental operation in the discrimination of membrane protein from another protein when an unknown protein sequence is obtained. *In particular, if a query sequence is predicted to have around seven TM helices, that sequence may be considered as a GPCR candidate.* When an amino acid sequence is translated into a numerical sequence of hydrophobicity indices (Kyte and Doolittle 1982, etc.) and the indices are plotted against the sequential number, the TM regions can be easily obtained because those regions become highly hydrophobic domains with 20–30 residues (corresponding to membrane thickness) and are connected by loop regions with low hydrophobicity. **TMFinder** (Charles et al. 2001) predicts TM regions using this physicochemical character.

The idea that "the TM helix domain has high average hydrophobicity" is the most fundamental one. However, using only this characteristic, a TM helix may be undistinguishable from several helices inside the hydrophobic core of globular proteins. By taking protein sequence length and amino acid amphiphilicity into consideration, **SOSUI** (Hirokawa et al. 1998) can predict TM helices by rejecting false-positive helices in globular proteins and therefore, highly accurate (97 %) discrimination between globular protein and membrane proteins can be accomplished.

Kim et al. developed a physicochemical algorithm for identifying TM proteins from genomic databases using a **quasi-periodic feature classifier (QFC)** (Kim et al. 2000). From a sequential plot of physicochemical parameters, several parameters of amino acids (amino acid usage index, log average of hydrophobicity periodicity, log average of polarity periodicity, and variance of first-order polarity derivative) are evaluated in a window of protein sequences and those values are then used in a linear discriminant function to separate GPCRs from non-GPCRs. The performance on their test dataset shows 96 % positive identification of known GPCRs.

In the analysis of TM helix proteins, the term '***topology prediction***' involves the simultaneous prediction of TM helical regions and the exposed side (extracellular or cytoplasmic side) of the loops that connect neighboring helices. The fundamental view of the inside-and-outside prediction of membrane protein is the ***positive inside rule*** (Andersson and von Heijne 1994) that the loops exposed to the cytoplasmic side have an abundance of basic residues, such as Lys and Arg, compared to the loops exposed to the extracellular side. **TopPred** is the first program to perform the inside-and-outside prediction of membrane protein (Elofson and von Heijne 2007) based on the *positive inside rule*. **SCANPI** (Bernsel et al. 2008) yields approximately 90 % predictive accuracy by combining the positive inside rule and the free energy calculation of protein folding from an amino acid sequence.

There are methods that model the feature of a TM helix with inside and outside loop information, and predict the topology of a query sequence by calculating the fitness score using the predefined profile or HMM. The loops at the extracellular/cytoplasmic side, the helix termini at the extracellular/cytoplasmic side, the central region of a TM helix, and sometimes the large globular domains exposed to water are incorporated into the TM helix HMM. **TMHMM** (Krogh et al. 2001) and its improved version (Viklund and Elofsson 2004), **HMMTOP** (Tusnády et al. 2001), and **MEMSAT** (Jones 2007) are typical programs that yield the highest accuracies.

7TMHMM (http://tp12.pzr.uni-rostock.de/~moeller/7tmhmm/submission.php) (Möller et al. 2001a) is a GPCR-specific predictor. This model is derived from TMHMM (Krogh et al. 2001) and contains sub-models for the seven TM helices and the cytoplasmic and extracellular loops, but is not trained on GPCR sequences. It uses a data-mining approach, combining pattern discovery with membrane topology prediction to find patterns of amino acid residues in the TM domains of GPCR.

GPCRHMM (http://gpcrhmm.cgb.ki.se) (Markus et al. 2006) is an HMM that is based on a large dataset representing the entire GPCR superfamily. GPCRHMM's sensitivity is approximately 15 % higher than that of the best TM helix predictors at comparable false-positive rates. When this method was applied to five proteomes, 120 sequences with no annotations were obtained as novel GPCRs. GPCRHMM strongly rejected a family of arthropod-specific odorant receptors believed to be GPCRs. Detailed analysis showed that the sequences were indeed very different from those of the other GPCRs.

There are other methods that do not take into consideration the TM model. NN-based method (**PHD-htm** (Rost et al. 1995)) and SVM-based methods (**SVMtm** (Yuan et al. 2004), **SPOCTOPUS** (Viklund et al. 2008), and **MEMSAT-SVM** (Nugent and Jones 2009)) directly learned parameters from sequence information to TM helix prediction. Those methods show high predictive accuracy.

The motif-based search method is also a popular tool for discriminating GPCR from other proteins. Cobanoglu et al. discriminated Class A GPCRs from other types (Cobanoglu et al. 2011) by discovering key receptor-ligand interaction sites selected by an original technique (Distinguishing Power Evaluation (DPE)). Application of this method to a GPCR sequence showed that it outperformed several other GPCR Class A subfamily prediction tools.

In general, sequence motif based methods are strongly dependent on the results of similarity of primary sequence alignments. In this regard, it is necessary to design a motif discovery and application method that is not strongly dependent on primary sequence similarity. The **gpcrmotif** (Gangal and Kumar 2007) is a method that uses a reduced alphabet representation (Eric et al. 2009) where similar functional residues (e.g., positively charged residues, negatively charged residues, etc.) have similar symbols. This reduced alphabet representation can accurately classify known GPCRs and the results obtained are comparable to those of PRINTS and PROSITE. For well-known GPCR sequences in SWISSPROT database, there are no false negatives and only a few false positives. This method is applicable to almost all known classes of GPCRs. It also predicts more than one class for certain sequences. This method has annotated 695 orphan receptors, 121 of which belong to Class A GPCRs (Gangal and Kumar 2007).

10.3.5 Tools for Hierarchical GPCR Family Classification

Proteochemometrics (Lapinsh et al. 2001, 2002, 2005) is an alignment-free approach for the discrimination and classification of proteins. It employs five "z values" derived from 26 physicochemical properties using principal component analysis. The five z values represent amino acid hydrophobicity, steric bulk/polarizability, polarity, electronic effects, and electrophilicity. Lapnish et al. translated amino acid sequences into these chemical parameters and applied this numerical sequence to the GPCR classification (Lapnish et al. 2002).

GPCRTree (Davies et al. 2007, 2008a, b) (http://igrid-ext.cryst.bbk.ac.uk/gpcrtree/) is a system that uses an alignment-free classification based on the physicochemical properties of amino acids. Using *proteochemometrics* (Lapinsh et al. 2001, 2005), the five physicochemical values are calculated for each amino acid in the sequence and are used to generate 15 attribute values. This system yields accuracies of 97 % at the class level, 84 % at the subfamily level, and 75 % at the sub-subfamily level when the system was applied to BIAS-PROFS GPCR dataset (Davies et al. 2007).

A better strategy would be to combine different classifiers to improve both specificity and sensitivity for the identification of a broader spectrum of GPCR candidates. **7TMRmine** (Lu et al. 2009) is a Web server that integrates alignment-free and alignment-based classifiers specifically trained to identify seven TM helix type receptors as well as TM helix prediction methods. This tool enables users to easily assess the distribution of GPCR families in diverse genomes or individual newly discovered proteins. Users can submit protein sequences for analysis or explore pre-analyzed results of multiple genomes. This server currently includes prediction results and a summary of statistics for 68 genomes.

Several GPCR classification methods are available. They are based on statistical models (HMM, etc.), machine learning methods (decision tree, SVM, kNN, CA, etc.), and methods that combine different types of algorithms.

Qian et al. generated a phylogenetic tree based profile HMM (**T-HMM**) (Qian et al. 2003) and showed its superiority in generating a profile for a group of similar proteins. When T-HMM was applied to the generation of the common features of GPCRs, the generated profile yielded high accuracy in GPCR function classification by ligand type and by coupled G-protein type.

Huang et al. used the decision tree method (bagging classification tree) (Huang et al. 2004) to predict GPCR subfamily and sub-subfamily based on the amino acid composition of a protein. They adopted the C4.5 algorithm for classification tree construction. To improve prediction accuracy, they used the bootstrap aggregating (bagging) procedure and the prediction accuracy actually became higher than that obtained by using a single classification tree. In a cross-validation test, they achieved a predictive accuracy of 91.1 % for GPCR subfamily classification and 82.4 % for sub-subfamily classification (Huang et al. 2004).

PRED-GPCR (Papasaikas et al. 2004) (http://athina.biol.uoa.gr/bioinformatics/PRED-GPCR/) uses family-specific profile HMMs to determine which GPCR family a query sequence resembles. The approach proposed by this method exploits the descriptive power of profile HMMs and uses an exhaustive discrimination assessment method to select only highly selective and sensitive profile HMMs for each family. The collection of these profile HMMs constitutes a signature library that is scanned for significant matches with a given query sequence.

GPCRsclass (Bhasin and Raghava 2005) (http://www.imtech.res.in/raghava/gpcrsclass/) is a tool for predicting amine binding receptors from amino acid sequences. For this purpose, SVM based tools were developed. The average accuracy of the method for two cases based on amino acid composition and dipeptide composition is 89.8 and 96.4 %, respectively, when evaluated using the fivefold cross-validation test.

GPCRpred (Bhasin and Raghava 2004) (http://www.imtech.res.in/raghava/gpcrpred/) uses an SVM-based method and the dipeptide composition of proteins to predict GPCR family and subfamily levels for a query sequence. The predictive power of this method is higher than that of the amino acid composition based method because dipeptide composition takes into account both the amino acid composition and the local consensus sequence of amino acids. This method can distinguish GPCRs from non-GPCR sequences with nearly 100 % accuracy. It can also predict classes and subfamilies of GPCRs with higher than 80 % accuracy.

GPCR-CA (Xiao et al. 2009) uses cellular automaton (CA) images to represent GPCRs through their pseudo amino acid composition (*PseuAA composition* (Chou 2001)), which was originally introduced to improve protein subcellular localization prediction and membrane protein type prediction. GPCR-CA has a two-layered prediction engine in which the first layer is used to determine whether a protein is a GPCR or a non-GPCR, and the second layer is used to classify the protein into the six functional classes. The overall success rates of the prediction for the first and second layers are higher than 91 and 83 %, respectively. The same research group also developed a tool that uses both adaptive kNN algorithm and CA images (Xiao and Qiu 2010). Based on CA images, complexity measure factors derived from each of the protein sequences are adopted for its PseuAA composition. GPCRs are categorized into nine subtypes. The overall success rate for nine class (rhodopsin-like receptor, peptide, hormones receptor, glutamate and calcium receptor, fungal mating pheromone receptor, cyclic AMP receptor, odorant receptors, gustatory receptor, frizzled/smoothened family, T2R family) identification is approximately 83.5 %. The high success rate indicates the high potential of the adaptive kNN algorithm and CA images for classifying GPCRs.

GPCR-SVMFS (Li et al. 2010) uses a novel three-layer predictor based on SVM and feature selection is developed for the prediction and classification of GPCRs directly from amino acid sequences. In this method, mRMR (maximum relevance minimum redundancy) algorithm is applied to pre-evaluate features with discriminative information, and genetic algorithm (GA) is utilized to find the optimized feature subsets. The overall accuracy of the three-layer predictor at the superfamily, family, and subfamily levels was determined by a cross-validation test on two non-redundant datasets and was approximately 0.5–16 % higher than those of GPCR-CA and GPCRPred.

GPCR-MPredictor (Muhammad and Asif 2012) (http://111.68.99.218/gpcr-mpredictor/) combines individual classifiers for the prediction of GPCRs and can efficiently predict GPCRs at five levels: GPCR or non-GPCR, family, subfamily, sub-subfamily, and subtype levels. This program analyzes the discriminative power of different feature extraction and classification strategies in the case of GPCR prediction and then uses an evolutionary ensemble approach to enhance prediction performance. Features are extracted using the amino acid composition, *PseuAA composition*, and dipeptide composition of receptor sequences. Different classification approaches, such as kNN, SVM, probabilistic neural network (PNN), decision tree method, naive Bayes method etc., are used to classify GPCRs at the five levels.

10.3.6 Tools for Predicting GPCR Function in View of Protein–Protein Interaction Network

To predict GPCR function in the cell, it is necessary to know the type of coupling G-protein as the signal transduction pathway is related to the G-protein type.

Enumerated (Qian et al. 2003; Cao et al. 2003; Möller et al. 2001b; Sreekumar et al. 2004) are the previously studied methods and tools for predicting coupling G-proteins. **T-HMM** of Qian et al. yielded high accuracy in the GPCR classification by coupled G-protein, as well as in the classification by ligand type (Qian et al. 2003). The method of Cao et al., which is based on the naive Bayes model, was able to predict G-protein with 72 % sensitivity from 55 GPCRs (Cao et al. 2003). Using pattern extraction, Möller et al. reported >90 % specificity with 30–40 % sensitivity (Möller et al. 2001b). Sreekumar et al. succeeded in reducing the prediction error rate to <1 % (Sreekumar et al. 2004) using HMM classification. Subsequent to those works, the following useful Web tools were published.

GRIFFIN (Yabuki et al. 2005) (http://griffin.cbrc.jp; G-protein and Receptor Interaction Feature Finding INstrument) is a Web tool that predicts GPCR and G-protein coupling selectivity by using SVM and HMM. Based on the assumption that whole structural segments of ligands, GPCRs, and G-proteins are essential to determine GPCR and G-protein coupling selectivity, various quantitative features are selected for ligands, GPCRs, and G-protein complex structures, and "*the most effective parameter set*" for predicting G-protein type are selected by evaluating predictive accuracy in SVM. The main part of GRIFFIN includes a hierarchical SVM classifier that utilizes feature vectors which is composed of "*the most effective parameter set*", and this classifier is useful for Class A GPCRs, the major GPCR family. For opsins and olfactory subfamilies of Class A and other minor families (Classes B, C, and frizzled/smoothened), the binding G-protein is predicted with high accuracy using HMM. Most importantly, this method is the first to predict G-protein type by inputting both ligand molecular weight and GPCR sequence information. It can predict G-proteins with sensitivity and specificity exceeding 85 % for the dataset with known coupling property.

PRED-COUPLE (Sgurakis et al. 2005) (http://athina.biol.uoa.gr/bioinformatics/PRED-COUPLE2/) is a Web method that predicts the coupling specificity of GPCRs to the four families of G-proteins (including $G_{12/13}$). This method can predict coupling to more than one family of G-proteins, as is experimentally determined. Similar to the first version, the PRED-COUPLE2 system implements a library of refined profile HMMs. The profiles are trained by the intracellular domain sequences of 188 GPCRs with known coupling properties. All HMMs are constructed and calibrated by the

HMMER (Eddy 1998) software package. In order to produce the final prediction, scores from individual profiles are combined by using an Artificial Neural Network. Multiple predictions should be made for receptors in the form of more than one output above the threshold.

Conventional research of GPCR function has taken into consideration the relationship between a ligand and an isolated GPCR. However, recently, new topics related to GPCR-GPCR complexes have emerged (Szidonya et al. 2008). In some cases, one GPCR activates another GPCR through the contact surface of their complex. In other cases, some GPCRs binding to specific G-proteins change the type of G-protein when the GPCR-GPCR complex is formed. The prediction of interactions between GPCRs and other proteins has attracted significant attention for next-generation drug design (Szidonya et al. 2008; Susan et al. 2002).

Filizola and Wenstein reviewed several bioinformatics approaches for GPCR oligomerization prediction (Flizola and Wenstein 2005). The correlated mutation algorithm (Govel et al. 1994; Oliveira et al. 1993) is a powerful tool for the accurate prediction of physical contact. The evolutional trace method (Lihitage et al. 1996) is also useful to determine family-specific functional regions. Using those methods, Filizola and Wenstein analyzed the occurrence probability of residues, which exposed to lipid surface, in predicting the oligomerization interface of GPCRs and reported the importance of the TMH4-TMH6 regions as the oligomerization interfaces (Flizola and Wenstein 2005).

GRIP (Nemoto et al. 2009) (http://grip.cbrc.jp/GRIP/) is a unique WEB tool that predicts the interface of oligomerized GPCRs by means of the spatial cluster detection (SCD) method (Cook et al. 2007). Here, the three procedures for query sequence analysis involve: (1) gathering of homologous sequence families, (2) pre-searching of template structure to which sequence families are mapped, and (3) detection of clusters of conserved residues around structural surfaces. The conserved residues within the detected sector are considered to correspond to the residues constituting the interface. Unfortunately, because of the shortage of the learning dataset, it is difficult to evaluate the accuracy of this system. Nevertheless, it works well at least for known complexes.

10.4 Conclusion

In this review, many bioinformatics tools that predict the functions of GPCRs from amino acid sequences were introduced. The tools have several classification layers that are divided according to GPCR family hierarchy, such as class, family, subfamily, and sub-subfamily, by using motif-based methods, alignment–free methods, statistical models, and machine learning methods. These methods have achieved prediction accuracies of around 90 % (see Sects. 10.3.4, 10.3.5, and 10.3.6).

Around 10 years ago, bioinformatics tools for predicting GPCR function saw a surge in number, but the number saturated thereafter. However, in the past 3–5 years, new bioinformatics tools for GPCR function prediction have emerged in succession. Why is this so? During the first surge that occurred approximately 10 years ago, protein sequences were determined rapidly from genome sequences, and it became possible to analyze a large number of GPCR sequences. However, although bioinformatics tools were required temporarily, it seemed that the demand for them reached saturation as experimental results and protein tertiary structures were restricted in those days.

On the other hand, in the past several years, the amount of biological information showed a marked increase due to remarkable advances in experimental techniques. For example, the genome sequences of many species have been determined by next-generation sequencers. Moreover, the tertiary structures of GPCRs are rapidly increasing in number (structures of approximately 13 families in 2013) with improved crystal analysis technology, and regions linked with a function can now be determined directly. Under such circumstances, the demand for bioinformatics tools has resurfaced.

In recent years, although the analysis of target proteins has seen much advancement, the number of de-orphanized receptors is still small. Therefore, GPCRs still maintain the position of the most important drug design target. In the environment where the competition for drug target search has intensified among pharmaceutical companies, the requirement of bioinformatics for function classification in GPCR proteome has naturally emerged. In this regard, what should be the purpose of developing bioinformatics tools from now on? If the purpose is to solve the problem of *'classification to fit a known category'* as described in Sect. 10.2, finding ways to increase prediction accuracy would be a good goal of development.

In order to classify sequences that have no similarity, bioinformatics researchers should set their sights on the development of optimal algorithms and parameters. Therefore, GPCR is positioned as a suitable target for bioinformatics research. In reality, however, it remains a difficult problem whether such bioinformatics research can meet the demand of researchers who actually design experiments. Of course, it is very important to predict to which family a novel receptor belongs as this information is expected to support the design of new experiments. It is also very important to predict, for a novel GPCR, the type of signal it transmits to a cell and the biological changes that take place as a result of signal transduction. Current GPCR function prediction tools cannot achieve this yet.

In bioinformatics, a vast quantity of data in a complicated system are needed as learning data. Fortunately, such a vast amount of experimental data would be obtained by scaling up experimental techniques. We also expect the development of new prediction tools using those data. Obviously, the close cooperation of researchers in the fields of biochemistry, structural biology, and bioinformatics is indispensable to this end.

Acknowledgement I appreciate deeply the insightful discussions with Dr. Minoru Sugihara (CBRC), Dr. Masami Ikeda (Aoyama Gakuin University), Dr. Wataru Nemoto (Tokyo Denki University), and Dr. Hiroyuki Toh (CBRC).

References

Altschul SF, Gish W, Miller W, Myers EW, Lipman DJ (1990) Basic local alignment search tool. J Mol Biol 215:403–410

Altschul SF et al (1997) Gapped BLAST and PSI-BLAST: a new generation of protein database search programs. Nucleic Acids Res 25:3389–3402

Andersson H, von Heijne G (1994) Membrane protein topology: effects of delta mu H+ on the translocation of charged residues explain the 'positive inside' rule. EMBO J 13:2267–2272

Attwood TK, Coletta A, Muirhead G, Pavlopoulou A, Philippou PB, Popov I, Roma-Mateo C, Theodosiou A, Mitchell A (2012) The PRINTS database: a fine-grained protein sequence annotation and analysis resource – its status in 2012. Database 2012: Article ID bas019 doi:10.1093/database/bas019

Bernsel A, Viklund H, Falk J, Lindahl E, von Heijne G (2008) SCAMPI: prediction of membrane-protein topology from first principles. Proc Natl Acad Sci USA 105:7177–7181

Bhasin M, Raghava GP (2004) GPCRpred: an SVM-based method for prediction of families and subfamilies of G-protein coupled receptors. Nucleic Acids Res 32(Web Server issue):W383–W389

Bhasin M, Raghava GP (2005) GPCRsclass: a web tool for the classification of amine type of G-protein-coupled receptors. Nucleic Acids Res 33(Web Server issue):W143–W147

Bishop CM (1994) Neural networks and their applications. Rev Sci Instrum 65:1803–1832

Brauner-Osborne H, Wellendorph P, Jensen AA (2007) Structure, pharmacology and therapeutic prospects of family C G-protein coupled receptors. Curr Drug Targets 8:169–184

Cao J, Panetta R, Yue S, Steyaert A, Young-Bellido M, Ahmad S (2003) A naive Bayes model to predict coupling between seven transmembrane domain receptors and G-proteins. Bioinformatics 19:234–240

Chapter MC et al (2010) Chemical modification of class II G protein-coupled receptor ligands: frontiers in the development of peptide analogs as neuroendocrine pharmacological therapies. Pharmacol Ther 125:39–54

Charles M et al (2001) TM Finder: a prediction program for transmembrane protein segments using a combination of hydrophobicity and nonpolar phase helicity scales. Protein Sci 10:212–219

Chou KC (2001) Prediction of protein cellular attributes using pseudo-amino acid composition. Proteins 43:246–255

Civelli O, Saito Y, Wang Z, Hans-Peter Nothacker H-P, Reinscheid RK (2006) Orphan GPCR and their ligands. Pharmacol Ther 110:525–532

Cobanoglu MC, Saygin Y, Sezerman U (2011) Classification of GPCRs using family specific motifs. Trans Comput Biol Bioinformatics 8:1495–1508

Cook AJ, Gold DR, Li Y (2007) Spatial cluster detection for censored outcome data. Bio Metrics 63: 540–549

Crasto C, Marenco L, Miller P, Shepherd G (2002) Olfactory receptor database: a metadata-driven automated population from sources of gene and protein sequences. Nucleic Acids Res 30:354–360

Cristianini N, Shawe-Taylor J (2000) An introduction to support vector machines: and other kernel-based learning methods. Cambridge University Press, Cambridge, xi, 189

Davies MN et al (2007) On the hierarchical classification of G protein-coupled receptors. Bioinformatics 23:3113–3118

Davies MN et al (2008a) Optimizing amino acid groupings for GPCR classification. Bioinformatics 24:1980–1986

Davies MN et al (2008b) GPCRTree: online hierarchical classification of GPCR function. BMC Res Notes. doi:10.1186/1756-0500-1-67

Dong M et al (2008) Insights into the structural basis of endogenous agonist activation of family B G protein-coupled receptors. Mol Endocrinol 22:1489–1499

Dong M, Cox RF, Miller LJ (2009) Juxtamembranous region of the amino terminus of the family BG protein-coupled calcitonin receptor plays a critical role in small-molecule agonist action. J Biol Chem 284:21839–21847

Dorota L et al (2012) G-protein coupled receptors-recent advances. Acta Biochem Pol 59:515–529

Eddy SR (1998) Profile hidden Markov models. Bioinformatics 14:755–763

Eichinger L, Noegel AA (2005) Comparative genomics of Dictyostelium discoideum and Entamoeba histolytica. Curr Opin Microbiol 8:606–611

Eilers M et al (2005) Comparison of class A and D G protein-coupled receptors: common features in structure and activation. Biochemistry 44:8959–8975

Elofson A, von Heijne G (2007) Membrane protein structure: prediction versus reality. Annu Rev Biophys Rev Biochem 76:125–140

Eric LP, Jane K, Julie AT, Rob P (2009) Reduced amino acid alphabets exhibit an improved sensitivity and selectivity in fold assignment. Bioinformatics 25:1356–1362

Finn RD et al (2010) The Pfam protein families database. Nucleic Acids Res 38(Database issue):D211–D222

Flizola M, Wenstein H (2005) The study of G-protein coupled receptor oligomerization with computational method and bioinformatics. FEBS J 272: 2926–2938

Fredriksson R, Schiöth HB (2005) The repertoire of G-protein-coupled receptors in fully sequenced genomes. Mol Pharmacol 67:1414–1425

Fredriksson R, Lagerström MC, Lundin LG, Schiöth HB (2003) The G-protein-coupled receptors in the human genome form five main families. Phylogenetic analysis, paralogon groups, and fingerprints. Mol Pharmacol 63:1256–1272

Fridmanis D et al (2007) Formation of new genes explains lower intron density in mammalian rhodopsin G protein-coupled receptors. Mol Phylogenet Evol 43:864–880

Gangal R, Kumar KK (2007) Reduced alphabet motif methodology for GPCR annotation. J Biomol Struct Dyn 25:299–310

Gaulton A, Attwood TK (2003) Bioinformatics approaches for the classification of G-protein-coupled receptors. Curr Opin Pharmacol 3:114–120

Gollery M (2008) Handbook of hidden Markov models in bioinformatics. Chapman & Hall/CRC Press, Boca Raton/London, xix, 156

Govel V, Sander C, Schneider R, Valencia A (1994) Correlated mutations and residue contacts in proteins. Proteins 18:309–317

HaitinaT, Fredriksson R, Foord SM, Schioth H, Gloriam DE (2009) The G protein-coupled receptor subset of the dog genome is more similar to that in humans than rodents. BMC Genomes 10:24. doi:10.1186/1471-2164-10-24

Harmar AJ et al (2009) IUPHAR-DB: the IU-PHAR database of G protein coupled receptors and ion channels. Nucleic Acids Res 37:D680–D685

Hill CA, Fox AN, Pitts RJ, Kent LB, Tan PL, Chrystal MA, Cravchik A, Collins FH, Robertson HM (2002) G protein-coupled receptors in Anopheles gambiae. Science 298:176–178

Hirokawa T, Boon-Chieng S, Mitaku S (1998) SOSUI: classification and secondary structure prediction system for membrane proteins. Bioinformatics 14:378–379

Hodges PE et al (2002) Annotating the human proteome: the human proteome survey database (HumanPSD™) and an in-depth target database for G protein-coupled receptors (GPCR-PD™) from incyte genomics. Nucleic Acids Res 30:137–141

Horn F et al (1998) GPCRDB: an information system for G protein – coupled receptors. Nucleic Acids Res 26:275–279

Horn F, Vriend G, Cohen FE (2000) Collecting and harvesting biological data: the GPCRDB and NucleaRDB information systems. Nucleic Acids Res 29:346–349

Horn F et al (2003) GPCRDB information system for G protein – coupled receptors. Nucleic Acids Res 31:294–297

https://www.ebi.ac.uk/chembl/sarfari/gpcrsafari

Huang Y, Cai J, Ji L, Li Y (2004) Classifying G-protein coupled receptors with bagging classification tree. Comput Biol Chem 28:275–280

Hunter S et al (2009) InterPro: the integrative protein signature database. Nucleic Acids Res 37(Database issue):D211–D215

Jones DT (2007) Improving the accuracy of transmembrane protein topology prediction using evolutionary information. Bioinformatics 23:538–544

Josefsson LG (1999) Evidence for kinship between diverse G-protein coupled receptors. Gene 239: 333–340

Karchin R, Karplus K, Haussler D (2002) Classifying G-protein coupled receptors with support vector machines. Bioinformatics 18:147–159
Karmik SS, Gogonea C, Patil S, Saad Y, Takezako T (2003) Activation of G-protein-coupled receptors: a common molecular mechanism. Trends Endocrinol Metab 14:431–437
Kerashvili G et al (2010) GPCR-OKB: the G protein coupled receptor oligomer knowledge base. Bioinformatics. doi:10.1093/bioinformatics/btq 264
Kim J, Moriyama EN, Warr CG, Clyne PJ, Carlson JR (2000) Identification of novel multi-transmembrane proteins from genomic databases using quasi-periodic structural properties. Bioinformatics 9:767–775
Kobilka BK (2007) G protein coupled receptor structure and activation. Biochim Biophys Acta 1768:794–807
Kobilka B, Schertler FX (2008) New G-protein-coupled receptor crystal structures; insights and limitations. Trends Pharm Sci 29:79–83
Kohonen T (1990) The self-organization map. Proc IEEE 9:1464–1480
Kojima M, Hosoda H, Date Y, Nakazato M, Matsuo H, Kangawa K (1999) Ghrelin is a growth-hormone-releasing acylated peptide from stomach. Nature 402:656–660
Krogh A, Larsson B, von Heijne G, Sonnhammer ELL (2001) Predicting transmembrane protein topology with a hidden Markov model: application to complete genomes. J Mol Biol 305:567–580
Kyte J, Doolittle RF (1982) A simple method for displaying the hydropathic character of a protein. J Mol Biol 157:105–132
Lapinsh M et al (2001) Development of proteochemometrics: a novel technology for the analysis of drug-receptor interactions. Biochimica Et Biophysica Acta- Gen Subj 1525:180–190
Lapinsh M et al (2005) Improved approach for proteochemometrics modeling: application to organic compound–amine G protein-coupled receptor interactions. Bioinformatics 21:4289–4296
Lapnish M, Gutcaits A, Prusis P, Post C, Lundstedt T, Wikberg JES (2002) Classification of G-protein coupled receptors by alignment-independent extraction of principle chemical properties of primary amino acid sequences. Protein Sci 11:795–805
Li Z, Zhou X, Dai Z, Zou X (2010) Classification of G-protein coupled receptors based on support vector machine with maximum relevance minimum redundancy and genetic algorithm. BMC Bioinformatics 11:325
Lihitage O, Bowrne HR, Chhen FE (1996) An evolutionary trace method defines binding surfaces common to protein families. J Mol Biol 257:342–358
Lu G, Wang Z, Jones AM, Moriyama EN (2009) 7TMR-mine: a Web server for hierarchical mining of 7TMR proteins. BMC Genomics 10:275. doi:10.1186/1471-2164-10-275
Madabushi S, Gross AK, Philippi A, Meng EC, Wensel TG, Lichtarge O (2004) Evolutionary trace of G protein-coupled receptors reveals clusters of residues that determine global and class-specific functions. J Biol Chem 279:8126–8132
Malbon CC (2004) Frizzleds: new members of the superfamily of G-protein-coupled receptors. Front Biosci 9:1048–1058
Markus W, Lukas K, Erik LL (2006) Sonnhammer, a general model of G protein- coupled receptor sequences and its application to detect remote homologs. Protein Sci 15:509–521
Menzies T, Hu Y (2003) Data mining for very busy people. IEEE Comput 36:22–29
Mizadegan T, Benkö G, Filipek S, Palczewski K (2003) Sequence analysis of G-protein-coupled-receptors: similarities to rhodopsin. Biochemistry 42: 2759–2767
Möller S, Vilo J, Croning MD (2001a) Supplementary material for the G-protein coupling prediction paper of Croning Prediction of the coupling specificity of GPCRs to their G proteins Bioinformatics (Suppl 1):174–181
Möller S, Vilo J, Croning MD (2001b) Prediction of the coupling specificity of G protein coupled receptors to their G proteins. Bioinformatics 17:174–181
Muhammad N, Asif UK (2012) GPCR-MPredictor: multi-level prediction of G protein-coupled receptors using genetic ensemble. Amino Acids 42(5):1809–1823
Muramatsu T, Suwa M (2006) Statistical analysis and prediction of functional residues effective for GPCR-G-protein coupling selectivity. Protein Eng 19:277–283
Nemoto W, Fukui K, Toh H (2009) GRIP: a server for predicting interfaces for GPCR oligomerization. J Recept Signal Transduct Res 29:312–317
Nugent T, Jones DT (2009) Transmembrane protein topology prediction using support vector machines. BMC Bioinformatics 10:159
Nygaard R, Frimurer TM, Holst B, Rosenkilde MM, Schwartz TW (2009) Ligand binding and micro-switched in 7TM receptor structures. Trends Pharmacol Sci 30:249–259
Okuno Y, Tamon A, Yabuuchi H, Niijima S, Minowa Y, Tonomura K, Kunimoto R, Feng C (2008) GLIDA: GPCR-ligand database for chemical genomics drug discovery-database and tools update. Nucleic Acids Res 36:D907–D912
Oliveira L, Paiva ACM, Vriend G (1993) A common motif in G-protein coupled seven transmembrane helix receptors. J Comp Aid Mol Des 7:649–658
Ono Y, Fujibuchi W, Suwa M (2005) Automatic gene collection system for genome-scale overview of G-protein coupled receptors in eukaryotes. Gene 364:63–73
Palczewski K et al (2000) Crystal structure of rhodopsin: a G protein- coupled receptor. Science 289:739–745
Papasaikas PK, Bargos PG, Litou ZI, Promponas VJ, Hamodrakas SJ (2004) PRED-GPCR: GPCR recognition and family classification server. Nucleic Acids Res 32:W380–W382
Parthier C et al (2009) Passing the baton in class B GPCRs: peptide hormone activation via helix induction? Trends Biochem Sci 34:303–310

Pearson WR, Lipman DJ (1988) Improved tools for biological sequence comparison. Proc Natl Acad Sci USA 85:2444–2448

Pin JP, Galvez T, Prezeau L (2003) Evolution, structure, and activation mechanism of family 3/C G- protein-coupled receptors. Pharmacol Ther 98:325–354

Prabhu Y, Eichinger L (2006) The Dictyostelium repertoire of seven transmembrane domain receptors. Eur J Cell Biol 85:937–946

Qian B, Soyer OS, Neubig RR, Goldstein RA (2003) Depicting a protein's two faces: GPCR classification by phylogenetic tree-based HMMs. FEBS Lett 554: 95–99

Rasmussen SG et al (2011) Crystal structure of the Beta 2 adrenergic receptor-Gs protein complex. Nature 477:549–555

Rosenbaum DM, Rasmaussen SGF, Kobilka BK (2009) The structure and function of G-protein-coupled receptors. Nat Insight 459:356–363

Rost B, Casadio R, Fariselli P, Sander C (1995) Transmembrane helices predicted at 95% accuracy. Protein Sci 4:521–533

Ruiz-Gómez A, Molnar C, Holguín H, Mayor F Jr, de Celis JF (2007) The cell biology of Smo signalling and its relationships with GPCRs. Biochim Biophys Acta 1768:901–912

Saito Y, Nothacker H-P, Wang Z, Steven HSL, Leslie F, Civelli O (1999) Molecular characterization of the melanin-concentrating-hormone receptor. Nature 400:265–269

Sakurai T, Amemiya A, Ishii M, Matsuzaki I, Chemell RM, Tanaka H et al (1998) Orexins and orexin receptors: a family of hypothalamic neuropeptides and G protein-coupled receptors that regulate feeding behavior. Cell 92:573–585

Satagopam VP, Theodoropoulou MC, Stampolakis CK, Pavlopoulos GA, Papandreou NC, Bagos PG, Schneider R, Hamodrakas SJ (2010) GPCRs, G-proteins, effectors and their interactions: human-gpDB, a database employing visualization tools and data integration techniques. Database doi:2010:baq019

Sgurakis NG, Bagos PG, Papasaikas PK, Hamodorakas SJ (2005) A method for the prediction of GPCRs coupling specificity to G-proteins using refined profile Hidden Markov Models. BMC Bioinformatics 6:104. doi:10.1186/1471-2105-6-104

Shakhnarovish-rovich G, Darrell T, Indyk P (ed) (2005) Nearest-neighbor methods in learning and vision, IEEE. The MIT Press, Cambridge, MA

Sigrist CJ et al (2010) PROSITE, a protein domain database for functional characterization and annotation. Nucleic Acids Res 38(Database issue):D161–D166

Smith TF, Waterman MS (1981) Identification of common molecular subsequences. J Mol Biol 147:195–197

Sreekumar KR, Huang Y, Pausch MH, Gulukota K (2004) Predicting GPCR–G-protein coupling using hidden Markov models. Bioinformatics 20:3490–3499

Susan R, George Brian FOD, Samuel PL (2002) G-protein coupled receptor oligomerization and it potential for drug discovery. Nature review 1:808–820

Suwa M, Ono Y (2009) Computational overview of GPCR gene universe to support reverse chemical genomics study. In: Koga H (ed) Reverse chemical genetics, Methods in Mol Biol 577, 1st edn. Springer, Tokyo

Suwa M, Ono Y (2010) A bioinformatics strategy to produce a cyclically developing project structure-comprehensive functional analysis of the drug design target genes. Synthesiology 3:1–12

Suwa M, Sugihara M, Ono Y (2011) Functional and structural overview of G-protein-coupled receptors comprehensively obtained from genome sequence. Pharmaceuticals 4:652–664

Szidonya L, Cserzo M, Hunyady L (2008) Dimerization and oligomerization of G-protein-coupled receptors: debated structures with established and emerging functions. J Endocrinol 196:435–453

Tatemoto K, Hosoya M, Habata Y, Fujii R, Kakegawa T, Zou MX et al (1998) Isolation and characterization of a novel endogenous peptide ligand for the human APJ receptor. Biochem Biophys Res Commun 251:471–476

Theodoropoulou MC, Bagos PG, Spyropoulos IC, Hamodrakas SJ (2008) gpDB: a database of GPCRs, G-proteins, effectors and their interactions. Bioinformatics 24:1471–1472

Thora KB, David EG, Sofia HH, Helena K, Fredriksson R, Helgi BS (2006) Comprehensive and phylogenetic analysis of the G protein-coupled receptors in human and mouse. Genomics 88:263–273

Troemel ER et al (1995) Divergent seven transmembrane receptors are candidate chemosensory receptors in C. elegans. Cell 83:207–218

Tusnády GE, Simon I, Jayasinghe S, Hristova K, White SH (2001) Bioinformatics 17:49–50

Vassilatis DK et al (2003) The G protein-coupled receptor repertoires of human and mouse. Proc Nat Acad Sci USA 100:4903–4908

Vauquelin G, Mentzer B (2007) G. Protein-coupled receptors. Wiley, West Sussex

Viklund H, Elofsson A (2004) PRO/PRODIV-TMHMM: best alpha-helical transmembrane protein topology predictions are achieved using hidden Markov models and evolutionary information. Protein Sci 13:1908–1917

Viklund H, Bernsel A, Skwark M, Elofson A (2008) SPOCTOPUS: a combined predictor of signal peptides and membrane protein topology. Bioinformatics 24:2928–2929

Wess J (1998) Molecular basis of receptor/G-protein – coupling selectivity. Pharmacol Ther 80: 231–264

Wise A et al (2002) Target validation of G-protein coupled receptors. Drug Discov Today 7:235–246

Wolfram S (1984) Cellular automata as models of complexity. Nature 311:419–424

Xiao X, Qiu WR (2010) Using adaptive K-nearest neighbor algorithm and cellular automata images to predicting G-protein-coupled receptor classes. Interdiscip Sci Comput Life Sci 2:180–184

Xiao X, Wang P, Chou KC (2009) GPCR-CA: A cellular automaton image approach for predicting G-protein–coupled receptor functional classes. J Comput Chem 30:1414–1423

Yabuki Y, Mutamatsu T, Hirokawa T, Mukai H, Suwa M (2005) GRIFFIN: a system for predicting GPCR-G-protein coupling selectivity using support vector machines and a hidden Markov model. Nucleic Acids Res 32:W148–W153

Yuan Z, Mattick JS, Teasdale RD (2004) SVMtm: support vector machines to predict transmembrane segments. J Comput Chem 25:632–636

Index

M. Filizola (ed.), *G Protein-Coupled Receptors - Modeling and Simulation*, Advances in Experimental Medicine and Biology 796, DOI 10.1007/978-94-007-7423-0,

F

G

H

I

J

K

L

M

MIX
Papier aus verantwortungsvollen Quellen
Paper from responsible sources
FSC® C105338

If you have any concerns about our products,
you can contact us on
ProductSafety@springernature.com

In case Publisher is established outside the EU,
the EU authorized representative is:
Springer Nature Customer Service Center GmbH
Europaplatz 3, 69115 Heidelberg, Germany

Printed by Libri Plureos GmbH
in Hamburg, Germany